Carmen Flury
Counting on Computers

Studies in the History of Education and Culture

—

Edited by
Meike Sophia Baader, Elke Kleinau, and Karin Priem

Volume 7

Carmen Flury

Counting on Computers

New Information Technologies and Curricular Change in East Germany, 1960s to 1990

DE GRUYTER
OLDENBOURG

Funding for the open access publication of this book was generously provided by the project group "Reforms in Education: A European Research Network," funded by the Swedish Research Council.

The research on which this book is based was made possible by the funding of the research project "Europe and the Digital Agenda" by the Swiss National Science Foundation (SNSF-Project no. 182217).

ISBN 978-3-11-144763-6
e-ISBN (PDF) 978-3-11-144864-0
e-ISBN (EPUB) 978-3-11-144890-9
ISSN 2748-9531
DOI https://doi.org/10.1515/9783111448640

This work is licensed under the Creative Commons Attribution-NonCommercial-NoDerivatives 4.0 International License. For details go to https://creativecommons.org/licenses/by-nc-nd/4.0/.

Creative Commons license terms for re-use do not apply to any content (such as graphs, figures, photos, excerpts, etc.) not original to the Open Access publication and further permission may be required from the rights holder. The obligation to research and clear permission lies solely with the party re-using the material.

Library of Congress Control Number: 2024945576

Bibliographic information published by the Deutsche Nationalbibliothek
The Deutsche Nationalbibliothek lists this publication in the Deutsche Nationalbibliografie; detailed bibliographic data are available on the internet at http://dnb.dnb.de.

© 2025 the author(s), published by Walter de Gruyter GmbH, Berlin/Boston
The book is published open access at www.degruyter.com.

Cover image: Delegation from the district of Erfurt at the parade through the Berlin city center on July 4, 1987 on the occasion of the city's 750th anniversary, © Federal Archives, Image 183-1987-0704-077 / Photographer: Thomas Uhlemann / License CC-BY-SA 3.0
Typesetting: Integra Software Services Pvt. Ltd.
Printing and binding: CPI books GmbH, Leck

www.degruyter.com

Acknowledgements

I am deeply grateful to my supervisor, Prof. Dr. Michael Geiss, for providing the impetus and inspiration to engage so intensively with the history of digitalization in education, and for his mentorship, guidance, and constructive criticism throughout the course of my dissertation project. I would also like to thank Prof. Dr Lucien Criblez as my second supervisor and Prof. Dr Marcelo Caruso for their support and their valuable feedback.

I thank the Swiss National Science Foundation for funding the research project "Europe and the Digital Agenda" (SNF-Project no. 182217), as part of which I conducted my research on the introduction of computer technology in education in the GDR. I also thank the project group "Reforms in Education: A European Research Network," funded by the Swedish Research Council, for the generous financial support that has made the open access publication of this book possible.

Many thanks go to my colleagues at the Education and Digital Transformation Research Center at the Zurich University of Teacher Education. I am especially grateful to my colleagues in the "Europe and the Digital Agenda" project team, Rosalía Guerrero and Fabian Grütter. It was truly a pleasure to work with you, and I couldn't have wished for a better team. I would also like to thank Silvana Flütsch Keravec, Liana Pirovino, Mirjam Nievergelt, and Angela Bonetti, who encouraged and motivated me during the meetings of our wonderful writing group.

Finally, I would like to express my deep gratitude to my family and closest friends, who were always there for me during this intense period of writing and working, and who eased my worries and frustrations when I felt stuck: Silvia and Balz Flury, Barbara, David and Mara Jansen, Elias Sprengel, and Sophia Gisler.

Contents

Acknowledgements —— V

1 Introduction —— 1
  1.1 Research Interest —— 6
  1.2 State of Research —— 10
  1.3 Methodology and Sources —— 26

2 A Sociotechnical Imaginary for "Socialist" Computerization —— 36
  2.1 Computers as a Tool Controlled by the Working People —— 38
  2.2 Computer-Aided Governance of the Economy and the State —— 40
  2.3 Computer Technology as an Engine of Economic Growth —— 43
  2.4 Computers as a Symbolic Arena in the Ideological Battle Between the Systems —— 44
  2.5 Computer Technology as a Driver of Social Progress —— 46
  2.6 The Computer as a Tool for the Development of the Socialist Personality —— 48

3 Cybernetics and Computer Science: Emergence of a New Discipline —— 50
  3.1 Cybernetics as an Intellectual Precursor of Computer Science and Informatics —— 50
  3.2 Institutionalization of Computer Science and Informatics in Higher Education —— 61
  3.3 Continuing Education and Training in Computing and Information Processing —— 87
  3.4 Chapter Conclusion —— 95

4 Training the Socialist Workforce for the Computer —— 98
  4.1 Building a Computer Industry in the GDR —— 99
  4.2 A Workforce for the Era of Mainframe Computers —— 102
  4.3 Envisioning the Computerization of Industrial Production and Offices —— 112
  4.4 Introducing Computer Technology in Vocational Education and Training —— 119
  4.5 The Computerization of the Socialist Workplace —— 133
  4.6 Chapter Conclusion —— 149

5 Informatics For All? Computer Literacy in Schools —— 155
- 5.1 Electronic Pocket Calculators and Early Curricular Reforms —— 161
- 5.2 A Curriculum for Computer Instruction in General Education —— 170
- 5.3 Teacher Training for Computer Education —— 194
- 5.4 A Dedicated Computer for Schools: The "Bildungscomputer" (BIC) —— 206
- 5.5 Chapter Conclusion —— 214

6 Extracurricular Computer Education —— 220
- 6.1 Popular Education and Extracurricular Learning about Computer Technology —— 222
- 6.2 Computer Technology Education in Popular Media —— 237
- 6.3 Computer Games in the Context of Socialist Education —— 242
- 6.4 Pedagogical Control and Guidance in the Development of Software in Schools —— 246
- 6.5 Chapter Conclusion —— 249

7 Conclusion and Outlook —— 253
- 7.1 Introducing Computer Education to Navigate Technological Change —— 254
- 7.2 A Multifaceted and Fluid Imaginary of Socialist Computerization —— 263

List of Abbreviations —— 267

List of Figures —— 269

List of Tables —— 271

Historical Sources —— 273

Literature —— 275

1 Introduction

Since its advent, computer technology was perceived as revolutionary.[1] Its development was accompanied by a sense of excitement about the future, a promise of endless possibilities and radical changes that the widespread use of computers would bring. The computer came to be envisioned as a powerful tool for information processing by promising a solution to almost any problem. From its origins in research and computing centers, it soon found its way into workshops, factories, offices, leisure centers, classrooms, and private homes. But for the computer to become established in so many areas of the economy and society, the wider population first had to make sense of this new technology and its role in society.[2]

In education, the establishment of the computer as, for example, a powerful calculator, a pluripotent information processor, or even a universal teaching and learning aid, focused much more on its ascribed utility value, the possibilities and expected effects of its use, rather than on the device itself as a piece of technical hardware.[3] Thus, as educational policymakers in most industrialized countries were faced with the challenge of responding to the technological innovation of modern computers, a fundamental concern was the question of how and for what purposes a future society would use them. In this situation, normative visions of a desirable computerized future provided decisionmakers with general principles for deciding whether and how to introduce computers into education, and how to design computer curricula to prepare people for the envisioned future. In this sense, the development of a shared understanding of a desirable computerized future was a prerequisite for the acceptance of the computer and its

---

[1] Michael S. Mahoney, "The History of Computing in the History of Technology," *Annals of the History of Computing* 10, no. 2 (April–June 1988): 123.

[2] For a history of how people made sense of the microcomputer by answering the question of what they *hoped* it could be used for, see Laine Nooney, *The Apple II Age: How the Computer Became Personal* (Chicago: The University of Chicago Press, 2023), 14.

[3] The rich and multifaceted history of the introduction of computers in education and how computers were imagined to support teaching and learning is illustrated by the contributions in the edited volumes of Carmen Flury and Michael Geiss, eds., *How Computers Entered the Classroom, 1960– 2000: Historical Perspectives* (Berlin/Boston: De Gruyter Oldenbourg, 2023), Arthur Tatnall and Bill Davey, eds., *Reflections on the History of Computers in Education: Early Use of Computers and Teaching about Computing in Schools* (Berlin/Heidelberg: Springer, 2014), and John Impagliazzo and John A. N. Lee, eds., *History of Computing in Education* (New York: Springer, 2004) illustrate. On the history of the PLATO system as an example of how computing was imagined and established as an educational technology in the USA, see Bill Cope and Mary Kalantzis, "A little history of e-learning: finding new ways to learn in the PLATO computer education system, 1959–1976," *History of Education* 52, no. 6 (2023): 905–36.

place in education. It enabled a common vision of who the future users would be, what they would use the computer for, and how they would use it. These shared visions of a desirable future involving computer technology, thus, provided a basis for educational policymakers to consider what kinds of knowledge, skills and values needed to be imparted to people in the present in order to prepare them for the imagined future.[4] At the same time, the political mission to bring computers into education can also be understood as an attempt of policymakers to actually help make the imagined future a reality – by setting the course of educational policy in the present to orient social and technological development towards the imagined social and technological order of the future.

This book is about the process of introducing computer technology into the education system of the German Democratic Republic (GDR) from the 1960s, when the first university courses in computer science and information technology were established, until 1990, when the GDR was reunified with West Germany. More specifically, it focuses on the imaginary of a desirable computerized future that guided this process. During the Cold War era, politicians, and educators in both the East and the West imagined possible roles, uses, and effects of computing in their respective education systems, and in doing so, they envisioned future utopias to strive for and dystopian futures to avert. As a case study of the GDR, it pursues, in particular, the historical reconstruction of a specific imaginary of socialist computerization propagated by the authoritarian socialist regime of the GDR, which guided policymaking and curricular reforms aimed at introducing computer technology into education.

The first modern computers were developed during the Second World War, motivated by the need to perform calculations at much higher speeds to solve practical problems. These early digital computers were massive, bulky machines made up of thousands of vacuum tubes and easily occupied the space of a whole room.[5] The first of its kind was the 'Z3' designed by the German engineer Konrad Zuse, who presented a fully functional model of the computer to the public in 1941. During the Cold War, the arms race between the USA and the Soviet Union fueled the development of more powerful computer technology, which played an important role in the military's development of sophisticated weapons and defence systems. The second generation of computers replaced vacuum tube technology with transistors, and the third eventually used integrated circuits, which proved to be faster and more reliable.[6] While the large mainframes and minicom-

---

[4] Lina Rahm, "Educational imaginaries: governance at the intersection of technology and education," *Journal of Education Policy* 38 no. 1 (2023): 46–68.
[5] Gerard O'Regan, *A Brief History of Computing* (London: Springer, 2012), 35.
[6] O'Regan, *A Brief History of Computing*, 54.

puters[7] developed in the period from the 1950s to the mid-1970s were highly specialized computers, developed for the use in science, business, and government, the mid-1970s saw the birth of the idea of the general-purpose home computer.[8] The year 1977 came to be known in the history of computer technology for the "1977 trinity," as it saw the launch of the first three ready-made personal computer models. Three American companies had each launched their product all in the same year, targeted at a wider market than just electronics hobbyists: Commodore with the PET 2001, Apple with the Apple II, and Tandy with the TRS-80.[9] These microcomputers were delivered fully built, and ready to be plugged in and switched on, rather than consisting of bare circuitry or having to be assembled from a do-it-yourself computer kit.[10] This development significantly lowered the barriers of entry into computing for the wider population. In addition, the creation and spread of simplified programming languages such as BASIC (Beginner's All-purpose Symbolic Instruction Code)[11] and later the availability of graphical user interfaces,[12] and off-the-self software[13] helped to make the computer more accessible to non-expert, casual computer users.

The "1977 trinity" had ushered in a new era of home computing in the 1980s, as the new microcomputers were now marketed to non-technical users as an affordable and more accessible consumer product. The development of the "micro" was based on the key invention of the microprocessor. The microprocessor has been at the heart of the development of ever smaller and more powerful consumer technology, from desktop computers to portable laptops, smartphones, and today's computing devices that can be carried in one hand, worn as a wristband,

---

[7] The label "minicomputer" can be misleading: While it was introduced in the 1960s to describe the new generation of much smaller, general-purpose computers compared to the large and expensive mainframes, these minicomputers were still about the size of a refrigerator.
[8] O'Regan, *A Brief History of Computing*, 53–54.
[9] Tom Lean, *Electronic Dreams: How 1980s Britain Learned to Love the Computer* (London/New York: Bloomsbury Sigma, 2016), 54; Nooney, *The Apple II Age*, 11.
[10] The first microprocessor-based computer was the "Altair 8800". It was released in 1975 and is often described as the first personal computer. While inexpensive (just under $400 US), it was sold by mail order as a self-assembly computer kit, consisting of "a single box containing the central processor, with a panel of switches and lights on the front; it had no display, no keyboard, and minimal memory" (Martin Campbell-Kelly et al., *Computer: A History of the Information Machine* (Boulder, CO: Westview Press, 2014), 235).
[11] Mark J. Lorenzo, *Endless Loop: The History of the BASIC Programming Language* (Philadelphia/Pittsburgh: SE Books, 2017).
[12] Thomas Haigh and Paul E. Ceruzzi, *A New History of Modern Computing* (Cambridge, MA: MIT Press, 2021), 243–61.
[13] Campbell-Kelly et al., *Computer: History of the Information Machine*, 186–188.

or slipped into a pocket.[14] Starting in the 1970s and 1980s, the innovation of the microprocessor affected industrialized countries all around the globe. In the GDR, the first microprocessor was manufactured in 1978, the "U808". Only two years later, it was followed by the much more powerful "U880," an 8-bit microprocessor and an unlicensed replica of the US-American Zilog's Z80.[15] The U880 was produced by the "VEB Mikroelektronik Karl Marx" in Erfurt and was the basis of most of the GDR's domestically produced 8-bit computers.

In the late 1970s, microelectronics and computer technology became a key element in the strategy of the GDR's authoritarian regime to overcome the permanent economic crises of the 1980s and to stabilize its political system.[16] In 1977, the leadership of the Socialist Unity Party (ger.: Sozialistische Einheitspartei der DDR, SED) decided to create a domestic microelectronics industry and devoted a staggering amount of money and human resources to the task of computerizing the socialist economy and society.[17] The centerpiece was the "Robotron" combine in Dresden, which became the most important hardware and software producer in the GDR.[18] Robotron was not only the heart of the GDR's computer industry, but also became a leader in the field of electronic computer technology within the Eastern Bloc.[19]

As a result of the SED's ambitious microelectronics program, political leaders soon recognized the need to train not just a few experts, but also the general public for the future use of the new computer technology. To this end, the Politburo and the Council of Ministers of the GDR passed a resolution in 1985 on the consequences that were to be drawn from technological developments for the education system. Educational policymakers and educators were instructed to initiate

---

[14] Nooney, *The Apple II Age*, 6.
[15] "Integrierte Schaltkreise," Robotrontechnik, last modified September 21, 2023, https://www.robotrontechnik.de/index.htm?/html/komponenten/ic.htm (accessed May 31, 2024).
[16] Frank Dittmann, "Microelectronics under Socialism," *Icon* 8 (2002): 49.
[17] Gerhard Barkleit, *Mikroelektronik in der DDR: SED, Staatsapparat und Staatssicherheit im Wettstreit der Systeme* (Dresden: Hannah-Arendt-Institut für Totalitarismusforschung, 2000), 35–53.
[18] Klaus Krakat, *Schlussbilanz der elektronischen Datenverarbeitung in der früheren DDR* (FS-Analysen, no. 5) (Berlin: Forschungsstelle für Gesamtdeutsche Wirtschaftliche und Soziale Fragen, 1990): 35; Gerhard Merkel, *VEB Kombinat Robotron, Sitz Dresden: Ein Kombinat des Ministeriums für Elektrotechnik und Elektronik der DDR* (Dresden: Förderverein für die Technischen Sammlungen Dresden, 2005), https://robotron.foerderverein-tsd.de/111/robotron111a.pdf (accessed May 31, 2024).
[19] Krakat, *Schlussbilanz*, 40.

the necessary curricular adjustments at all levels and in all areas of the GDR's educational system.[20]

This ambitious plan to computerize the GDR, together with the rapid pace of technological change, posed a major challenge to education policymakers and educators in the GDR. They were faced with the task of preparing new generations for an uncertain future: What skills would be needed in tomorrow's world, shaped by a technology with a yet unknown trajectory? What skills and knowledge would be rendered obsolete by technological change? At a time when ground-breaking scientific and technological advances were turning the gaze of political leaders and the general public towards the future, the role of education systems became focused on preparing children not only for the present, but increasingly for an anticipated world to come. As Arthur J. Lewis put it in 1981: "To educate children for the status quo is to educate them for obsolescence. Education for a prefigurative culture needs to build bridges between the present and the future."[21]

Today, historians might look back at the mid-20th century and call it the start of the computer age or refer to the 1980s as the early days of home computing and the cradle of the digital society. However, for the politicians and educators back then, it was by no means clear where technological change would lead them. In his book *Die Geschichte der Zukunft* (eng.: The History of the Future), Joachim Radkau highlights the fallacy of retrospective coherence and the limitations of our understanding of history. He suggests that historical events and transformations appear conclusive, explainable, and even meaningful only in hindsight, when they are reconstructed with the knowledge and perspective of our present time. In this sense, the inclination to impose current understandings and biases onto the past leads to the creation of a narrative that aligns with present perspectives. Radkau makes a compelling argument about the inherently unpredictable nature of the future, as our attempts to forecast or shape the future are often based on flawed assumptions and a limited understanding of the intricacies at play. Rather than rooted in factual knowledge, our understanding of the future is often rooted in hopes and fears — or in other words— in fictional expectations and collective imaginaries of what the future world could look like.

---

20 SAPMO DY 30/J IV 2/2/2138, "Standpunkte zu Konsequenzen aus der Entwicklung der Informatik und informationsverarbeitenden Technik für das Bildungswesen."
21 Arthur J. Lewis, "Education: Bridging Past, Present, and Future," *Journal of Thought* 16, no. 3 (1981): 62.

1.1 Research Interest

The research interest that motivated this study stems from the following consideration: Faced with an unknown trajectory of technological change and an uncertain future, political leaders, and educational policy makers in the 1970s and 1980s had few other options than to draw upon on compelling collective imaginaries of how they anticipated, hoped for, and desired the computerized future. In the interdisciplinary field of Science and Technology Studies (STS), the concept of "Sociotechnical Imaginaries" was coined in 2009 by Sheila Jasanoff and Sang-Hyun Kim to describe what particular communities or societies envision as both achievable and desirable in terms of the intertwined dynamics of technological innovation and social change in the future.[22]

In the introduction to their edited volume *Dreamscapes of Modernity*, Jasanoff and Kim define sociotechnical imaginaries as "as collectively held, institutionally stabilized, and publicly performed visions of desirable futures, animated by shared understandings of forms of social life and social order attainable through, and supportive of, advances in science and technology."[23] The authors point out that sociotechnical imaginaries can develop from the visions formed by small groups of people or even individuals, and subsequently "gain traction through blatant exercises of power or sustained acts of coalition building. Only when the originator's 'vanguard' vision comes to be communally adopted, however, does it rise to the status of an imaginary."[24] This applies, for example, to setting a common agenda within a political party — whether through a democratic process of consensus-building or by imposing a particular vision of the party leadership on its members. The latter would apply to the SED's authoritarian style of leadership.

Sociotechnical Imaginaries are thus a suitable analytical concept for investigating the collectively shared hopes, desires, and expectations about the future co-evolution of computer technology and the socialist society among the party members and the leadership of the SED, and, by extension, the powerful visions that guided the decisions and actions of the SED's educational policymakers in response to technological change.

The aim of this study is to investigate and understand how educational policies and curricular changes in response to the advent of computer technology are

---

[22] Sheila Jasanoff and Sang-Hyun Kim, "Containing the Atom: Sociotechnical Imaginaries and Nuclear Power in the United States and South Korea," *Minerva* 47, no. 2 (2009): 119–46.
[23] Sheila Jasanoff, "Future Imperfect: Science, Technology, and the Imaginations of Modernity," in *Dreamscapes of Modernity: Sociotechnical Imaginaries and the Fabrication of Power*, ed. Sheila Jasanoff and Sang-Hyun Kim (Chicago/London: University of Chicago Press, 2015), 4.
[24] Jasanoff, "Future Imperfect," 4.

intertwined with particular political agendas and ideologies regarding a computerized future, cast in the form of specific sociotechnical imaginaries. This perspective acknowledges that the integration of new technology into education is not a neutral process, but crucially shaped by broader social, cultural, and political contexts. By exploring these dynamics, this book intends to contribute to a deeper understanding of the complex relationships between education, technology, politics, and ideology. My research objective comprises of two elements: The reconstruction of an ideologically charged sociotechnical imaginary on the one hand, and an analysis of the educational policy response to technological change on the other. The first of two research questions is thus aimed at the sociotechnical of the GDR's political leadership, which dominated during the introduction of computer technology in the GDR:

1) How was the development and use of new computer technology envisioned by the SED, and what effects on the future socialist society and economy of the GDR were desired and anticipated?

In this sense, the aim of my research is to identify a possible future that has been normatively asserted as desirable within the framework of a specific imaginary in an institutionalized and powerful way. My second research question, then, asks about the concrete changes in education policy and curricula, such as the introduction of new subjects or the integration of new information technology into teaching and learning:

2) What was the SED's education policy response to the advent of computer technology, and in what way was it shaped by the dominant sociotechnical imaginary of this new technology?

In this regard, the focus lies on the interactions between sociotechnical imaginaries and education policymaking: What was the role assigned to education in bringing about the desired sociotechnical future the SED imagined for the GDR? And in turn, in what way was the education system expected to help propagate, stabilize, and reinforce the SED leadership's sociotechnical imaginary of computer technology among the people of the GDR?

My research aims to explore the entanglement between sociotechnical imaginaries, educational policymaking, and curricular development from a historical perspective, focusing on the case of the GDR. There are several reasons why the GDR makes for an appropriate and interesting case study with this research interest in mind. For one, the GDR was governed by an authoritarian regime with the power to translate its sociotechnical imaginaries into policy without the "interference" of democratic processes of consensus building. In contrast to democratic political systems, in which a multitude of stakeholders and interest groups can put forward and garner public support for their different sociotechnical imaginaries, the development of alternative imaginaries was suppressed by the SED's authoritarian regime. While individuals and small groups may have certainly had their own visions of a computerized future in the GDR, these imaginaries could not achieve the status of a sociotechnical imaginary as defined by Jasanoff (2015), that is, as a "collectively held, institutionally stabilized and publicly performed vision of a desirable future,"[25] without the backing of the government and the wider public. Unlike the SED, individuals and smaller opposition groups had neither the political power nor the democratic means to turn their visions and ideas into concrete policies – or they were simply silenced by the oppressive regime.[26] The case of the GDR, with one dominant sociotechnical imaginary proffered and promoted by its authoritarian leadership, thus allows for a more direct analysis of how a specific sociotechnical imaginary was translated into concrete policy, rather than necessarily having to shift focus to the power struggles and discursive practices of a multitude of groups competing for the primacy of their own sociotechnical imaginaries. But even within the SED apparatus there were power struggles and hidden rivalries, which, however, were not allowed to come out in the open.[27] Therefore, the sociotechnical imaginary of computerization in the GDR, as it was presented to the people of the GDR and enacted through policy changes in education, had to be brought forward as a unitary consensus, backed by an ostensibly unified party.

The GDR leadership's capacity to enforce and realize its sociotechnical imaginary of a "socialist" computerization in the GDR, however, was constrained by economic and technical feasibility, as well as the will and capacity of the people to support the cause of the leadership. While economic and technical constraints were more difficult to change through political intervention, a more promising

---

25 Jasanoff, "Future Imperfect," 4.

26 There certainly was popular discontent, opposition, and political activism in the GDR, expressed through acts of industrial sabotage, strikes, defacing of party symbols and depictions of SED politicians, as well as more subtle acts of protest and insubordination (Mary Fulbrook, "Popular Discontent and Political Activism in the GDR," *Contemporary European History* 2, no. 3 (1992): 269). However, as Mary Fulbrook notes, "for the greater part of the GDR's history [. . .] élites were able to exert effective control over the rumblings of discontent below, which remained for the most part easily isolated and uncoordinated" (Fulbrook, "Popular Discontent," 282).

27 On party discipline within the SED, see, for example, Thomas Ammer, "Die Machthierarchie der SED," in *Materialien der Enquete-Kommission "Aufarbeitung von Geschichte und Folgen der SED-Diktatur in Deutschland"* (12. Wahlperiode des Deutschen Bundestages), vol. II, 2 (1995), 813–15.

approach was to turn to education to mobilize public support and to develop the necessary technical skills among the working population. The GDR thus makes for an interesting case to investigate the nexus of sociotechnical imaginaries, changes in educational policy, and curricular development. Microelectronics and computer technology were high on the SED's political agenda since the late 1970s, and education was assigned a paramount importance in realizing the party's social, political, and economic vision for the computerization of the GDR.

Furthermore, the unique combination of the GDR's geographical, political, and economic situation in the 1970s and 1980s make it a particularly fascinating case to explore its educational response to technological change in the context of the Cold War. For one, the GDR was a small country on a coherent territory with a relatively homogenous population in terms of its ethnicity and language. This allowed the SED leadership to roll out changes in education policy and curricula much faster and comprehensively than, for example, in the Soviet Union, where the equipment of schools with computers and teaching materials for computer education was mostly concentrated on larger metropoles such as Moscow or Novosibirsk.[28] Wedged between the capitalist countries of Western Europe and the Socialist Bloc in the East, the GDR was neither geographically, nor politically isolated. It had political and economic ties to the Soviet Union and other socialist countries in Eastern Europe, but also to the capitalist West – most notably to West Germany.[29] Although the GDR was economically, politically, and militarily dependent on the Soviet Union, it did not have to follow Moscow's orders unconditionally. Especially under Honecker's rule during the 1980s, it gained more room for manoeuvre and opened the possibility of rapprochement with West Germany.[30] Finally, the GDR was technologically more advanced than most other countries in the Socialist Bloc, perhaps with the exception of the Soviet Union. It was a highly industrialized country where the challenge of the emergence and development of new computer technology and its impact on the education system was felt immediately and with full force, comparable to the situation in other in-

---

[28] Richard Judy and Jane M. Lommel, "The New Soviet Computer Literacy Campaign," *Educational Communication and Technology* 34, no. 2 (1986): 108–23; Iveta Kestere and Katrina Purina-Bieza, "Computers in the Classrooms of an Authoritarian Country: The Case of Soviet Latvia (1980s–1991)," in *How Computers Entered the Classroom, 1960–2000: Historical Perspectives*, ed. Carmen Flury and Michael Geiss (Berlin/Boston: De Gruyter Oldenbourg, 2023), 75–98.

[29] On GDR foreign policy and its relations with the Soviet Union and West Germany in the 1970s and 1980s, see Hermann Wentker, *Außenpolitik in engen Grenzen: Die DDR im internationalen System 1949–1989. Veröffentlichungen zur SBZ-/DDR-Forschung im Institut für Zeitgeschichte* (München: Oldenbourg Wissenschaftsverlag, 2007), in particular Part B, Chapters II and III.

[30] Wentker, *Außenpolitik*, 477–500.

dustrialized capitalist countries in the West.³¹ For this reason, the findings of this study on the case of the GDR will also be highly relevant for other individual country case studies, as well as comparative research on the educational policy consequences that were drawn in various socialist and capitalist countries from the rapid development of computer technology in the second half of the 20th century.

1.2 State of Research

Historical research on the development of computer technology has highlighted the key role of electronic data processing and computers within broader socioeconomic transformation processes. In particular, it has focused on the analysis of its effect as a pacemaker of change in workplaces and in the organization of the economy and social relations, shaped by political and societal actors, institutions, and discursive formations.³² The history of digital technology is a highly interdisciplinary field of research, as it is not just a history of technical developments of the computer. It is media and social history, encompassing cultural-historical approaches as well as elements of the history of economics and informatics.³³

For the most part, the histories of computers in education and computer education have been studied by historians of technology rather than historians of education. Most publications focus on Western countries with capitalist economies and democratic systems, most prominently the USA³⁴ and the UK.³⁵ However, over the

---

31 Wolfgang Hörner, "Informationstechnische Bildung," in *Vergleich von Bildung und Erziehung in der Bundesrepublik Deutschland und in der Deutschen Demokratischen Republik*, ed. Oskar Anweiler et al. (Köln: Verlag Wissenschaft und Politik, 1990), 620.
32 Martin Schmitt, Julia Erdogan, Thomas Kasper and Janine Funke, "Digitalgeschichte Deutschlands: Ein Forschungsbericht," *Technikgeschichte* 83, no. 1 (2016): 33.
33 Schmitt et al., "Digitalgeschichte Deutschlands," 34.
34 Larry Cuban, *Oversold and Underused: Computers in the Classroom* (Cambridge, MA: Harvard University Press 2001); Joy Lisi Rankin, *A People's History of Computing in the United States* (Cambridge, MA: Harvard University Press, 2018); Katie Day Good, *Bring the World to the Child. Technologies of Global Citizenship in American Education* (Cambridge, MA: MIT Press, 2020); Victoria Cain, *Schools and Screens. A Watchful History* (Cambridge, MA: MIT Press, 2021).
35 Neil Selwyn, "Learning to Love the Micro: The Discursive Construction of 'Educational' Computing in the UK, 1979–89," *British Journal of Sociology of Education* 23, no. 3 (2002): 427–43; Tilly Blyth, *The Legacy of the BBC Micro: Effecting Change in the UK's Cultures of Computing* (London: Nesta, 2012); Tilly Blyth, "Computing for the Masses? Constructing a British Culture of Computing in the Home," in *Reflections on the History of Computing*, ed. Arthur Tatnall (Berlin/Heidelberg: Springer, 2012), 231–42; Tom Lean, "Mediating the Microcomputer: The Educational Character of the 1980s British Popular Computing Boom," *Public Understanding of Science* 22, no. 5 (2012):

past decade, a growing number of historical case studies have also been published on the introduction of computers into the education systems of socialist countries. Noteworthy contributions on this subject have been published in edited volumes, for example on the cases of Hungary,[36] Latvia[37] and Poland.[38] In the recently published book by Victor Petrov on computing in Bulgaria,[39] chapter six covers state efforts to introduce computer education in universities, vocational training, schools and to establish after-school youth clubs and computer hobby magazines. The case of the Soviet Union was studied by Gregory Afinogenov, who focuses on Andrei Ershov, a prominent Soviet computer scientist and vocal advocate of universal computer literacy in the Soviet Union and abroad.[40] In the mid-1980s, Ershov's appeals to the Ministry of Education and the Politburo led to the introduction of school informatics in the USSR. Afinogenov describes how Ershov sought to persuade the socialist leadership of the USSR to introduce a far-reaching computer literacy program, arguing that such an initiative was in line with the goals of perestroika. However, the computer literacy campaign he imagined was also meant to help realize his own utopian vision of "democratizing the information structure of society".[41] Ksenia Tatarchenko tells the story of the Soviet computer literacy campaign through Ershov's correspondence, highlighting how the state failed to provide classrooms with computers in time to teach them 'algorithmic thinking', essentially turning the large-scale initiative into a dry run in many places, where programming had to be taught using blackboards and chalk rather than new computer technology.[42] Zbigniew Stachniak looks beyond the large-scale state programs for computer literacy, focusing instead on everyday cultural practices within the computer hobby scene in the Soviet

---

546–58; Neil Selwyn, "Making the Most of the 'Micro': Revisiting the Social Shaping of Microcomputing in UK Schools," *Oxford Review of Education* 40, no. 2 (2014): 170–88; Tom Lean, *Electronic Dreams*).

36 Lajos Somogyvári, Máté Szabó, and Gábor Képes, "How Computers Entered the Classroom in Hungary: A Long Journey from the Late 1950s into the 1980s," in *How Computers Entered the Classroom, 1960–2000: Historical Perspectives*, ed. Carmen Flury and Michael Geiss (Berlin/Boston: De Gruyter Oldenbourg, 2023), 39–74.

37 Iveta Kestere and Katrina Elizabete Purina-Bieza, "Computers in the Classrooms of an Authoritarian Country."

38 Sysło, Maciej M. "The First 25 Years of Computers in Education in Poland: 1965 – 1990." In *Reflections on the History of Computers in Education*, ed. Arthur Tatnall and Bill Davey (Berlin/Heidelberg: Springer, 2014), 266–90.

39 Victor Petrov, *Balkan Cyberia* (Cambridge, MA: MIT Press, 2023).

40 Gregory Afinogenov, "Andrei Ershov and the Soviet Information Age," *Kritika: Explorations in Russian and Eurasian History* 14, no. 3 (2013): 561–84.

41 Afinogenov, "Andrei Ershov," 579.

42 Ksenia Tatarchenko, "Why We Should Remember the Soviet Information Age," in *Vertical atlas*, ed. Leonardo Dellanoce et al. (Arnhem: ArtEZ Press, 2022), 264–68.

Union.[43] In a similar vein, the book *Gaming the Iron Curtain* by Jaroslav Švelch tells the fascinating history of amateur programmers and the emerging computer game culture in Czechoslovakia.[44] In both countries, the socialist regime responded to the grassroots movement of computer enthusiasts with benign indifference or sometimes even selective support for the creation of computer clubs and magazines to distribute instructions for building microcomputers or to swap homebrew games and programs.

With regard to the history of computer technology and computing in the GDR, five broad strands of research can be distinguished: (1) The economic and political history of the microelectronics and computer industry, (2) imaginaries of the future and sociotechnical visions of technology-led progress of the GDR's socialist authoritarian regime (3) the social history of the computerization of workplaces, (4) the cultural history of computing, and (5) the history of computer education and the formation of the academic discipline of informatics.

The bulk of historical research on computer technology in the GDR is concerned with the political and economic context, in which the country's microelectronics and computer industry developed. Among them are historical accounts by contemporary witnesses who were themselves involved in the development and establishment of computer technology in the GDR, and who subsequently explored their first-hand experiences from a historical perspective and shared their insights.

Erich Sobeslavsky and Nikolaus Joachim Lehmann recount the SED's technology policy from 1957 to 1968 as part of both the economic and political history of the GDR, describing in detail the political decisions of the SED leadership in building up an electronic data processing industry in the GDR.[45] The first part of the book tells the history of how in the 1960s, the GDR leadership attempted to reduce the technological gap between the GDR and the western industrialized countries in the field of computer technology through the targeted and substantial deployment of financial and human resources. In 1964, the GDR's Politburo and Council of Ministers adopted a "data processing program," which was a comprehensive political and economic program to build up the domestic electronic data processing sector and to promote the use of this technology in state planning and eco-

---

43 Zbigniew Stachniak, "Red Clones: The Soviet Computer Hobby Movement of the 1980s," *IEEE Annals of the History of Computing* 37, no. 1 (2015): 12–23.
44 Jaroslav Švelch, *Gaming the Iron Curtain: How Teenagers and Amateurs in Communist Czechoslovakia Claimed the Medium of Computer Games* (Cambridge, MA: MIT Press, 2018).
45 Erich Sobeslavsky and Nikolaus Lehmann, *Zur Geschichte von Rechentechnik und Datenverarbeitung in der DDR 1946–1968* (Dresden: Hannah-Arendt-Institut für Totalitarismusforschung, 1996).

nomic production.⁴⁶ The program also included the training of skilled personnel in the production and use of data processing machines, and for the management and supervision of the implementation of electronic data processing projects on site. However, Sobeslavsky and Lehmann argue that the inefficiency of the socialist planning system and the shortcomings of the centrally planned economy only further widened the technological gap in comparison with the capitalist West.⁴⁷

It is worth noting that within the historiography of the GDR microelectronics industry, the search for the reasons for its "decline" or diagnoses and analyses of the "failures" of the SED's technology policy seem to be of particular interest to authors. Gerhard Barkleit also studied the SED's economic and technology policy in the development of a domestic microelectronics industry, examining in detail the policy decisions and arguments of the SED's Politburo in the late 1970s that led to the establishment of a largely self-sufficient microelectronics industry in the GRD.⁴⁸ Attending more closely to the global economic and political context, Barkleit points out the key importance of international relations in analyzing and explaining the trajectory of the GDR's technology policy and development. A key driver behind the Politburo's decision in 1977 to develop a domestic microelectronics industry was the concern to ensure the exportability of the GDR's mechanical and electrical engineering products, which was of great importance to the competitiveness of the national economy in the global market. However, creating a largely self-sufficient domestic microelectronics industry, which attempted to produce almost the entire range of microelectronic components and devices, proved to be a risky and ultimately unwise decision for a relatively small country like the GDR. But as Barkleit argues, this path was taken in response to the specific predicament the GDR found itself at the time: On the one hand, the capitalist West had imposed an embargo on high-technology goods against the Socialist Bloc,⁴⁹ and on the other, the socialist

---

46 Sobeslavsky and Lehmann, *Geschichte von Rechentechnik und Datenverarbeitung*, 66–70.
47 Sobeslavsky and Lehmann, *Geschichte von Rechentechnik und Datenverarbeitung*, 7.
48 Barkleit, *Mikroelektronik in der DDR*.
49 The Coordinating Committee (COCOM) was a multilateral regime initiated by the USA to control trade with the Socialist East. The COCOM embargo was intended to prevent the transfer of military equipment. In addition to strategic goods, this also included the export of so-called "dual-use goods," which could serve both military and civilian purposes. Most of the computer technology and microelectronic components fell into the latter category (Simon Donig, "Die DDR-Computertechnik und das COCOM-Embargo 1958–1973. Technologietransfer und institutioneller Wandel im Spannungsverhältnis zwischen Sicherheit und Modernisierung," in *Informatik in der DDR – eine Bilanz*, ed. Friedrich Naumann and Gabriele Schade (Bonn: Gesellschaft für Informatik, 2006), 254; Christopher Leslie, "From CoCom to Dot-Com: Technological Determinisms in Computing Blockades, 1949 to 1994," in *Histories of Computing in Eastern Europe*, ed. Christopher Leslie and Martin Schmitt (Cham: Springer, 2019), 196–225.

partner states in the East had either proved unreliable or refused to cooperate meaningfully in the development of computer technology. The Soviet Union, in particular, was unwilling to cooperate in the development of high technology because it was part of the Soviet military-industrial complex to which the GDR was denied access.[50] Olaf Klenke makes a similar argument by foregrounding the challenge of globalization to frame the GDR's economic and political struggles with its domestic computer industry.[51] Klenke argues that the failure of building an innovative and successful computer industry cannot be explained with system-immanent factors alone. Instead, a crucial factor in explaining these struggles was the GDR's exclusion from the processes of internationalization: free trade of high-technology goods, the international division of labor, and cooperation in technological development with the economies of the Western world.

Raymond Stokes examined East German technology policy from 1945 to 1990, with a particular focus on high-tech industries such as semiconductors and electronics, petrochemicals, and automation.[52] The author departs from the question of why the socialist system and economy of the GDR failed and, in particular, what GDR-specific factors might explain its persistent technological lag behind the FRG, given that both countries had inherited a rich German technological tradition and a vibrant culture of innovation. In the GDR, earlier German efforts in field of microelectronics and computing[53] were resuscitated under Honecker in the 1970s with massive material, financial, and human resources, accompanied by a strategy to copy Western developments. Despite significant technological breakthroughs and successful imitation of innovations from the West, the author paints a picture of steady economic and political decline. He explains this decline as much in terms of the political failures of the SED and the inherent contradictions of the socialist system, as in geopolitics and wider changes in the global economy. Stokes concludes that the SED's research and development strategy of closely shadowing the Western leaders in computer technology and focusing on cutting-edge technology was "utterly inappropriate for its own demographic, economic, and technological conditions" and contributed to the country's eventual

---

50 Barkleit, *Mikroelektronik in der DDR*, 32.
51 Olaf Klenke, *Ist die DDR an der Globalisierung gescheitert? Autarke Wirtschaftspolitik versus internationale Weltwirtschaft – Das Beispiel Mikroelektronik* (Frankfurt am Main: Peter Lang, 2001).
52 Raymond G. Stokes, *Constructing Socialism: Technology and Change in East Germany 1945–1990* (Baltimore/London: Johns Hopkins University Press, 2000).
53 Here, Stokes refers to the work of German engineer and pioneering computer scientist Konrad Zuse, who build the world's first programmable computer Z3 in 1941 (Stokes, *Constructing Socialism*, 179–80).

collapse by starving other sectors of the economy of ideas and funding.[54] James Cortada adds to this research by highlighting how the SED leadership rhetorically elevated microelectronics and computer technology to a field of decisive importance for the social and economic prosperity of the GDR.[55] In no other socialist country, he claims, had a single technology been so closely tied to national intentions and public policy by its political leaders.[56] In Cortada's account, it was not least the failure of the political leadership to deliver on its promises and effectively leverage new computer technology, that contributed to the discrediting of the SED party leadership in the late 1980s and its eventual downfall.[57]

The SED's strategy for the development to combine domestic innovation with both legal and illegal technology transfer to develop its computer industry is the focus of publications by Simon Donig (2010), Rüdiger Bergien (2019) and Martin Schmitt (2019). Donig points out that, despite the SED's assertions of computer technology's ideological "neutrality," the appropriation of Western technology by the socialist states was only possible because fundamental ideas such as efficiency, cost-benefit relations or rationality were shared on both sides of the Iron Curtain.[58] An example of this was the strong influence of IBM[59] system concepts on computer development in the GDR. IBM's "System/360" also became the technical model for the COMECON-wide development of mainframe computers in the "Ryad" series (German: ESER), at the behest of the Soviet Union and the GDR. The clandestine import of semiconductors for copying, industrial espionage, and the smuggling of Western technology for reverse engineering were the domain of the "Stasi" (Ministerium für Staatssicherheit, eng.: Ministry for State Security), which played an important role in the SED's strategy for advancing the domestic microelectronics and computer industry. However, Western computer manufacturers, too, profited from and played their part in circumventing the embargo, as Bergien shows: In 1970, the West German company Siemens sold four mainframe computers to the Central Institute for Information and Documentation (ger.: Zentralinstitut für Information

---

54 Stokes, *Constructing Socialism*, 194.
55 James W. Cortada, "Information Technologies in the German Democratic Republic (GDR), 1949–1989," *IEEE Annals if the History of Computing* 34, no. 2 (2012).
56 Cortada, "Information Technologies," 44.
57 Cortada, "Information Technologies," 45.
58 Simon Donig, "Appropriating American Technology in the 1960s: Cold War Politics and the GDR Computer Industry," *IEEE Annals of the History of Computing* 32, no. 2 (2010): 32–45.
59 International Business Machines Corporation (IBM) is an American multinational technology corporation which came to dominate the global computer market in the 1960s and 1970s. For the history of IBM as an influential player in the global information technology market, see James W. Cortada, *IBM: The Rise and Fall and Reinvention of a Global Icon* (Cambridge, MA: MIT Press, 2019).

und Dokumentation, ZIID), a branch of the GDR's Ministry for State Security, for 20 million Marks.[60] Already in the second half of the 1960s, the Stasi had acquired computers of the French manufacturer Bull and sent staff and unofficial collaborators to Paris and Munich to attend computer courses at Bull and Siemens and pass on their knowledge to other employees of the GDR's secret service.

More recently, Martin Schmitt provided a detailed technological and economic history of the digitalization in the financial sector from the late 1950s to the early 1990s, comparing the developments in the two Germanies with regard to the use of computers in savings banks.[61] In contrast to other scholarship on the history of digital computing which foregrounds its origins in mathematics and engineering,[62] Schmitt draws attention to the fact, that computing also had roots in the spheres of clerical work and office machinery such as tabulating and accounting machines. Furthermore, while the research discussed previously focused predominantly on the economic and technology politics of the SED leadership, Schmitt also accounts for the concrete effects of technological change in the financial sector; namely the changes in how financial services were provided, how payment transactions were administered, or how the new technology altered the work of bank clerks and the experience of clients. The first to encounter computer technology in the 1960s in this branch of the economy was the back-office staff without customer contact. Bank clerks were introduced to computer terminals in the 1970s, but it was not until the 1980s that customers came into contact with computer technology in the banking sector, with the introduction of ATMs and bank statement printers. Because the SED leadership prioritized industry over non-manufacturing sectors in the computerization of the GDR's economy, the impact of new information technology unfolded slower and with a delay in comparison with the financial sector in West Germany. Despite striking structural parallels between technological change in the GDR and West Germany, Schmitt identifies a number of systemic and cultural characteristics of the pathway to a "socialist digital era."[63] While West Germany took a more decentralized approach, with individual banks going to great lengths to migrate their own control systems into the digital age, the GDR took the opposite approach, seeking to create a centralized collective system for all. In West

---

[60] Rüdiger Bergien, "Programmieren mit dem Klassenfeind: Die Stasi, Siemens und der Transfer von EDV-Wissen im Kalten Krieg," *Vierteljahrshefte für Zeitgeschichte* 67, no. 1 (2019): 1–2.

[61] Martin Schmitt, *Die Digitalisierung der Kreditwirtschaft: Computereinsatz in den Sparkassen der Bundesrepublik und der DDR, 1957–1991* (Göttingen: Wallstein, 2021).

[62] See, for example, Christine Pieper, *Hochschulinformatik in der Bundesrepublik und der DDR bis 1989/90* (Stuttgart: Franz Steiner Verlag, 2009); Martin Campbell-Kelly, "Origin of Computing," *Scientific American* 301, no. 3 (2009): 62–69.

[63] Schmitt, *Digitalisierung der Kreditwirtschaft*, 608.

Germany, optical readers were deployed to make receipts computer readable. In banks in the GDR, on the other hand, data was entered manually by mostly female, low-paid workers. Schmitt explains this difference in the approach to data input not simply with the GDR's "technological backwardness," but also refers to systemic differences in wage costs, workers' interests and the centralized political system of the GDR.[64] Regarding the question of a distinctly "socialist" computerization, Schmitt argues that the historical context of socialism in the GDR did not produce "socialist" computer technology, but rather structured the spatial expansion of new information technology and the practice of its use in the form of software. The author thus calls for historians of computerization to consider the programming code of software as a source worthy of historical investigation. Schmitt goes as far as to argue that "traces of socialism" can indeed be found in the software of the GDR's savings banks: The software of savings banks contained the banks' logics and modes of operation and were thus documenting their workflows and routines.[65] In addition, the program code was designed to run smoothly on the specific hardware available at the time when the software was created. In the form of "legacy systems," old routines dating back to the days of punch-card machines were migrated and converted to new systems – constituting a digital form of path dependency within software systems.

A second strand of historical research on the GDR, which is closely related to the literature on economic and technological policy, is concerned with the discourses and visions of automation[66] as well as future expectations[67] propagated by the SED. As these discourses, visions, and expectations are closely related to

---

64 Schmitt, *Digitalisierung der Kreditwirtschaft*, 473.
65 Schmitt, *Digitalisierung der Kreditwirtschaft*, 609.
66 Uwe Fraunholz, "'Revolutionäres Ringen für den gesellschaftlichen Fortschritt': Automatisierungsvisionen in der DDR," in *Technology Fiction: Technische Visionen und Utopien in der Hochmoderne*, ed. Uwe Fraunholz and Anke Woschech (Bielefeld: transcript, 2012), 195–219; Martin Schwarz, "'Zauberschlüssel zu einem Zukunftsparadies der Menschheit': Automatisierungsdiskurse der 1950er- und 1960er-Jahre im deutsch-deutschen Vergleich" (PhD diss., TU Dresden, 2015), https://nbn-resolving.org/urn:nbn:de:bsz:14-qucosa-191580 (accessed May 31, 2024).
67 Martin Sabrow, "Zukunftspathos als Legitimationsressource: Zu Charakter und Wandel des Fortschrittsparadigmas in der DDR," in *Aufbruch in die Zukunft. Die 1960er Jahre zwischen Planungseuphorie und kulturellem Wandel. DDR, CSSR und Bundesrepublik Deutschland im Vergleich*, ed. Heinz-Gerhard Haupt and Jörg Requate (Weilerswist: Velbrück Wissenschaft, 2004), 165–84; Annette Schuhmann, "Die Zukunft der Arbeit in der Übergangsgesellschaft: Überlegungen zur Produktion von (Zukunfts-)Erwartungen in der DDR," in *Vergangene Zukünfte von Arbeit: Aussichten, Ängste und Aneignungen im 20. Jahrhundert*, ed. Franziska Rehlinghaus and Ulf Teichmann (Bonn: Dietz, 2019), 157–78.

"sociotechnical futures" in the broader sense, this strand of research is particularly relevant for informing my own research.

Martin Sabrow coined the term of the "socialist expectation society" (ger.: Sozialistische Erwartungsgesellschaft) to describe the SED's constant invocation of imaginaries, hopes, and promises for the future. Imaginaries of the future, of social progress and thus the permanent production of high expectations – albeit often detached from reality – were a central part of the SED's political propaganda to legitimize its rule.[68] Sabrow analyzes the SED's rhetoric of social progress and pathos of the future from the 1950s until the end of the 1980s. While the concept of progress in the GDR retained its lexical meaning unchanged, its significance in the political culture of rule in the GDR changed from a utopian concept of political awakening with totalitarian features in the 1950s to a scientifically calculable planned variable in the 1960s. In the 1970s, the concept of progress increasingly became an externally determined challenge, and finally lost the rest of its legitimizing binding force along with its cultural development function over the course of the 1980s.[69]

Uwe Fraunholz emphasizes that visions of the future, which extrapolated current trends and developments, were essential to stabilize the socialist system and the SED's claim to power. At the same time, these visions served to postpone the realization of an imagined society to a more or less distant future.[70] Fraunholz shows how the shared understanding of the transformative power of computerization contributed to a new sense of planning optimism in both East and West Germany. In particular, the use of computer and automation technology in production served as the basis for this new-found optimism. The ability to automate various aspects of production processes was seen as a significant advance and a potential catalyst for social progress. Socialist visions of automation were fuelled by the belief that automated production would eventually lead to the triumph of communism over capitalism. Accordingly, computer and automation technology were heralded as the basis for the desired future of communism. Simultaneously, computerization and automation would inevitably expose the contradictions of the capitalist system, which was portrayed as inherently incapable of uniformly controlling entire sectors of the economy and thus of fully realizing the potential of these new technologies.[71]

Martin Schwarz found that there were attempts in both the GDR and West Germany to de-couple rationalization from the connotation of economic crisis

---

68 Sabrow, "Zukunftspathos als Legitimationsressource," 169.
69 Sabrow "Zukunftspathos als Legitimationsressource," 184.
70 Fraunholz, "Revolutionäres Ringen," 196.
71 Fraunholz, "Revolutionäres Ringen," 199.

and unemployment. In the GDR, the SED's rhetoric juxtaposed a positive "socialist" rationalization to a negative "capitalist" rationalization.[72] Automation and microelectronics technology was promoted by the political leadership as a "revolutionary" means to accelerate social and economic progress. However, Schwarz also shows how this rhetoric increasingly lost in appeal and persuasiveness over the course of the 1980s, since there was less and less hope for the future to be drawn from the technical present.[73]

Annette Schuhmann studied the imaginaries of the future, expectations, and "technocratic illusions" that underpinned three of the SED's political-economic reform projects: the chemistry program in 1958, the economic reform of the new economic system in 1963, and the microelectronics program in 1977.[74] The development of microelectronics was part of a broader program of economic rationalization in response to a shortage in labor and low work productivity.[75] The technocratic utopia of automating industrial production and computerizing administrative work was deemed visionary and unavoidable both in the East and West already in the 1950s and 1960s. While in the West, public discourse in the 1970s turned to questions of whether the social impact of microelectronics was a blessing or a curse, in the GDR, the SED's rhetoric of social progress and positive expectations remained unchanged. However, the tangible impact of technological innovation in the GDR was limited, and yet, the practical implementation of new technologies in workplaces led to social conflict and resistance on part of the workforce.[76] Schuhmann concludes that the doubts about the attainability of the ambitious goals in the field of microelectronics did not, however, lead to the kind of pessimism about technology in the GDR that could be observed at least in part in West Germany. Rather, the economic and technical stagnation increasingly counteracted the official political rhetoric of progress.[77]

A third strand of research on the computerization in the GDR is concerned with the social history of how it impacted workplaces and workers. Peter Hübner tells the history of work, workers, and technology in the GDR between 1971 and 1989, albeit not focusing primarily on perspective of workers themselves, but rather the SED's labor policies in response to challenges such as the use of computer technology to rationalize economic production and the social effects of computerization

---

72 Schwarz, "Zauberschlüssel zu einem Zukunftsparadies."
73 Schwarz, "Zauberschlüssel zu einem Zukunftsparadies," 274.
74 Schuhmann, "Zukunft der Arbeit."
75 Schuhmann, "Zukunft der Arbeit," 174.
76 Schuhmann, "Zukunft der Arbeit," 175.
77 Schuhmann, "Zukunft der Arbeit," 176.

on the workforce.[78] Throughout the 1970s and 1980s, the SED leadership launched campaigns to increase productivity and quality of production, in particular in key sectors such as the microelectronics and computer industry. The extensive use of automation technology and industrial robots was seen as a solution to alleviate the chronic shortage of labor. In addition, a system of performance- and skill-based wage differentiation was introduced. Financial incentives were introduced to encourage workers to accept shift work to fully exploit the potential of scarce and expensive computer technology by running the machines around the clock. However, the initiatives largely failed to achieve their goals. Shortcomings in the provision of materials and interruptions in the production process due to technical failures led to unplanned breaks, thus hampering rationalization efforts.[79] Olaf Klenke describes how such downtimes, in turn, led to periodic peaks of stress for workers trying to compensate for lost time. Klenke's study focuses on the social impact of technology-driven rationalization on workers in the GDR's microelectronics industry. The analysis also examines the reactions and forms of resistance on the part of the socialist workforce. In doing so, he highlights the blatant contradictions between information technology modernization rhetoric of the party leadership and the work reality of people working in microelectronic production.

The research of Dolores Augustine, in contrast, is concerned with the "educational elite" in the domain of computer technology, namely the software experts and computer scientists of the GDR.[80] Augustine's work is based on oral history and aims to reconstruct the gendered socialization and professional biographies as well as the self-image and work habitus of GDR engineers and mathematicians in the field of software development. Augustine's findings expose the myth of socialist meritocracy, as party membership was often a prerequisite for career advancement, and elites with the SED's approval to travel had access to foreign technology and international experiences. Women were at a disadvantage in entering the field of computer science and technology compared to men, who were more likely to have had access to electronics kits in their childhood and youth and were often less involved in domestic work and child rearing. Despite being at

---

[78] Peter Hübner, *Arbeit, Arbeiter und Technik in der DDR 1971 bis 1989. Zwischen Fordismus und digitaler Revolution* (Bonn: J.H.W. Dietz Nachf., 2014).
[79] Hübner, *Arbeit, Arbeiter und Technik*, 184–85.
[80] Dolores L. Augustine, "Berufliches Selbstbild. Arbeitshabitus und Mentalitätsstrukturen von Software-Experten in der DDR," in *Eliten im Sozialismus. Beiträge zur Sozialgeschichte der DDR*, ed. Peter Hübner (Köln: Böhlau, 1999), 405–433; Dolores L. Augustine, "The Socialist 'Silicon Ceiling': East German Women in Computer Science" in *International Symposium on Technology and Society – Women and Technology: Historical, Societal, and Professional Perspectives, New Brunswick NJ, USA, 29–31 July 1999. Proceedings*, 347–355.

a disadvantage in terms of a successful career in the field of new information technology, women made up the majority of students in software-related majors in the GDR.[81] Augustine attributes this largely to "pronatalist state policies" aimed at encouraging women to "combine motherhood and work outside the home," and the SED's efforts to bring more women into computer science and engineering professions.[82] However, the author also points out the importance of gendered socialization: While women accounted for more than half of the students enrolled in the software-focused major "Information Processing," they were outnumbered by male students in the more hardware-oriented major "Information Technology." Software development appealed to women with an aptitude for mathematics and science, who did not want to work directly with machines. Moreover, computing was portrayed as a field that would thrive in the future, and a career in this area promised an opportunity to advance economically and socially. However, the persistence of traditional gender roles in reproduction meant that women still bore the main responsibility for children and the household also in the socialist GDR. Thus, despite being highly qualified, female computer experts often found themselves in non-supervisory and non-managerial jobs more readily compatible with family responsibilities.[83]

A fourth strand of research focuses on the cultural history of computing in the GDR. Jens Schröder analyzes the computer games scene in the GDR in the context the country's economic history and the SED's efforts to develop a domestic microelectronics industry.[84] Schröder shows that the state-supported development of a computer game scene in the GDR was politically motivated and intended to play a system-stabilizing role: According to Honecker's maxim of the "unity of economic and social policy," the microelectronics industry was not only to develop high technology for research and industry, but also to contribute to providing the GDR population with modern electronic consumer goods that were already widely available in the capitalist West.[85] From the mid-1980s, the availability of domestically produced microcomputers, which could be connected to a joystick, formed the basis for the development of a computer games scene in the GDR. However, as computers were mostly distributed to educational institutions and companies, they hardly ever found their way into private households. Thus, the computer game scene took place mainly in state-run youth and leisure cen-

---
81 Augustine, "Socialist 'Silicon Ceiling'," 348.
82 Augustine, "Socialist 'Silicon Ceiling'," 348.
83 Augustine, "'Socialist Silicon Ceiling'," 352–53.
84 Jens Schröder, *Auferstanden aus Platinen: Die Kulturgeschichte der Computer- und Videospiele unter besonderer Berücksichtigung der ehemaligen DDR* (Stuttgart: ibidem, 2010).
85 Schröder, *Auferstanden aus Platinen*, 65.

ters. Even before the advent of digital networks such as the Internet, social networks of computer-enthusiastic young people who met in computer clubs or formed pen friendships to exchange computer games, made it possible to build up extensive game collections, which often included self-programmed games or pirated copies of Western computer games.[86] As Schröder points out, the SED's cultural policy in the area of computer gaming also served another important purpose, namely the cultivation of a new generation of skilled computer users for the computerization of the GDR's economy. Educational policymakers and educators hoped that computer games would spark young people's enthusiasm for computers and informatics and that, in the medium term, they would be able to recruit from a pool of qualified young people for the domestic computer industry and the computerized working world of the future.[87]

More recently, Julia Gül Erdogan published her comprehensive study comparing the social phenomenon of "hackers" or computer hobby cultures in the GDR and West Germany.[88] Her work highlights the role of the computer as a means of self-empowerment and curious exploration, by recounting the history of the self-determined appropriation of computer technology by young, and predominantly male computer enthusiasts in the late 1970s and 1980s. Erdogan (2021) shows that "hacking" and the creative use of computer technology was a community-building practice – not only in the GDR, where the lack of private computers made working together on a single device even more common than in West Germany. The hackers in West Germany built decentralized networks for both local and cross-border communication, using bulletin board systems. While the West German scene was dominated by the Chaos Computer Club which also pursued a subversive political agenda, computer hobbyists in the GDR were supported by the regime through socialist youth organizations, state-sponsored computer clubs and the state-run technical press – but were also to some extent controlled and monitored. The inadequate telecommunication infrastructure and state surveillance hampered the establishment and use of private digital networks.[89] In-person meet-ups in computer clubs thus became a decisive element of community-building among young computer enthusiasts in the GDR: Computer hobbyists could meet like-minded peers, share experiences, and learn from each other. Especially in the GDR, where comparatively few you people had their own computer at home, it was only the computer clubs that gave them the opportunity to

---

86 Schröder, *Auferstanden aus Platinen*, 93–94.
87 Schröder, *Auferstanden aus Platinen*, 101–102.
88 Julia G. Erdogan, *Avantgarde der Computernutzung: Hackerkulturen der Bundesrepublik und der DDR* (Göttingen: Wallstein, 2021).
89 Erdogan, *Avantgarde der Computernutzung*, 97–98.

use computers for their very own purposes and even get to know Western computer models. Computer clubs thus allowed young people to use computers in an alternative way to the pedagogically guided computing in schools and other educational institutions and offered a different way of learning about computer technology through creative exploration.[90]

Finally, a fifth strand of research is concerned with the history of the establishment and development of informatics as an academic discipline, and computer education in the educational system of the GDR.

Already in 1990, in the midst of the political upheavals of the German reunification, an edited volume was published under the authority of the West German Federal Ministry of Intra-German Relations, providing a comprehensive comparison of the educational systems of West Germany and the GDR.[91] One chapter in the volume focuses on the particular issue of education in information technology and was contributed by Wolfgang Hörner, a West German educational scientist who later became professor for comparative education at the University of Leipzig. Hörner provides a brief overview of the changes in education policy and curricula in response to the development and introduction of new information technology in the two Germanies.[92] Hörner notes that despite ideologically motivated demarcation attempts, the conceptualization of the phenomenon of new information technology and its consequences in the East and West are strikingly similar. This seems particularly evident in the case of the two Germanies, which, due to their similar levels of industrialization, were confronted with analogous problems at about the same time – not least in view of the conclusions that were drawn for the development of the education system.[93] Hörner highlights that the GDR, in contrast to West Germany, chose the direct way of a mandatory "informatics instruction for all." This general computer education was strongly focused on computer use, despite the policymakers' assertion that not user skills, but problem-solving skills were foregrounded.[94]

Between 2006 and 2010, three edited volumes on the history of informatics in the GDR were published, in the form of the proceedings of the German Informatics Society's symposia on informatics in the GDR, which provide a rich collection of historical accounts on the broad field of research, development, teaching and

---

90 Erdogan, *Avantgarde der Computernutzung*, 265.
91 Oskar Anweiler et al., eds., *Vergleich von Bildung und Erziehung in der Bundesrepublik Deutschland und in der Deutschen Demokratischen Republik* (Köln: Verlag Wissenschaft und Politik, 1990).
92 Hörner, "Informationstechnische Bildung," 620–37.
93 Hörner, "Informationstechnische Bildung," 620.
94 Hörner, "Informationstechnische Bildung," 635.

learning with and about new computer technology.⁹⁵ To a considerable extent, this literature represents a history written by contemporary witnesses, consisting of first-hand accounts of their own experiences as pupils, teachers, mathematicians, engineers and computer scientists in the GDR. The contributions cover diverse topics from hard- and software development in the GDR to the use of computers in science and industry, as well as the institutionalization of informatics in higher education. Overall, the topic of computer education in schools is covered in only two chapters in one of the edited volumes. In one, Werner Schmidt recounts his own experience as a student in the early 1960s, long before schools were equipped with computers or taught programming.⁹⁶ As a participant in the first mathematics Olympiad in the GDR in 1961/62, he had the opportunity to see the "OPREMA" computer in Jena.⁹⁷ The other chapter pertaining to education in the same volume is by Immo Kerner, who had been involved in the development of a computer science curriculum for grade 11 in general education and co-authored several textbooks, teaching and learning materials both for pupils and for the training of mathematics teachers in informatics and programming. His contribution consists of a recollection of his own memories of the introduction of calculators in schools and, soon after, the introduction of computer education in general education, the equipping of schools with computer technology and the preparation of teachers to teach computer science.⁹⁸

Christine Pieper compared the establishment and further development of informatics and computer science in higher education in West Germany and the GDR until 1989/90.⁹⁹ The detailed description of the development of higher education structures and study programs in the field of higher education informatics, which stays close to historical sources and institutional records, provided a valuable source of information for my own research. Pieper traces the origins of informatics

---

95 Friedrich Naumann and Gabriele Schade, eds., *Informatik in der DDR – eine Bilanz. Tagungsband zu den Symposien 7. bis 9. Oktober 2004 in Chemnitz und 11. bis 12. Mai 2006 in Erfurt* (Bonn: Gesellschaft für Informatik, 2006); Birgit Demuth, ed., *Informatik in der DDR: Grundlagen und Anwendungen. Drittes Symposium 'Informatik in der DDR,' 15. und 16. Mai 2008 in Dresden, Deutschland* (Bonn: Gesellschaft für Informatik, 2008); Wolfgang Coy and Peter Schirmbacher, eds., *Informatik in der DDR – Tagung Berlin 2010. Tagungsband zum 4. Symposium 'Informatik' in der DDR am 16. und 17. September 2010 in Berlin* (Berlin: Humboldt-Universität zu Berlin, 2010).
96 Werner H. Schmidt, "Informatikausbildung an Schulen der DDR," in *Informatik in der DDR – eine Bilanz*, ed. Friedrich Naumann and Gabriele Schade (Bonn: Gesellschaft für Informatik, 2006), 378–81.
97 Schmidt, "Informatikausbildung an Schulen," 378.
98 Immo O. Kerner, "Vorbereitung des Informatik-Unterrichts an den Schulen der DDR," in Friedrich Naumann and Gabriele Schade (Bonn: Gesellschaft für Informatik, 2006), 422–31.
99 Pieper, *Hochschulinformatik*.

as an academic discipline back to its beginnings in the fields of mathematics and engineering. As computer technology and informatics gained both in scientific and political relevance, the different departments argued over the shape of the disciplinary core and institutional affiliation of the emerging field.[100] Eventually, the field of informatics and computer science developed into and was institutionalized as an academic discipline in its own right. Pieper describes in detail the individual development of information technology research and education at the six universities of the GDR in Leipzig, Halle-Wittenberg, Berlin, Jena, Rostock, and Dresden, as well as at the more vocationally oriented technical colleges and engineering schools. The author focuses on the multiple processes of negotiation between the Ministry of Higher Education, universities, and industry in defining and shaping the academic research field and subject of informatics.

While changes in curricula and educational policy in response to the development of new information technologies in higher education in the GDR have been the focus of historical research, very little has so far been written about the introduction of computer technology in other areas of the GDR's system of education. Moreover, a comprehensive view of the changes in education in response to the development of modern computer technology in the GDR is still lacking, one that brings these threads together, fills in the missing pieces, and places them in a broader historical and technological context. I address this research gap in two ways: 1) by situating my historical investigation of curricular changes and educational policy responses within the economic and political context GDR at the time; and 2) by linking these changes and responses to the sociotechnical imaginaries of computer technology which were propagated by the SED to pave the way for an anticipated computerized future under socialism. Thus, this research not only contributes to enriching the history of education in the GDR regarding modern computer technology, which was a key area of economic and technological policy in the 1970s and 1980s. It also sheds light on the role of sociotechnical imaginaries and future expectations in educational policymaking, particularly with regard to technology education. Against the background of ongoing debates about the role of computers, digital tools, and artificial intelligence in education today, it is worthwhile to explore how present expectations, hopes, and fears of our digital future impact education policies and curricula – or in other words, how sociotechnical imaginaries shape education. From a methodological perspective, this study seeks to demonstrate that the concept of sociotechnical imaginaries can be fruitfully applied to the historical study of education and to the analysis of educational policymaking.

---

100 Pieper, *Hochschulinformatik*, 9.

1.3 Methodology and Sources

This study draws upon the theoretical concept of sociotechnical imaginaries, which originated in the field of Science and Technology Studies and was first suggested by Jasanoff and Kim in 2009,[101] and subsequently elaborated by Jasanoff in 2015.[102] The concept of sociotechnical imaginaries focuses on "shared understandings of forms of social life and social order attainable through, and supportive of, advances in science and technology."[103] Based on the fundamental conviction that technological development is decisively shaped by social processes – driven by collectively shared imaginaries – the concept of sociotechnical imaginaries also brings into focus how people imagine and hope for technologically induced social change. The notion of sociotechnical imaginaries, thus, integrates both the idea of socially driven technological change and the idea of technologically driven social change. At its core is the idea of a "mutual shaping," which emphasizes that the social and technological must be seen as interdependent and intertwined, as mutually constituting each other.[104]

In the following, I will focus in detail on four components which Jasanoff includes in the definition of sociotechnical imaginaries, namely the premise that they are "collectively held," "institutionally stabilized," "publicly performed," and that they constitute "visions of desirable futures."[105]

The definition of sociotechnical imaginaries as "collectively held" raises the question as to what kind of collective stands behind any identified imaginary. In an earlier version of their definition, Jasanoff and Kim associated sociotechnical imaginaries with nation-states, as "collectively imagined forms of social life and social order reflected in the design and fulfillment of *nation-specific* and/or technological projects."[106] However, the unit of the nation state, taken as a more or less homogenous "collective" is inherently problematic. Carol Bacchi, for example, argues that "any suggestion that 'sociotechnical imaginaries' find their origins in national political cultures, therefore, raises a question about the way in which

---

[101] Sheila Jasanoff and Sang-Hyun Kim, "Containing the Atom: Sociotechnical Imaginaries and Nuclear Power in the United States and South Korea," *Minerva* 47, no. 2 (2009): 119–46.
[102] Sheila Jasanoff, "Future Imperfect: Science, Technology, and the Imaginations of Modernity," in *Dreamscapes of Modernity: Sociotechnical Imaginaries and the Fabrication of Power*, ed. Sheila Jasanoff and Sang-Hyun Kim (Chicago/London: University of Chicago Press, 2015), 1–33.
[103] Jasanoff, "Future Imperfect," 4.
[104] Donald MacKenzie and Judy Wajcman, "Introductory essay: the social shaping of technology," in *The social shaping of technology*, ed. Donald MacKenzie and Judy Wajcman (Buckingham/Philadelphia: Open University Press, 1999), 23.
[105] Jasanoff, "Future Imperfect," 4.
[106] Jasanoff and Kim, "Containing the Atom," 120 (emphasis added).

the term may suppress recognition of contestation of the assumed norms in any selected imaginary."[107] When Jasanoff further elaborated their definition in 2015, she stated that actors other than nation-states can also create sociotechnical imaginaries, namely corporations, social movements, or professional societies.[108] However, Bacchi criticizes that while this perspective acknowledges that multiple imaginaries may coexist in a society, it tends to neglect the examination of any tensions that may arise within these imaginaries.[109]

In response to these concerns, my work focuses not on an assumed *national* sociotechnical imaginary of the GDR, but rather on the imaginary produced and propagated by the country's authoritarian political leadership, the SED. This conception leaves room for the possible presence of alternative sociotechnical imaginaries shared by other, non-state groups and social collectives in the GDR, which, however, did not have the political means and power to assert their visions and imaginings among the broader public. In addition, I conceptualize the SED's imaginary not as a homogenous, monolithic imaginary, but as a multifaceted one, which opens up the possibility to investigate how different aspects of the same imaginary led to both symbiotic effects as well as tensions among these aspects of the same imaginary. This particular point will be further elaborated in the second chapter, which discusses the various aspects of the SED's dominant sociotechnical imaginary of computer technology.

The notion of sociotechnical imaginaries as "institutionally stabilized" refers to the idea that sociotechnical imaginaries are often associated with and backed by powerful institutions such as state authorities, expert bodies, foundations, or companies. Institutions often have the means and resources at their disposal to be effective creators and disseminators of sociotechnical imaginaries, by generating the necessary public attention and support for their own visions of a desirable future. Political institutions, media organizations, professional communities of practice, or schools can help to reinforce and legitimize a particular sociotechnical imaginary by incorporating it into their policies, practices, and programs. Educational institutions can serve as powerful agents in propagating a specific imaginary, by shaping the way how society thinks about and relates to a technology.[110] In the GDR, all of these institutions were under state control and subject to

---

107 Carol Bacchi, "Sociotechnical Imaginaries and WPR: Exploring connections," posted November 29, 2022, https://carolbacchi.com/2022/11/29/sociotechnical-imaginaries-and-wpr-exploring-connections (accessed May 31, 2024).
108 Jasanoff, "Future Imperfect," 4.
109 Bacchi, "Sociotechnical Imaginaries."
110 Annika B. Rensfeldt and Catarina Player-Koro, "'Back to the future': Socio-technical imaginaries in 50 years of school digitalization curriculum reforms," *Seminar.net* 16, no. 2 (2020).

the SED's authoritarian regime. The SED leadership thus had virtually unbridled power to determine how computer technology was to be developed, used, and regulated – and by extension, the power to influence the people's beliefs, values, and assumptions related to computer technology. Furthermore, the SED stabilized its sociotechnical imaginary of computer technology by aligning it with socialist ideology, and the party's political values and priorities, which had been embedded within and reinforced through the GDR's institutions long before the advent of new computer technology.[111] Sociotechnical imaginaries develop within the historical context of already established social frameworks of technology and society, and draw on existing sociotechnical ideas, problematizations, and values.[112] They are shaped by and embedded in the local cultural, social, political, and institutional context of their time, which in turn, can serve to stabilize a sociotechnical imaginary.

Closely related to the institutional stabilization is the public performance of sociotechnical imaginaries. On the one hand, this can refer to the affirmation of the imaginary itself, its public reuse and endorsement by authority figures, or quite literally, public performances and spectacles to showcase a country's innovativeness and technological prowess (see Figure 1).

On the other hand, public performance of sociotechnical imaginaries refers to how they are deployed and shape the present. Once imaginaries become widely accepted and integrated into a society, they have the power to shape research and technology development, as well as the allocation of public and private resources. Sociotechnical imaginaries thus promote collective narratives and facilitate a shared understanding of social reality. But most importantly, they have significant normative consequences. While they may initially serve as descriptions of potentially attainable futures, they can quickly turn into prescriptions for the desired futures that ought to be attained.[113]

In this sense, sociotechnical imaginaries constitute a "crucial reservoir of power and action [that] lodges in the hearts and minds of human agents and in-

---

[111] See, for example, Martin Sabrow's account of how the paradigm of "progress" was established by the SED as a guiding concept in political discourse in the 1950s and retained this role until the end of the GDR but was interwoven with changing political priorities and conceptions of the future and thus assumed different functions in political discourse over time (Sabrow, "Zukunftspathos als Legitimationsressource").
[112] Uli Meyer, "The Institutionalization of an Envisioned Future. Sensemaking and Field Formation in the Case of 'Industrie 4.0' in Germany," in *Socio-Technical Futures Shaping the Present*, ed. Andreas Lösch et al. (Wiesbaden: Springer VS, 2019), 116.
[113] Paul Waller, *Nightmare of the Imaginaries. A Critique of Socio-technical Imaginaries Commonly Applied to Governance* (May 17, 2020), http://dx.doi.org/10.2139/ssrn.3605494 (accessed May 31, 2024).

Figure 1: The delegation of the district of Erfurt during the big parade through the city center of Berlin to celebrate the 750th anniversary of Berlin on 4th July 1987. The photo shows the presentation of desktop computers from Sömmerda (the Robotron PC 1715) as part of the celebrations.
Source: BArch, Image no. 183-1987-0704-077 / Photographer: Thomas Uhlemann. Licenced under CC-BY-SA 3.0, Wikimedia Commons.

stitutions."[114] Lösch et al. point out that "any imagination of new technologies is necessarily coupled with the shaping or even inventing of novel kinds of users and usages, of images of altered social relations or even societal regimes."[115] These imagined usages and users as well as the anticipated social effects of technology use provide an orientational framework, for example, to guide the design of new curricula, by outlining who would use computer technology in this collectively imagined future, in what ways and for what purposes, and in what kind of social context. In this sense, sociotechnical imaginaries are enacted and performed by the developers and users of a technology, but also by educational policy makers, the designers of new curricula, and the educators who teach their students about technology. Crucially, they provide orientation and a framework

---

114 Jasanoff, "Future Imperfect," 17.
115 Andreas Lösch et al., eds., *Socio-Technical Futures Shaping the Present* (Wiesbaden: Springer VS, 2019), 2.

for the coordination and motivation of action.[116] Uli Meyer argues that the credibility of an anticipated sociotechnical future is not the decisive factor in determining whether it is successful.[117] Instead, sociotechnical imaginaries provide useful scripts to guide political decision-making and action, as well as professional practice, especially under conditions of uncertainty. Thus, despite often telling a general and rather vague story, an imaginary can still effectively achieve the goal of reducing uncertainty: "It does so by creating an impression of urgency, which is connected to existing debates, and at the same time providing a solution for this problem."[118]

Finally, the specification of sociotechnical imaginaries as "visions of *desirable* futures" highlights the normative character of imaginaries, which are typically grounded in "positive visions of social progress."[119] It also suggests that any sociotechnical imaginary is an expression of the community's preferences for what is considered "desirable" and worthy of pursuit in the first place, that is, the community's shared understanding of which goals technology is meant to help accomplish, and how it should or should not be used.[120] The concept of the sociotechnical imaginary thus emphasizes that "political agendas are driven by culturally-specific belief and value systems that produce different forms of technopolitical order. [. . .] In this sense, the notion of sociotechnical imaginaries invites a close reading of the various expectations and concerns, the diverse norms, mores, and ideologies that guide and inform the articulation of national policies."[121]

The concept of the sociotechnical imaginary is a suitable tool to identify and discuss computer technology-related normative visions of the future in the SED's official policy discourse, as well as in the discourse among pedagogues and computer scientists, and in the public media. Focusing on the case of the GDR and its authoritarian political leadership, the notion of sociotechnical imaginaries serves to highlight the role of the SED in how a "global" technology such as the computer was discursively and imaginatively appropriated and adapted to specific sociopolitical system of the GDR – its established social order and relations, practices, values, and beliefs. Investigating the political leaderships' imagination and political propaganda of a "desirable future" involving computer technology seems par-

---

116 Meyer, "The Institutionalization of an Envisioned Future, 116.
117 Meyer, "The Institutionalization of an Envisioned Future."
118 Meyer, "The Institutionalization of an Envisioned Future, 116.
119 Jasanoff, "Future Imperfect," 4.
120 Jasanoff, "Future Imperfect," 4.
121 Gernot Rieder, "Tracing Big Data Imaginaries through Public Policy: The Case of the European Commission," in *The Politics and Policies of Big Data: Big Data, Big Brother?*, ed. Ann R. Sætnan, Ingrid Schneider, and Nicola Green (New York/London: Routledge, 2018), 3.

ticularly pertinent in the case of the GDR, as previous research on its history has pointed out that the constant production of anticipated futures, filled with promises of social progress, was a prominent feature of the SED's political discourse throughout its rule.[122] The concept of sociotechnical imaginaries, thus, seems well suited to investigate how a specific vision of a sociotechnical future mediated and shaped policies to introduce computers into education in the GDR.

The SED's sociotechnical imaginary of new information technology was performed, enacted, and manifested in technical artefacts, public and specialist discourse, curricular materials, and policymaking. Methodically, my approach combines discourse analysis with the analysis of historical policy documents and curricula. The historical sources analyzed include GDR newspaper articles, books and magazines on computing, as well as policy documents, and school and university curricula regarding the introduction of computer technology into the GDR's economy and society, and more specifically, of informatics and computing into education.

With regard to archival sources, this study draws upon government policy papers, drafts, political decisions, minutes of negotiations, and correspondence of state ministries and central institutes for education,[123] the Council of Ministers, as well as party documents, decisions and programs from the SED's Central Committee and its Politburo, which are all kept at the German Federal Archive in Lichterfelde, Berlin. With regard to the introduction of computer technology and curricular development in schools, the study draws upon archival materials stored at the Research Library for the History of Education at DPF (ger.: Bibliothek für Bildungsgeschichtliche Forschung des DIPF, BBF) in Berlin, in particular the documents, reports, working papers, draft curricula, and correspondence of the GDR's Academy of Pedagogical Sciences (APW), as well as the BBF's collection of GDR school curricula and the collection of Pedagogical Lectures.[124]

---

122 Sabrow "Zukunftspathos als Legitimationsressource;" Schuhmann, "Zukunft der Arbeit."
123 Ministry for General Education (ger.: Ministerium für Volksbildung), State Secretary for Vocational Education and Training (ger.: Staatssekretariat für Berufsbildung), Ministry for Higher Education (ger.: Ministerium für Hoch- und Fachschulwesen), Central Institute for Vocational Education and Training (ger.: Zentralinstitut für Berufsausbildung), Central Institute for the Continuing Education of Teachers and Educators (ger.: Zentralinstitut für die Weiterbildung der Lehrer und Erzieher).
124 On the fascinating historical source genre of pedagogical lectures from the GDR, see Josefine Wähler and Maria-Annabel Hanke, "'Pacemakers Report': GDR Pedagogical Innovators and the Collection of Pädagogische Lesungen, 1952–1989," *Paedagogica Historica* 58, no. 1 (2022): 66–83; Katja Koch and Felix Linström, "Die Pädagogischen Lesungen im Rahmen der DDR-Lehrer*innenweiterbildung – Eine Systematisierung," in *DDR-Unterricht im Spiegel der*

The monthly journal *Einheit – Zeitschrift für Theorie und Praxis des Wissenschaftlichen Sozialismus* (eng.: Unity – Journal for Theory and Practice of Scientific Socialism) was included in the analysis of historical discourse of computer technology in the GDR, as it is a rich and informative source for the study of sociotechnical imaginaries shared among high party functionaries of the SED. The articles in the journal cover predominantly political, theoretical, and philosophical deliberations on state socialism and the socialist society, as well as economic matters. Of particular interest for this study were articles pertaining to new information technologies and computing and its expected or desired effects on the socialist society and economy of the GDR. As the official mouthpiece of the SED, the journal was aimed at party functionaries at all levels of the party hierarchy, to support them in their practical day-to-day work and to convey the party leadership's position on virtually every topic of interest to the SED.[125] The authors were leading SED functionaries or carefully selected senior employees of central party institutes and academies. Between 1967 and 1989, the journal *Einheit* was the responsibility of the SED Central Committee Secretary Kurt Hager. Hager headed the ideological commission of the SED's Politburo and was secretary for science, education, and culture, playing a decisive role in shaping GDR education and cultural policy in the 1970s and 1980s.[126]

While the above-mentioned sources document policymaking processes and curricular development, they also contain useful information regarding the discourse among SED functionaries, computer technology experts, and pedagogues regarding computer technology and its integration into education. They reveal little about the public discourse on computer technology in the GDR, or put differently, how the SED broadly communicated their sociotechnical imaginary to the people. For this reason, I have also included two SED-controlled daily newspapers in my analysis, *Neues Deutschland* and *Berliner Zeitung*, as well as popular science and technology magazines, books published by the GDR's state press on comput-

---

Pädagogischen Lesungen, ed. Katja Koch and Tilman von Brand (Baltmannsweiler: Schneider Verlag Hohengehren, 2021), 35–54.

125 Siegfried Lokatis, "Falsche Fragen an das Orakel? Die Einheit der SED," in *Zwischen 'Mosaik' und 'Einheit': Zeitschriften in der DDR*, ed. Simone Barck, Martina Langermann, and Siegfried Lokatis (Berlin: Christoph Links Verlag, 1999), 592.

126 Bernd-Rainer Barth and Helmut Müller-Enbergs, "Hager, (Leonhard) Kurt," in *Wer war wer in der DDR? Ein Lexikon ostdeutscher Biographien*, ed. Helmut Müller-Enbergs et al. (Berlin: Christoph Links Verlag, 2010), https://www.bundesstiftung-aufarbeitung.de/de/recherche/kataloge-datenbanken/biographische-datenbanken (accessed May 31, 2024).

ing, and children's and youth magazines such as *technikus*,[127] *Jugend + Technik*,[128] and *Frösi*.[129] Children's and youth magazines, played an important role in educating the younger generation about new computer technology beyond the confines of formal schooling, thus extending the reach of the SED's sociotechnical imaginary into the sphere of children's leisure activities and interests.

For the purposes of this study, all quotations from historical sources in German have been faithfully translated into English. In cases where the English translation does not adequately capture the German meaning, or in the case of specific GDR idioms, the original German expression has been added to the approximate English translation.

An important limitation of this study, both due to its methodological framework and the selected source materials, concerns the future expectations, imaginations, and visions of the less powerful, less organized, and less articulate members of society. While this study focuses on the sociotechnical imaginary put forward and propagated by party leaders and high ranking party members in their role as SED functionaries, the available historical sources in the form of written documents do neither contain information on possible alternative imaginaries shared among certain communities or groups of people in the GDR, nor do they provide indication of whether the people of the GDR, including SED party members, believed in or endorsed the dominant sociotechnical imaginary offi-

---

127 *technikus* was a monthly popular science magazine aimed at school-aged children in grades 5 to 10, covering the topics of mathematics, natural sciences, and technology. The magazine was aimed at stimulating an interest in science and technology among children and support polytechnic school education (Klaus Pecher, "Kinderzeitschriften in der DDR – erziehungsstaatliche Okkupation der Kindheit," in *Kinderzeitschriften in der DDR*, ed. Christoph Lüth and Klaus Pecher (Bad Heilbronn: Verlag Julius Klinkhardt, 2007), 36.

128 The magazine *Jugend und Technik* (eng.: Youth and Technology) was published by "Junge Welt," which was associated with the SED's youth organization 'Free German Youth' (FDJ) (Pecher, "Kinderzeitschriften in der DDR," 182). The magazine was dedicated to the topics of technology and popular science, and contained reports on new scientific and technological advances, and the use of new technologies in the GDR and other socialist countries. The magazine also contained practical suggestions and instructions for smaller scientific and technical hobby projects. Starting in 1987, the magazine published a series of instructions for the self-built "Ju-Te" home computer.

129 *Frösi* was a monthly children's magazine for members of the socialist "Ernst Thälmann Pioneer Organization," a subdivision of the "Free German Youth" for ages 6 to 13. It contained educational comics and short picture stories, easy to understand articles on nature and technology, as well as suggestions and templates for handicrafts (Dieter Wilkendorf, "Was bleibt? Die Kinderzeitschrift 'Fröhlich sein und singen – Frösi' im Erinnern und Nachdenken ihres Chefredakteurs," in *Kinderzeitschriften in der DDR*, ed. Christoph Lüth and Klaus Pecher (Bad Heilbrunn: Verlag Julius Klinkhardt, 2007), 139–51).

cially propagated by the SED. What this study can do, however, is to explore the powerful sociotechnical imaginary of the SED's upper echelons. Regardless of whether this imaginary was able to persuade and win over the hearts and minds of lower-level SED officials, party members and the rest of the population, it was effective in shaping economic and educational policy in the GDR, and both triggered and guided the development of new curricula and courses of study.

Furthermore, this study does not aim to capture the practice of teachers and learners in relation to computer technology. It focuses on how prescriptive policies and documents, such as curricula and teaching materials, envision educational practice, rather than whether and how a particular sociotechnical imaginary of computer technology was realized in teaching and learning.

The study begins with a chapter, that unravels and explores the SED leadership's sociotechnical imaginary of computer technology, by drawing on articles published by party functionaries in the journal *Einheit* and the state-controlled newspapers *Neues Deutschland* and *Berliner Zeitung*. Different aspects of the imaginary are identified, which have each been activated and operationalized to varying degrees in the different educational sectors within the framework of educational policy reforms and curriculum innovations. How this process of educational and curricular reforms played out in the different sectors of the GDR's educational system, as well as the role of the SED's dominant sociotechnical imaginary in these processes, is the subject of the following empirical chapters.

The third chapter follows the process of the institutionalization of computing and informatics in higher education. It traces the origins of the computer imaginary back to an earlier discourse on cybernetics and tells the story of how the socialist leadership of the GDR, after initial ideological reservations, appropriated cybernetic theory and, soon after, embraced computer technology. The emergence of computer technology in the GDR was accompanied by the creation of dedicated study courses in higher education. The new field of knowledge found fertile ground in various disciplines such as mathematics, engineering, and economics, which took on the task of training computer experts.

The fourth chapter focuses on developments in vocational education and training (VET) in response to the advent of computer technology. The development of a domestic computer industry and the SED's plans to computerize the GDR's economy required the education and training of computer users, developers, maintenance, and repair personnel. The party leadership recognized the need for a substantial transformation of the GDR's occupational system and the qualification structure of its workforce to accommodate for the SED's sociotechnical imaginary of socialist computerization. VET policymakers were thus faced with the challenge of undertaking a major curriculum reform project, despite the unknown trajectory of computer technology and its uncertain impact on jobs and skill requirements. The chapter also shows, how the SED's sociotechnical imaginary of socialist computerization in some instances conflicted with the reality in workplaces, threatening to spark discontent among workers and a loss of trust in the political leadership and its promises of a desirable computerized future.

The fifth chapter tells the story of how and why computers were introduced into general education in the second half of the 1980s. A new compulsory computer course was to be introduced for all schoolchildren, prompting a debate about the value, importance, and content of computer education in general education among educators, computer experts and education policymakers, but also about possible cooperation with other socialist countries in this matter. Similarly, the development of a dedicated national school computer caused dissatisfaction among educators and pedagogues because of its technological conservatism, which seemed at odds with the SED's promises and sociotechnical visions of progress.

The sixth chapter takes a closer look at how children and the youth learned about computers and experimented with their use outside of formal education. While the SED encouraged and supported extracurricular activities involving computer technology, it was wary of giving the younger generation too much leeway in how they appropriated this new technology and for what purposes they would use it. Thus, extracurricular computer use was to some extent controlled by the SED by largely restricting access to computers to educational institutions and youth centers, as well as by providing pedagogical (and ideological) guidance to computer-savvy children and young people. Books on computing and magazine articles were meant to provide young people with the necessary knowledge and suggestions for computer projects and programming activities that were in line with the values, purposes, and uses of computer technology as envisioned by the SED.

The final chapter summarizes the findings on the GDR's pedagogical and educational policy response to the advent of the computer and relates them to the SED's sociotechnical imaginary. It highlights the ways in which, on the one hand, the SED's imaginary of computer technology was propagated and reinforced through the educational system and thus aimed at shaping people's understanding of computers and the purposes for which they should (and should not) be used. On the other hand, the dominant sociotechnical imaginary guided the work of educational policymakers and pedagogues in drafting new curricula, skill profiles and descriptions for new jobs involving computers in an envisioned computerized future in the socialist GDR.

2 A Sociotechnical Imaginary for "Socialist" Computerization

In 1988, the GDR computer expert Gerhard Saeltzer published a reference book for workers with the title *Kollege Personalcomputer* (eng.: My colleague, the personal computer).[130] The book promised an easy-to-understand introduction to the world of using personal computers. In the book, Saeltzer emphasized the historically unprecedented scale and speed with which computers were entering society and the economy. He painted a vision of the near future, in which hundreds of thousands, if not millions, of workers would be using computers in their workplaces and would need to be trained to do so. Computer education, Saeltzer argued, was becoming a decisive key to sustained and strong growth in productivity, quality, and efficiency of work, products, and services in the GDR. Importantly, he stressed that it was not the computer itself, but in fact the "education of its users," that was the key to unlocking the envisioned future of social and economic progress.[131]

Saeltzer's book reflects on a number of contemporary expectations and hopes for the future computerized society of the GDR, and the role assigned to education in bringing about this particular desired future, which were propagated at the time by the SED leadership. Crucially, the SED's sociotechnical imaginary contained a variety of political, ideological, economic, and social aspects. For the purposes of my research, I therefore conceptualize the sociotechnical imaginary of computer technology in the GDR as a single, albeit multi-faceted imaginary which became dominant during the 1970s and 1980s within the context of the authoritarian SED regime. Rather than speaking of a plurality of contending sociotechnical imaginaries, I argue that the concept of multiple aspects of a single, more or less coherent imaginary is a more accurate reflection of the political landscape in the GDR.

The SED leadership had good reasons not to allow for multiple imaginaries to compete in a free market of ideas, but rather to assert a single, comprehensive imaginary in line with its political, technological, and economic agenda. A multiplicity of imaginaries would have held the potential for conflict and dissent that could have threatened the party's claim to power. Through its monopoly of power, the SED was able to put forward a single sociotechnical imaginary, politically legitimized by the party leadership, rather than having to contend with a

---

[130] Gerhard Saeltzer, *Kollege Personalcomputer* (Leipzig: VEB Fachbuchverlag Leipzig, 1988), 219.
[131] Saeltzer, *Kollege Personalcomputer*, 219 (translated from German).

multiplicity of different, perhaps even conflicting, imaginaries that would potentially divide the population over the ideal realization of computer technology's potential. Promoting a single monolithic imaginary, however, might have been disadvantageous, as it could prove to be too rigid to be adapted over time and in response to specific challenges without losing popular support. A multifaceted sociotechnical imaginary, on the other hand, allowed the SED leadership to present a variety of interconnected aspects, which could be dynamically highlighted or promoted in response to specific target groups, current political issues or to call upon fundamental sociocultural beliefs and political ideologies. For example, the vision of computer technology as a tool to reduce the burden on workers could be foregrounded to mobilize the workforce in support of a policy program to computerize workplaces, or the envisaged positive effects of computer technology on productivity and economic competitiveness could be emphasized in response to prevailing economic crises. However, these facets can form part of the same sociotechnical imaginary, promoting a common vision of a desirable future in which computer technology is used for the advancement of the common good of the people living and working in the GDR.

While sociotechnical imaginaries can be invoked to garner support for a political program or intervention, they also contain certain "blind spots," whether intentional or not.[132] This is particularly evident in the case of the GDR, where the authoritarian regime of the SED did not allow for the emergence of alternative sociotechnical imaginaries that might have provided a corrective, pointing out certain aspects that were omitted by the imaginary promoted by the party leadership. For example, the SED's dominant sociotechnical imaginary of computer technology completely concealed any negative social effects of computer technology under socialism, as well as contradictions or conflicts that might arise between different aspects of the imaginary. For example, the imagined future impact and use of computers as a means of increasing economic productivity and competitiveness might run counter to the expected relief of workers from stress and heavy workloads as illustrated in the chapter on vocational training. A multifaceted imaginary can serve to deliberately obfuscate the potential conflicts between these different aspects, which could potentially threaten the harmonious larger picture of the ideal sociotechnical future envisioned. Certain individual facets of the imaginary could be invoked by party leaders depending on the situation

---

[132] Renan G. Leonel Da Silva and Larry Au, "The Blind Spots of Sociotechnical Imaginaries: COVID-19 Scepticism in Brazil, the United Kingdom and the United States," *Science, Technology and Society* 27, no. 4 (2022): 611–29.

and target group at hand, without necessarily having to confront the contradictions and gaps in the larger vision of the future that they presented as desirable.

The SED used a variety of media and communication channels to propagate and consolidate its sociotechnical imaginary among the people of the GDR – not least through the means of the educational system. In what follows, I will unravel the sociotechnical imaginary of computer technology that was prevalent in popular science and technology books, political speeches, policy documents, and newspaper and magazine articles. At least six defining aspects of this politically dominant imaginary can be identified in the historical sources. They are often not present in the sources in isolation but are intertwined with one or more of the other aspects. The six aspects are: (1) The computer as a tool controlled by the working people, (2) Computer-aided governance, (3) Computer technology as an engine of economic growth, (4) Computers as a symbolic arena in the ideological battle between the systems, (5) Computer technology as a driver of social progress, and (6) The computer as a tool to further the development of the socialist personality.

2.1 Computers as a Tool Controlled by the Working People

The cornerstone of the SED's sociotechnical imaginary was the notion of a "computerization for the benefit of all people," which was considered a hallmark of the socialist vision of the development and use of computer technology. A first characteristic of the dominant sociotechnical imaginary propagated by the SED was thus the idea of computers as tools in the hands – and under the control of – the working people. Humans were regarded as the masters of production and as the masters of technology.[133] Workers were considered active shapers of the computerization of the world of work and society. It would not be some technocratic elite, but the broad masses of computer users themselves, who decisively influenced the economic and social effects of the development and use of this new technology.[134] The role of people in shaping technological change, and in determining the aspirations and values that underpin this process, has thus been consistently emphasized by the authorities: "Technology can only ever be the object of human, social interests. The most complicated and meaningful computers and

---

[133] Werner Gerth, "Jugend und wissenschaftlich-technische Revolution," *Einheit*, no. 6 (1987): 528.
[134] Kurt Koopmann and Karl-Heinz Thieme, "Moderne Technologien als hoher moralischer Anspruch," *Einheit*, no. 8 (1987): 708.

robots are, outside of this reference to human goals and purposes, simply a meaningless accumulation of pieces of metal and glass."[135]

This emphasis on human control and shaping of technological innovation and change – presented as an innate value of the socialist society – was contrasted with the alleged technological determinism prevalent in capitalist societies: "The theory, so much strained by bourgeois philosophers of technology today, about an alleged independence of technology from humans, the assertion that technology is following increasingly its own laws, which are moving further and further away from the goals and purposes desirable from a human and social point of view, is wrong in its entirety. It is a false reflection of the fact that capitalist society is not able to use the progress of technology unhindered and directly for the enrichment of human life."[136]

Despite the strong emphasis placed on human agency, technology was still seen as a key driver of social change in its own right. Technology was conceptualised by the SED leadership as a guiding force that in and of itself triggers societal change and progress – albeit mediated by the sociopolitical context. This kind of "soft" techno-determinism leaves space for human choice and intervention, in contrast to purely or "hard" techno-deterministic visions of computer technology.[137] According to the latter, society organizes itself in order to meet the needs of technology, the societal effects of which are beyond society's control. Instead, the political authorities of the SED emphasized the importance of social context and human agency. In reference to Marx, political authorities of the GDR pointed out the role of political, social, and organizational aspects intertwined with material-technological affordances. In this sense, the development and use of technology were not to be understood and treated as a purely technical process, but always in connection with the productive activity of human beings, and thus, as an interaction between technology, economy, and society.[138]

As a consequence of this conception, the participation and necessary inclusion of all people in the GDR in the "scientific-technological revolution" was considered to be of great importance. According to the authoritarian leadership, the tasks at hand could and should not be left solely to specialists, as the socialist soci-

---

135 Harry Nick, "Informationsverarbeitende Technik erschliesst neue Quellen des ökonomischen und sozialen Fortschritts," *Einheit*, no. 7 (1986): 617 (translated from German).
136 Nick, "Informationsverarbeitende Technik," 617 (translated from German).
137 Leo Marx and Merritt R. Smith, "Introduction," in *Does Technology Drive History? The Dilemma of Technological Determinism*, ed. Merritt R. Smith and Leo Marx (Cambridge, MA/London: MIT Press, 1994): ix–xv.
138 Claus Krömke, "Ökonomische Strategie weist Wege zur Steigerung der Arbeitsproduktivität," *Einheit*, no. 9 (1987): 784.

ety required and enabled the mass participation of the working people.[139] Education and training for the broader masses was perceived as an imperative in face of an envisioned future in which computer technology would gain such fundamental importance in the lives of the working people of the GDR, that dealing with computers "will be as much a part of daily life as reading and writing."[140] In order to make working people the masters of this new technology and to realize its ascribed potential, the GDR's education system was tasked with familiarizing the imagined future users with computers, as well as providing them with political and ideological guidance in their mental appropriation of computer technology and conscious engagement with its use.[141]

2.2 Computer-Aided Governance of the Economy and the State

The SED's sociotechnical imaginary of computers also covered the idea of efficient centralized state planning and economic management with the aid of computer technology and large-scale information processing. This aspect was particularly prominent in the early days of the development of computer technology and was part of the debates on cybernetic theory.[142] Accordingly, the economy, or the state were considered as complex organisms controlled and regulated by a plethora of information and communication processes. Computers, then, were perceived as powerful tools to capture, process, and manage the flow of information in governing the planned economy and the socialist society of the GDR. As such, computers were imagined as a remedy to the shortcomings of planned economies in the Socialist Bloc, driven by the idea that state planning and the socialist economy could operate far more efficiently, if planners were provided with large amounts of relevant information, and the technical means to process and make use of it.[143]

Visions of computer-aided governance of the socialist economy were also apparent in other countries, and historical attempts to make them a reality bear witness to this fact. In the 1960s, Soviet cyberneticist Viktor Glushkov proposed a

---

[139] Gerd Rossow, "Wissenschaftlich-technischer Fortschritt und Kulturniveau der Arbeiterklasse," *Einheit*, no. 8 (1987): 760.
[140] Karl Hartmann, "Schlüssel für kräftiges Wachstum zum Wohle des Volkes," *Einheit*, no. 7 (1986): 596 (translated from German).
[141] Horst Enders, "CAD-Technologie meistern – eine Herausforderung an die politisch-ideologische Arbeit," *Einheit*, no. 7 (1986): 605.
[142] See chapter 2 on computer technology in higher education.
[143] Eugen Loebl, "Computer socialism," *Studies in Soviet Thought* 11, no. 4 (1971): 295.

National Automated System for Computing and Information Processing (OGAS), which aimed to create a nationwide computer network for economic planning and management. OGAS sought to connect various sectors of the economy, allowing for centralized data collection, analysis, and decision-making. However, the project faced significant technical and bureaucratic challenges and was ultimately abandoned in 1970 after failing to secure extensive government support. Another notable example is the project Cybersyn in Allende's Chile.[144] Project Cybersyn was an early computer network developed between 1970 and 1973 under Chile's socialist president Salvador Allende. Led by the British cyberneticist Stafford Beer, a multidisciplinary team in Chile drafted cybernetic models of factories and created a computer network to transmit data between government and the factory floor. Cybersyn sought to enable real-time decision-making and enhance centralized control in the management of the socialist economy, by providing decision-makers with up-to-date information on production levels, resource allocation, and market conditions.[145]

Despite high hopes for making socialist planned economies and societal governance more effective, cybernetic theories of societal governance were by no means perceived as unproblematic by the authoritarian leadership of the GDR, as the opening of the chapter on higher education in this book will show. Ideas of a technocratic and potentially emancipatory way to govern the socialist society and economy threatened the established power structures and the administrative status quo. From an ideological viewpoint, computer technology as a tool of cybernetic control posed the danger of machines ruling over humans – or in other words, that computers, "the supposed tools of man,"[146] would become the masters of state planning. Against this dystopian vision, the SED's sociotechnical imaginary of computer technology as a tool of governance thus foregrounded the paramount role of political leaders and economic policymakers in setting the goals for the economic system. The party remained the unquestioned political authority, while computers were merely imagined to support their decision-making, and a powerful tool to control and regulate the planned economy more meticulously to reach the goals set by the SED leadership. Therefore, what the SED leadership envisioned in terms of computer-aided governance was no reorganization of society and political power structures, but merely an adjustment of its governing techni-

---

144 Eden Medina, *Cybernetic Revolutionaries: Technology and Politics in Allende's Chile* (Cambridge, MA: MIT Press).
145 Eden Medina, "Designing Freedom, Regulating a Nation: Socialist Cybernetics in Allende's Chile," *Journal of Latin American Studies* 38, no. 3 (2006): 571–606; Medina, *Cybernetic Revolutionaries*.
146 Loebl, "Computer socialism," 299.

ques aimed at stabilizing the status quo of autocratic party leadership. Against this background, computer-aided governance of society and the economy was seen as a way to further support the legitimacy of party decisions through the rationality of the computer, and to increase the efficacy of political interventions to reach the SED's goals. It was envisioned that computers could not only support decision-making, but that through simulation, it would become possible to foresee the consequences of decisions to an unprecedented extent. This, in turn, would allow for political decisions and actions based on pure rationality.[147]

In particular, computers were imagined as indispensable tools for the effective coordination and control within the GDR's system of "democratic centralism,"[148] and the hierarchical chain of command in governing the economy. Accordingly, computers could serve in planning and management in state-owned enterprises and government, as well as in the administrative-command system of reporting economic indicators to higher authorities and coordinating production according to the decisions and the national economic plans of the party leadership. In 1962, the State Planning Commission acquired a French "Bull Gamma 3" vacuum-tube computer for the purpose of economic-mathematical modelling of the economy, which was not possible with the domestically available punch card technology.[149] In 1965, GDR cybernetician Klaus Dieter Wüstneck attempted to develop a cybernetic model of the new economic system of planning and management of the national economy. In his work, he conceived of the entire economy of the GDR as a system that is in a constant state of adaptation and can be adapted through feedback loops of information.[150] Two years later, Wüstneck became a candidate for the SED Central Committee and was appointed head of the newly formed Commission for Cybernetics at the Ministry of Higher Education, with the task of establishing cybernetics in research and higher education.[151] The honor bestowed on Wüstneck can be understood as an expression of the party's acceptance of cybernetic theories for the control of the socialist economy and society – as long as they did not challenge the party's claim to power.

---

[147] Horst Denzer, "Kybernetische Planung und Politische Ordnungsform: Ein Aspekt der Kybernetikdiskussion in der DDR," *Zeitschrift Für Politik* 15, no. 1 (1968), 70–71.
[148] Mary Fulbrook, "Democratic Centralism and Regionalism in the GDR," in *German Federalism*, ed. Maiken Umbach (London: Palgrave Macmillan, 2002), 146–71.
[149] Martin Schmitt, "Socialist Life of a U.S. Army Computer in the GDR's Financial Sector: Import of Western Information Technology into Eastern Europe in the Early 1960s," in *Histories of Computing in Eastern Europe*, ed. Christopher Leslie and Martin Schmitt (Cham: Springer, 2019), 159.
[150] Wüstneck 1965.
[151] Jérôme Segal, "Kybernetik in der DDR: Begegnung mit der marxistischen Ideologie," *Dresdener Beiträge zur Geschichte der Technikwissenschaften*, no. 27 (2001): 62–63.

2.3 Computer Technology as an Engine of Economic Growth

The SED leadership and economic policymakers in the GDR were convinced of the broad applicability of microelectronics, robotics, and computer technology in the economy, and their potential to serve as powerful means to modernize the material basis of production.[152] Increasing economic productivity was a key promise of the SED's sociotechnical imaginary of computer technology use in workplaces. As computer technology increasingly permeated the production and reproduction process, it was envisaged that it would enable new levels of automation, a sharp increase in labor productivity, a reduction in costs, and an improvement in product quality.[153] In the late 1970s and the 1980s, microelectronics and computer technology became keywords among politicians and economists of the GDR for boosting economic performance and productivity by using computers and automation technology.[154] The sociotechnical imaginary promoted by the SED promised not only a higher quality and precision that was beyond human capabilities, but also a higher productivity and a reduction in labor cost. Automation technology promised that human work functions would be taken over by the machine, especially the control of the production process. This, in turn, was expected to result in an increase of labor productivity by leaps and bounds, even while maintaining the previous processing speeds for the individual work operations, because considerably fewer workers would be needed for the production process.[155] Notably, the anticipated economic growth through the use of computer technology was not perceived simply as a pleasant benefit, but as a necessity. The development and application of new technologies were considered decisive in whether or not the GDR's economy would be able to satisfy the people's needs and compete successfully on the world market: "It is in this area that decisions are made about the rate of growth of labor productivity, which determines how our economy will meet the needs of our people, the diverse internal requirements of our country's development and how it will hold its own in the world."[156] The SED saw economic growth, based on the mastery of new technologies, as vital to meeting the growing demands of consumers both domestically and on the world

---

152 Hartmann, "Schlüssel für kräftiges Wachstum," 592.
153 Hans Modrow, "Schlüsseltechnologie – politische Bewährung auf dem Hauptkampffeld," *Einheit*, no. 8 (1987): 733.
154 Hartmann, "Schlüssel für kräftiges Wachstum," 591.
155 Krömke, "Ökonomische Strategie," 783.
156 Zentralkomitee der SED, *Bericht des Zentralkomitees der Sozialistischen Einheitspartei Deutschlands an den XI. Parteitag der SED, Berichterstatter: Genosse Erich Honecker* (Berlin: Dietz Verlag, 1986), 49 (translated from German).

market.[157] According to the SED's sociotechnical imaginary, which posited computer technology as an engine of economic growth, the development and use of this new technology was therefore necessary not only to appease its own people, but also to gain the upper hand in the ideological conflict of the Cold War, which in the 1980s was to a large extent also fought in the economic arena. For this reason, this facet of the imaginary, which posited computer technology as an engine of economic growth, was closely intertwined with another facet related to the ideological conflict between capitalist and socialist systems.

2.4 Computers as a Symbolic Arena in the Ideological Battle Between the Systems

In political speeches and papers covering the topic of computer technology, SED leaders often referred to the ideological and political competition between the systems, as well as a "race against time." In this regard, the dominant sociotechnical imaginary positioned new information technologies as a decisive arena in the political battle between the systems, and consequently, a challenge for the socialist leadership of the GDR to prove their political mettle.[158] The rhetoric of a symbolic or ideological battle both against capitalism and time reflects a sense of urgency behind the ambition to stay ahead of the curve. The SED leadership claimed that the speed and dynamism of the "scientific and technological revolution" in all industrialized countries would only bring tangible productivity advantages to those who determined the international state of the art themselves, at least in some areas.[159] According to the SED narrative, however, the pace of the necessary increase in labor productivity was set by the increasingly rapid international scientific and technological progress.[160] In this regard, the leading global competitors were Western capitalist countries. The SED leadership posited that the USA and NATO were seeking to strengthen their economic, political, and military position in the confrontation with socialist states through excellence in the field of key technologies. Accordingly, the socialist community also had to achieve

---

[157] Gerd Friedrich, "Beschleunigung des Reproduktionsprozesses in den Kombinaten," *Einheit*, no. 7 (1986): 599.
[158] Modrow, "Schlüsseltechnologie."
[159] Günter Kröber, "Individualität und Kollektivität bei der Meisterung der wissenschaftlich-technischen Revolution," *Einheit*, no. 9 (1987): 817.
[160] Friedrich "Beschleunigung des Reproduktionsprozesses," 598.

scientific and technological excellence to consolidate its own economic position worldwide.[161]

Computer technology, in particular, played a paramount role in this competition between capitalist and socialist systems, which was primarily conceived as a matter of economic competitiveness. The SED leadership was convinced that socialism, to prove its superiority and to establish its economic and political weight on the world stage, needed a higher labor productivity than capitalism.[162] The positive sociotechnical imaginary, envisioning a desirable future of computerization under socialism, was contrasted with an imagined dystopia of high technology under capitalism. In this capitalist dystopia, as envisioned by the SED, high technologies were linked to growing exploitation of workers, social cuts, and mass unemployment. Moreover, scientific, and technological innovations were allegedly subordinated to the "imperialist pursuit of power and profit," and thus misused to create "ever more diabolical weapons of mass destruction."[163] In this sense, the SED's imaginary posited that computer technology under capitalism would be turned into a threat to the existence of humanity, while by contrast, it would open up to humanity the perspective of high productivity and a betterment of living and working conditions under socialism. However, the political leadership of the GDR also noted that capitalist countries nevertheless succeeded in driving forward technological progress at a rapid pace, which made them serious contenders in the technological and economic competition of the two political systems.[164]

While the sociotechnical imaginary propagated by the SED painted a peaceful picture of socialist computerization for social and economic progress, it linked the development and use of computer technology in capitalist countries to the "military-industrial complex." In the SED's imaginary, computers played an important role in the Cold War between the systems, too, albeit in a "peaceful competition" of technological and scientific innovation and economic power. In the eyes of the authoritarian leadership, the mastery of the "modern key technologies" became even more important in the struggle to secure peace and strengthen socialism. It was considered an "objective necessity" to rule out war in the confrontation between socialism and capitalism, by instead resorting to a peaceful competition, in which making effective use of the newest technologies and scientific findings in economic production was of utmost importance.[165] According to the SED's sociotechnical

---

161 Hartmann, "Schlüssel für kräftiges Wachstum," 593.
162 Krömke, "Ökonomische Strategie," 788.
163 Modrow, "Schlüsseltechnologie," 734.
164 Modrow, "Schlüsseltechnologie," 734.
165 Modrow, "Schlüsseltechnologie," 734.

imaginary, computer technology was thus propagated as a means of securing peace and strengthening socialism in the conflict with capitalist systems. As illustrated by one of the political slogans of the SED in the second half of the 1980s – "My workplace, my battleground for peace" (ger.: Mein Arbeitsplatz – mein Kampfplatz für den Frieden)[166] –, every worker in the GDR was called upon to do his or her best to develop and use the new technology in the most effective way in order to gain the upper hand in the ideological conflict between socialism and capitalism. Only in this way, the SED leadership declared, could the GDR maintain its place among the advanced countries, further strengthen the "forces of peace," secure its achievements, and continue its social policy for the benefit of the working people.[167]

2.5 Computer Technology as a Driver of Social Progress

According to the principle of "unity of economic and social policy" under Honecker,[168] the SED leadership attempted to incentivize higher economic productivity, while at the same time fulfilling the party's promise of higher living standards. With this strategy, the SED aimed not only to ensure the long-term economic viability of the GDR, but also to rally popular support for the party's rule by promising a more "consumer-oriented socialism"[169] and introducing comprehensive social welfare measures.[170] Consequently, the SED's sociotechnical imaginary conceived of the computerization of the GDR not only as a contribution to economic growth, but crucially also as a sociopolitical program aimed at improving the living and working conditions of people in the GDR. Contrasting their own imaginary with the capitalist West, the political authorities claimed that in the socialist GDR, unlike capitalism, there was no reason for workers to fear for their jobs in the process of introducing new technologies. The provision of high-quality education and training for all would not only guarantee a job for everyone in the future but would also

---

[166] "Erfolgreich im Kampf um hohen Tempogewinn durch Mikroelektronik," *Neues Deutschland*, June 24, 1986, 3.
[167] Modrow, "Schlüsseltechnologie," 734.
[168] Mark Allinson, "More from Less: Ideological Gambling with the Unity of Economic and Social Policy in Honecker's GDR," *Central European History* 45, no. 1 (2012): 102–27.
[169] Andreas Malycha, *Die SED in der Ära Honecker: Machtstrukturen, Entscheidungsmechanismen und Konfliktfelder in der Staatspartei 1971 bis 1989* (München: De Gruyter Oldenbourg, 2014). See in particular chapter IV: "Der 'Konsumsozialismus' der Honecker-Ära" (pp. 177–256).
[170] Allinson "More from Less."

make workers the true masters of modern productive forces, giving them job satisfaction and the feeling of being needed and useful.[171]

In this line, the SED's sociotechnical imaginaries of computer technology promised the improvement of working conditions, as computers and computerized machines would free workers from doing repetitive, or dirty and strenuous work, allowing instead for more creative, mentally stimulating activities. It was envisioned that the "key technologies," and especially microelectronics, would pave the way for "fully realizing the advantages of socialism, bringing the fountains of social wealth to full flow, liberating human labor more and more from monotony and drudgery, enhancing its creative character, and promoting the development of socialist personalities to a greater extent."[172] A powerful image to illustrate this vision was the replacement of loud and dirty traditional industrial plants with modern, clean and emission-free factories in the budding field of microelectronics. In particular, the manufacture of electronic components was heralded as a pioneering example of the new high-tech production style, which was praised for radically improving working conditions. It was reported that production took place in bright and very clean, air-conditioned rooms, which allegedly increased the workers' enthusiasm and willingness to work.[173] Furthermore, the broad introduction of new technologies in economic production was promised to make it possible to increase the share of intellectually demanding, lively work by replacing simple manual labor and routine intellectual activity with technical means.[174] Crucially, this sociotechnical imaginary did not envision a replacement of human workers by machines, but on the contrary posited that the effective use of new computer technology would both enable and necessitate human creativity and responsibility to a greater extent than ever before. Therefore, an elevated role of workers and their level of qualification was anticipated. The SED asserted that new computer technology would not diminish the role of the human being in economic production, but instead even increase it: "Superficially, it might seem that the opposite is the case, that by transferring human functions also in relation to the control of production, the human worker is being 'pushed out' of the production process by automation. Examples of unmanned factory halls visually reinforce this impression. [. . .] Indeed, as a result of automation, fewer workers are needed in the production process, but they have to be more highly qualified. The workers, foremen and engineers who were previously responsible for a nar-

---

171 Wolfgang Rudolph, "Weiterbildung als ein erstrangiges Erfordernis," *Einheit*, no. 12 (1986): 1111.
172 Modrow, "Schlüsseltechnologie," 733 (translated from German).
173 Koopmann and Thieme, "Moderne Technologien," 712.
174 Hartmann, "Schlüssel für kräftiges Wachstum," 595.

rowly defined sub-section now bear responsibility for the mastery of an entire technological process."[175]

2.6 The Computer as a Tool for the Development of the Socialist Personality

Closely related to the imagination of computers as liberators of the workforce from monotonous and inhumane work was the aspect of envisioning the new technology as a means to expand the potential of humans beyond their physical and mental limitations,[176] and to contribute to the free development of the individual socialist personality.[177] The potential risk posed by the widespread use of new technologies, that humans would no longer remain the masters of technology and would instead be dominated by it, and that human thinking and feeling have to be increasingly subordinated to technology, only figured in the SED's dystopian sociotechnical imaginaries that was attributed to the capitalist countries. In socialism, on the other hand, technology was to serve the development of human personality; and humans must always remain its masters. This particular aspect of the SED's sociotechnical imaginary thus consisted in the promise that humanistic goals would always take precedence in the application of key technologies would under socialism: Technology should and would never be able to surpass or replace the human being as a total personality, but could only support and multiply his or her abilities.[178] The transfer of mental and manual work functions from humans to machines would therefore not equate to a loss, but was considered a prerequisite for a higher level of "humanity" in the work process, in which thinking as a typically human characteristic was supported, promoted, and positively challenged by technical means. In the SED's imaginary, the personal computer was envisioned as an instrument for amplifying and improving human capabilities – and crucially, it would always remain a mere instrument, and nothing more.[179]

However, the sociotechnical imaginary of computer technology to enhance individual personality development was not detached from the SED's strong em-

---

175 Krömke, "Ökonomische Strategie," 785 (translated from German).
176 Nick, "Informationsverarbeitende Technik," 611.
177 Lothar Hummel and Rudi Rosenkranz, "CAD/CAM im Dienste des Menschen," *Einheit*, no. 12 (1986): 1103.
178 Hartmann, "Schlüssel für kräftiges Wachstum," 595.
179 Hartmann, "Schlüssel für kräftiges Wachstum," 597.

phasis on socialist collectivism.[180] Thus, it needed to be coupled with the imperative of balancing individual desires and collective needs.[181] The interests and desires of the envisioned future users of computers needed to be aligned with societal – that is, the SED's political and economic goals. SED officials posited that the "modern productive forces." such as new computer technology, require a type of worker who is highly educated and skilled, who identifies personally with their work task, and who relates to the product of their labor and to the means of production as if they were their own. This, in turn, was thought to be possible only under social conditions in which both the product and the means of production belonged to the workers – that is, under socialism.[182] Against this background, the paramount task of the socialist education system was emphatically stressed, namely to educate people to become "socialist personalities" whose actions and behavior are characterized by a sense of responsibility for society as a whole and for themselves.[183]

In their work, Jasanoff and Kim point out that calling upon a certain sociotechnical imaginary can serve to "create the political will or public resolve to attain it."[184] In the GDR, the six aspects of the SED's sociotechnical imaginary of computer technology described above, were invoked to garner the support of the workers, teachers, and learners of the GDR to realize the political leadership's goals for the computerization of the country's economy and society. Additionally, they also informed the work of education policymakers and educators in the introduction of computers into schools and universities, as well as in developing new curricula, teaching, and learning materials for computer education. Together, the different aspects of this sociotechnical imaginary which became to dominate the political discourse around computer technology in the GDR, shaped both state policies and interventions, as well as the everyday practices of workers, teachers, and learners in their dealings with computer technology. It was under the guidance of the SED's powerful imaginary that the people of the GDR made up their minds about how and what computers should be used for, what needed to be learned about this new technology and for what purpose, and finally what kind of broader societal goals it should help to fulfill.

---

180 Angela Brock, "Producing the 'Socialist Personality'? Socialisation, education, and the Emergence of New Patterns of Behaviour," in *Power and Society in the GDR, 1961–1979*, ed. Mary Fulbrook (New York/Oxford: Berghahn Books, 2009), 220–52.
181 Koopmann and Thieme, "Moderne Technologien," 712.
182 Harry Nick, "Über Wesen, Effekte und Tragweite flexibler Automatisierung," *Einheit*, no. 9 (1987): 795.
183 Kretzschmar 1987, 140.
184 Jasanoff and Kim, "Containing the Atom," 123.

3 Cybernetics and Computer Science: Emergence of a New Discipline

The development of computer science and informatics in higher education in the GDR was built upon, and closely related to earlier efforts in the field of cybernetics. Cybernetics, as a theory of the (self-)regulation and control of systems, established itself in the socialist GDR in the 1950s and 1960s against initially fierce ideological reservations, and not entirely fading criticism over the next decade. In this process, as will be shown in the following, a specific sociotechnical imaginary of automation and information processing technology gained form and political persuasiveness. While cybernetics fell into disrepute again at the end of the 1960s (especially its application in the humanities and social sciences) and corresponding research eked out a niche existence, computer science was able to gain a foothold as a new scientific discipline in research and teaching over the course of the 1970s and 1980s.

With a similar ambivalence that characterized the adoption of cybernetics in the GDR, and as a continuation of this process, computer science initially struggled to establish itself as a field of research and development that fitted into the ideological foundations of the socialist political and ideological system. To gain a foothold in the GDR's higher education system, the young and still underdetermined field of computer science and technology had to be supported by a strong conviction that it served the interests and values of the socialist society and its political leadership. Therefore, the early phase of establishing computer science as an academic discipline and its institutionalization in the GDR's higher education system is closely linked to the creation of a powerful and durable sociotechnical imaginary of computer science and technology within the GDR's socialist system, which will be explored in this chapter.

3.1 Cybernetics as an Intellectual Precursor of Computer Science and Informatics

The field of cybernetics has played an important role in the historical development of modern information and communication technologies. The term cybernetics refers to the study of communication, control, and the processing of information in complex systems, which may be biological, social, mechanical or electronic in nature.¹⁸⁵ A key figure in the history of cybernetics is Norbert Wiener, who coined the term "cybernetics" in the 1940s and wrote a book of the same title in 1948.¹⁸⁶ Wiener's work laid the foundations for the field of cybernetics and had a significant influence on the development of computer science and technology. In fact, much of the research and teaching later carried out in the GDR under the label of "electronic data processing" or "computer science" had its origins in the cybernetics research institutes and research groups founded in the late 1950s and 1960s.¹⁸⁷ The invention of electronic computers made it possible to process and communicate information much faster and more efficiently than ever before and laid the foundations for the automated control of industrial processes. As a result, the early history of computing and the disciplines of computer science and computer engineering in the GDR was shaped by cybernetic concepts and methods and, what is of particular importance here, by their historically uneasy relationship with socialist ideologues in both the Soviet Union and the GDR.

The stance of the Socialist Bloc towards the field of cybernetics appears as somewhat paradoxical. On the one hand, there was a strong ideological opposition towards the "imperialist" theory of cybernetics, which was initially outright rejected in the Soviet Union and the GDR as a "pseudoscience."¹⁸⁸ The opponents of cybernetics – above all the representatives of the philosophy of materialist dialectics – criticized that cybernetics transferred technical concepts to biology and society, i.e., that it pursued a mechanistic-materialist approach. Conversely, by transferring biological concepts to the inorganic world or society, it attempted to "biologize" the entire system

---

185 Cybernetics was defined by Norbert Wiener as "the science of control and communication, in the animal and the machine" (Norbert Wiener, *Cybernetics: or Control and Communication in the Animal and the Machine* (Cambridge, MA: MIT Press, 1948)). However, cybernetics is less concerned with what systems are made of than with how they function. As a science dedicated to the study of abstract principles of organization in complex systems, cybernetics is not limited to communication and control processes in living beings and mechanical systems, but can also be applied to physical, technological, biological, ecological, psychological, or social systems (Francis Heylighen and Cliff Joslyn, "Cybernetics and Second Order Cybernetics," in *Encyclopedia of Physical Science and Technology (3ʳᵈ Edition), Vol. 4*, ed. Robert A. Meyers (New York: Academic Press, 2003), 155–70.
186 Wiener, *Cybernetics*.
187 Wissenschaftsrat, *Stellungnahmen zu den außeruniversitären Forschungseinrichtungen der ehemaligen Akademie der Wissenschaften der DDR in den Fachgebieten Mathematik, Informatik, Automatisierung und Mechanik* (Köln: Wissenschaftsrat, 1992), 17–67, http://www.adw-zki.de/docs/evaluierung_B051_4-92_AdW.pdf (accessed May 31, 2024).
188 David Holloway, "Innovation in Science – The Case of Cybernetics in the Soviet Union," *Science Studies* 4, no. 4 (1974): 299–337; Segal, "Kybernetik in der DDR," 50; Slava Gerovitch, *From Newspeak to Cyberspeak: A History of Soviet Cybernetics* (Cambridge, MA: MIT Press, 2002).

of science.[189] On the other hand, however, there was also serious interest in a theory of regulation and automation among scientists in the Soviet Union and the GDR, who carried out outstanding research work in the field. As an area of scientific research concerned with the regulation and control of processes, cybernetics was closely linked to the discourse on "electronic brains" and robots, which served the purposes of information processing and automatic control of production processes.[190] Its proponents, both in the GDR and abroad, discursively linked a compelling sociotechnical imaginary to cybernetics and cybernetic machines, which posited that their deployment in the socialist economy would allow for a significant boost in productivity, as well as freeing workers from physical and mental routine work, who would then be enabled to engage in more creative work on scientific and technical problems.[191]

Towards the end of the 1950s, the theory of cybernetics was rehabilitated to a certain degree both in the Soviet Union and the GDR.[192] Wiener, who had coined the term cybernetics, had openly criticized the American society and its educational system and engaged in opposition to nuclear weapons. He was thus no longer considered an "imperialist" researcher. Moreover, the Soviet military had recognized the significance of cybernetics for its purposes.[193] Already in 1956, Khrushchev had explicitly mentioned cybernetics as an important means to automate the Soviet industry. The program of the Soviet Union's Communist Party of 1961 stated that cybernetics, electronic computers, and automated control systems

---

189 Georg Klaus, "Die Kybernetik, das Programm der SED und die Aufgaben der Philosophen," *Deutsche Zeitschrift für Philosophie* 11, no. 6 (1963): 693.
190 Mike Hally, *Electronic Brains: Stories from the Dawn of the Computer Age* (Washington, DC: Joseph Henry Press, 2005).
191 "Sind Elektronengehirne klüger als der Mensch?," *Berliner Zeitung*, May 3, 1957, 6.
192 In "From Newspeak to Cyberspeak," Slava Gerovitch shows that in the Soviet Union, too, cybernetics was initially rejected, but later officially embraced under Khrushchev. This parallel suggests that the socialist regime of the GDR orientated itself strongly towards the Soviet Union in its official ideological assessment of cybernetics. In the early 1950s, the Soviet press ran campaigns against cybernetics because of what Gerovitch calls a "self-perpetuating Cold War propaganda discourse," which seriously diminished the range of applications for the first Soviet computers. To avoid ideological complications, computer scientists had to rid their language of cybernetic metaphors (Gerovitch, *From Newspeak to Cyberspeak*, 7). Under Nikita Khrushchev, however, the official political attitude to cybernetics changed fundamentally. Now officially embraced by the socialist regime, cybernetics was placed at the center of an ambitious project to transform Soviet science, as Soviet cyberneticians sought to establish cybernetics as a metascience that would provide a common language across different disciplines (Gerovitch, *From Newspeak to Cyberspeak*, chapter 5).
193 Segal, "Kybernetik in der DDR," 52.

3.1 Cybernetics as an Intellectual Precursor of Computer Science and Informatics — 53

were to be applied in industry, research, and administration.[194] In the GDR, the First Secretary of the Socialist Unity Party, Walter Ulbricht, announced at the 6th Party Congress in 1963 that cybernetics ought to be vigorously promoted and that cybernetic teachings should be applied in the comprehensive construction of socialism.[195]

It was the East German philosopher Georg Klaus who campaigned for the acceptance of cybernetics against the widespread belief in the Socialist Bloc in the 1950s that it was a "bourgeois pseudoscience."[196] In 1958, his lecture "Zu einigen Problemen der Kybernetik" (eng.: On some problems of cybernetics) was printed in the SED journal *Einheit*.[197] This was a significant step in the political acceptance of cybernetics in the GDR, essentially providing Klaus' standpoint, that cybernetics was not only compatible with socialist ideology, but could even contribute to its advancement, with the party's official zeal of approval.[198] Yet the article was understandably not sufficient to break all resistance to cybernetics in the GDR and establish cybernetics as a legitimate science. Strong objections continued against the new interdisciplinary theory by Marxist-Leninist philosophers, but also natural scientists, who feared the softening of firmly institutionalized disciplinary boundaries.[199] In 1961, Klaus published his book *Die Kybernetik in philosophischer Sicht* (eng.: Cybernetics from a Philosophical Perspective),[200] providing the first significant East German contribution to the field of cybernetics. Its aim was to point out the ways in which cybernetics could prove useful to Marxist philosophical thought, and to demonstrate that it was in fact compatible with dialectical materialism rather than contradicting it. Klaus called for a reappropriation of cybernetics by Marxist philosophy and argued that the notion of control was deeply rooted in dialectical materialism. In essence, the book developed both a cybernetic interpretation of Marxism, as well as a Marxist reading of cybernetics.[201]

In addition, the further acceptance of cybernetics as a new field of science in the GDR was helped by practical technological advances in computer technology, such as the development of the relay calculator "OPREMA," the first working com-

---

194 Segal, "Kybernetik in der DDR," 56.
195 Walter Ulbricht, *Das Programm des Sozialismus und die geschichtliche Aufgabe der Sozialistischen Einheitspartei Deutschlands. Schlusswort des Genossen Walter Ulbricht* (Berlin: Dietz Verlag, 1963), 345.
196 Holloway, "Innovation in Science."
197 Georg Klaus, "Zu einigen Problemen der Kybernetik," *Einheit 13, no. 7* (1958): 1026–40.
198 Segal, "Kybernetik in der DDR," 54–55.
199 Segal, "Kybernetik in der DDR," 55.
200 Georg Klaus, *Kybernetik in Philosophischer Sicht* (Berlin: Dietz Verlag, 1961).
201 Segal, "Kybernetik in der DDR," 56–58.

puter built in the GDR in 1955,[202] and the vacuum-tube based Programmable Calculator "D1" ('Dresden 1') built in 1956 at the Technical University of Dresden.[203] Klaus argued that cybernetics could provide a means to scientifically investigate and meet the GDR's social and economic challenges by extending the potential of Marxist thought and providing a new kind of cybernetics machines: "Computing machines are obviously instruments of production of a completely new kind. Their existence is equally interesting for epistemology and historical materialism. In view of these new developments, the concept of productive forces, which is fundamental to historical materialism, undergoes essential extensions and new differentiations. Learning machines, adaptive machines, and Markov machines (random machines) are the starting points of an evolving dialectical machine world, whose first initial elements were machines with built-in feedback loops."[204] As a new productive force, these new kinds of cybernetic machines were expected to help automate production processes, thus alleviating problems like the dramatic labor shortage that the GDR was struggling with.

As Klaus elaborated in a newspaper article in 1965, automation and computer technology would lead to a new relation between humans and machines, and thus create new roles for workers in the production process, shifting qualification needs from unskilled or low skilled manual labor to skilled occupations: "Whereas man has hitherto been coupled with the machine, a component of the machine as it were, he now steps beside the machine. The automatic self-regulation and self-optimisation of modern systems of machine production increasingly requires humans to be constructors, supervisors, maintenance workers or at least helping repairmen of the machines."[205] In line with Marxist philosophy, Klaus concluded that under socialism, a higher level of professional qualification for all workers was necessary to elevate humans above the new cybernetic machines and avoid relegating workers to a position subordinate to the machine, as was allegedly the case under the capitalist system: "The position of man in relation to automata is dialectically contradictory: either he does not acquire the new scientific and technical knowledge, in which case he really becomes a mere helper, a button-pusher, or, if he refuses to accept this, he is forced to acquire the qualifications of a technician, engineer, programmer, and so on. He MUST ascend to a level that makes him the

---

[202] Jürgen F. H. Winkler, "Oprema – The Relay Computer of Carl Zeiss Jena," August 26, 2019, https://arxiv.org/pdf/1908.09549.pdf (accessed May 31, 2024).
[203] Cortada, "Information Technologies," 37.
[204] Klaus, "Kybernetik, Programm SED und Aufgaben Philosophen," 698 (translated from German).
[205] "Marxistische Philosophie und technische Revolution," *Neues Deutschland*, March 16, 1965, 4 (translated from German).

3.1 Cybernetics as an Intellectual Precursor of Computer Science and Informatics — 55

controller of the machine, of the electronic calculating machines and control devices, the master of the technical revolution."[206] This implied that the appropriate response of a socialist society to scientific advances in cybernetics and technological change was a broad educational intervention that would enable all workers to acquire the necessary skills to become masters, not slaves, of the new machines.

By the early 1960s, the debate on cybernetics in the GDR had not yet been finally settled, but at least it had been defused. As a response to the growing importance of cybernetics, the German Academy of Sciences at Berlin (1972 renamed the Academy of Sciences of the GDR) set up a commission on cybernetics, with Georg Klaus as its chair.[207]

Simon Donig has pointed out the significance of the political and economic climate at the time for assigning a role to cybernetics in socialist society.[208] Notably, the erection of the Berlin wall on 13th August 1961 to stop the mass exodus of people from the GDR to the West heralded the start of a phase of increased political demarcation from the capitalist West and a focus on internal reforms for political stabilization and economic growth. In addition, the New Economic System of Planning and Management (ger.: Neues Ökonomisches System der Planung und Leitung, NÖSPL) was introduced in 1963 to reform the planned economy and increase economic growth by introducing free market elements, albeit to a very limited extent. Within this context, electronic data processing was perceived as a decisive instrument in realizing the planner's vision of systematic control over the GDR's economy.[209] The role of cybernetics in socialist governance and management was further substantiated by scientific publications such as the book by Heinz Liebscher, a former student of Georg Klaus, on "Kybernetik und Leitungstätigkeit" (eng.: Cybernetics and Management).[210] The book introduced its readers to cybernetic theories and methods, and highlighted the significance of cybernetics for socialist management and governance, which was explained to be an informational process that could be modelled and optimized using cybernetic theories and methods. A year later, a newspaper article boldly stated: "By the year 2000 at

---

206 "Marxistische Philosophie und technische Revolution," *Neues Deutschland*, March 16, 1965, 4 (translated from German).
207 Klaus Fuchs-Kittowski, "Zur Herausbildung von Sichtweisen der Informatik in der DDR unter Einfluss der Kybernetik I. und II. Ordnung," in *Kybernetik steckt den Osten an. Aufstieg und Schwierigkeiten einer interdisziplinären Wissenschaft in der DDR*, ed. Frank Dittmann and Rudolf Seising (Berlin: Trafo, 2007), 323–324.
208 Simon Donig, "Informatik im Systemkonflikt – Der Technik- und Wissenschaftsdiskurs in der DDR," in *Informatik in der DDR – eine Bilanz*, ed. Friedrich Naumann and Gabriele Schade (Bonn: Gesellschaft für Informatik, 2006), 462–78.
209 Donig "Informatik im Systemkonflikt," 469.
210 Heinz Liebscher, *Kybernetik und Leitungstätigkeit* (Berlin: Dietz Verlag,1966).

the latest, there will hardly be any jobs that do not require a minimum of cybernetic knowledge. In a few years, no senior executive will be able to do without cybernetics."[211] Virtually identical statements would be made 20 years later but replacing "cybernetics" with computer knowledge and programming skills.[212]

The discourse on cybernetics in the GDR in the 1960s was influenced by the reception of research output from both the Soviet Union and Western capitalist countries. However, the "capitalist West" played an ambivalent role in the political and scientific discourses on cybernetics and computer technology in the GDR. On the one hand, the imaginary of the "West" appeared as a threat and a competitor, providing a consistent benchmark in terms of scientific and technological progress as well as the perceived social and economic implications of the development and use of new technologies. On the other hand, the "West" served as a supplier of ideas and cultural transfers from which it was necessary either to strictly distance oneself or to ideologically appropriate the received ideas and inputs and integrate them into the edifice of socialist philosophy and politics.[213]

However, most important for the political acceptance and promotion of cybernetics was the conviction that its theory, methods, and machines would prove vastly more efficient and economically useful within the socialist planned economy than within capitalist systems. In an article dated March 1964, the GDR newspaper *Neues Deutschland* wrote: "The capitalists also use electronic calculators. But [. . .] modern calculating machines (lack) the space to develop fully if they cannot process the data of the entire national economy under uniform planning aspects. In this field, too, technology has outgrown capitalist private property. [. . .] One must want the right thing. One must want the prosperity of the people, the strengthening of the people's power, the consolidation of peace. All this shows that socialism can and will make more fruitful use of cybernetic machines than is

---

211 "Was gibt uns die Kybernetik?," *Neues Deutschland*, May 17, 1967, 2 (translated from German).
212 See, for example, the statement made in 1987 by GDR researcher and educator Gerd Hutterer: "As early as 1990, half a million working people in the GDR will be using computers at CAD/CAM workstations. Other workers will be masters of 170,000 office and personal computers, whose effective use will bring about economic and social results of great dimensions in all areas of society. To successfully solve the tasks at hand, there is a movement of learning in combines and enterprises whereby the working people are educated in the use of information processing technology." (Gerd Hutterer, "Orientierungen für eine erziehungswirksame Gestaltung des inhaltlichen und methodischen Herangehens an die Beschäftigung mit dem Computer in der außerunterrichtlichen Tätigkeit," in *Computer in der außerunterrichtlichen Tätigkeit. Standpunkte und Anregungen*, ed. Gerd Hutterer (Halle: Pädagogische Hochschule Halle N. K. Krupskaja, 1987), 6 (translated from German)).
213 Donig "Informatik im Systemkonflikt," 475.

ever possible under capitalism."²¹⁴ The SED leadership shared this belief in the superiority of the socialist system in fully exploiting the potential of cybernetics to gain the upper hand in the systemic conflict with the capitalist West.²¹⁵

As the acceptance of cybernetics in the GDR grew during the 1960s, it was perceived as less problematic when applied as a purely technical science, for example to technical systems such as electronic computers, but raised more ideological concerns when applied to the social sphere, such as society or the educational system.²¹⁶

One example of this was the application of cybernetics in educational research. This involved the idea that the processes of education and the formation of the socialist personality could be understood as a cybernetic system that could be made more efficient and better controlled. In the GDR, the application of cybernetics in educational research quickly became virtually synonymous with research into programmed instruction, which began in 1962.²¹⁷ Initially, these endeavours were focused on the study of approaches and experiences in other (including Western capitalist) countries and attempts to transfer them to the socialist educational system of the GDR.²¹⁸ From an ideological viewpoint, a heavy focus on algorithmic modelling of instruction as an operational process was rejected, as were references to the works on programmed instruction by the American Psychologists Burrhus Frederic Skinner and Norman Crowder.

In practice however, educational researchers in the GDR heavily drew on Skinner's and Crowder's methodology, albeit not explicitly mentioning doing so. Likewise, the algorithmic modelling of instructional processes was central to theoretical research on programmed instruction in the GDR.²¹⁹ By the mid-1960s, educational researchers in the "Forschungsgemeinschaft Programmierter Unterricht" (eng.: Re-

---

214 "Rebellion der Tatsachen – gegen wen?," *Neues Deutschland*, March 27, 1964, 3 (translated from German).
215 Sobeslavsky and Lehmann, *Geschichte von Rechentechnik und Datenverarbeitung*, 65.
216 For an overview of research undertaken in the GDR on cybernetics in the fields of philosophy, pedagogy, biology and medicine, natural sciences, and engineering as well as economics, see Frank Dittmann and Rudolf Seising, eds., *Kybernetik steckt den Osten an. Aufstieg und Schwierigkeiten einer interdisziplinären Wissenschaft in der DDR* (Berlin: trafo, 2007).
217 Nicole Zabel, "Die Lehrmaschinen und der Programmierte Unterricht – Chancen und Grenzen im Bildungswesen der DDR in den 1960er und 1970er Jahren," in *Jahrbuch für Historische Bildungsforschung 2014, Vol. 20*, ed. Sektion Historische Bildungsforschung der DGfE together with Bibliothek für Bildungsgeschichtliche Forschung des Deutschen Instituts für Internationale Pädagogische Forschung (DIPF) (Bad Heilbrunn: Verlag Julius Klinkhardt, 2015), 131.
218 BBF APW 9396.1, "Forschungsmaterial zur Programmierung von Lehr- und Lernprozessen in der allgemeinbildenden Schule."
219 Zabel, "Lehrmaschinen und der Programmierte Unterricht," 132.

search Group on Programmed Instruction) at the DPZI ("Deutsches Pädagogisches Zentralinstitut," the German Pedagogical Central Institute and the predecessor of the Academy of Pedagogical Sciences of the GDR) began to develop and test their own programmed instruction materials for the use in general schools, primarily in mathematics and natural sciences. The efforts culminated in a large-scale pilot test to assess the effectiveness of programmed instruction between 1966 and 1968, involving roughly 5000 students in 160 school classes. But the mixed, if not disappointing results of the test, the high cost of developing teaching machines and programmed instruction materials, as well as persistent ideological concerns about the de-ideologization of socialist education, increasing individualization and a compromise of the teacher's role as a leader in the classroom led to a marked decline in enthusiasm for cybernetic pedagogy in the GDR.[220]

As a result, cybernetic pedagogy was increasingly relegated to a niche existence over the course of the 1970s. In 1980, research on programmed instruction was abandoned for good, as the advances in computer technology seemed to promise a more effective way to increase the efficiency of educational processes.[221]

Similarly, in 1967, the GDR cyberneticist Klaus-Dieter Wüstneck was elected as a candidate for the Central Committee of the SED and soon after was appointed head of the Commission for Cybernetics in the Ministry of Higher Education. The commission had the task of developing and implementing curricula on cybernetics in higher education. However, its proposals were never acted upon, and Cybernetics was never able to establish itself as an independent course of study in the GDR.[222] From 1969 onwards, cybernetics disappeared from the SED's educational policy agenda. The new focus in the 1970s and 1980s was on microelectronics, automation technology, computers, and programming. Electronic data processing and informatics were now the center of political attention, and as a result, corresponding research centers and study courses were created.

Both Segal and Donig identify an "anti-cybernetic turn" in the GDR in the year 1969.[223] The reasons for this were the increasingly harsh criticism of SED ideologues over referencing Western capitalist cyberneticians, as well as growing fears that cybernetics could challenge the party's leadership role. The latter was a reaction to

---

220 Zabel, "Lehrmaschinen und der Programmierte Unterricht." On technocratic visions of the replacement of the teacher that underpinned the development and introduction of teaching machines and programmed instruction, see Marcelo Caruso, *Geschichte der Bildung und Erziehung: Medienentwicklung und Medienwandel* (Stuttgart: utb, 2019), 196–98.
221 BBF APW 9396.1, "Forschungsmaterial zur Programmierung von Lehr- und Lernprozessen in der allgemeinbildenden Schule."
222 Segal, "Kybernetik in der DDR," 62–63.
223 Donig, "Informatik im Systemkonflikt," 470–71; Segal, "Kybernetik in der DDR," 64.

the growing body of cybernetic research on the self-regulation of social and economic systems. At its core, the cybernetics movement of the 1960s sought to create a unified theory of all systems, namely a theory of the principles and patterns that govern the behavior and interactions of all systems that would be applicable to all fields and disciplines – from engineering to biology, psychology, and the social sciences. This goal was potentially dangerous to an authoritarian political leadership that claimed sole power over the governance of society and the national economy, since such knowledge could empower individuals and groups to challenge its authority and question its methods of control. Furthermore, the cybernetic approach of bringing together different disciplines under the umbrella of a unifying theory could also reveal the interconnections and interdependencies between different systems, which had the potential to challenge disciplinary boundaries as well as traditional power structures and hierarchies. For these reasons, certain political and scientific elites in the GDR, who feared that it could threaten their established position in society fiercely opposed cybernetics. When Erich Honecker officially took power at the 8th Party Congress in 1971, he crushed the universalist aspirations of cyberneticians in the GDR and their remaining hopes of transforming scientific disciplines and, more broadly, the socialist system. The new General Secretary of the GDR solemnly announced that it was now finally proven that cybernetics and systems research were "pseudosciences."[224] Cybernetics once again fell out of political favor and disappeared for good from the SED leadership's scientific policy agenda. Nevertheless, the concepts of cybernetics had already permeated many disciplines and fields of academic study, including computer science and information theory, robotics, economics, and organizational theory. It did not achieve the status of a universal science in the GDR but was rather fragmented into separate disciplines – similar to the fate of cybernetics in the Western academic establishment.[225]

However, the same year of the "anti-cybernetic turn" also saw the founding of the Central Institute for Cybernetics and Information Processes (ger.: Zentralinstitut für Kybernetik und Informationsprozesse, ZKI), which existed until the end of the GDR. In contrast to the application of cybernetics in the social sciences and humanities, or even its conception as a universal theory, the ZKI focused primarily on technical cybernetics. Its research was mainly concerned with technical cybernetic systems such as information processing machines and the development of automated control systems, later also with artificial intelligence and software

---

[224] Segal, "Kybernetik in der DDR," 64.
[225] Ronald R. Kline, *The Cybernetics Moment: Or Why We Call Our Age the Information Age* (Baltimore: Johns Hopkins University Press, 2015), 182–83.

development for computer-integrated manufacturing.[226] This rather narrow technical focus of cybernetics seemed less problematic for the party's ideologues and lent itself to integration into the state technology policy of the late 1970s and 1980s, with its focus on automation and computer technology.

Thus, by the late 1960s, the sociotechnical imaginaries of what could be achieved with cybernetics and automation were enshrined in the imaginary of computer technology, which consolidated the utopian hopes and ideals for the future of socialist society and economy that had emerged from cybernetic ideas: increasing productivity and freeing workers from boring and repetitive tasks. This imaginary gained popularity at a time marked by a political shift that Morandi (1998) describes as the "economization of the GDR's national mission."[227] Economic productivity was now foregrounded by the SED as a decisive factor in the conflict between capitalist and socialist systems, and in turn the increased importance of technological and scientific progress in achieving this goal was emphasized. Thus, in the 1970s and 1980s, the "scientific-technological revolution" became a common catchword on the SED's political agenda.[228] In this context, the computer was particularly well received as a cybernetic machine capable of processing information and automatically controlling production processes.

Subsequently, the political discourse on cybernetics in the GDR was gradually replaced by microelectronics, electronic data processing, and computers. Notably, in the Soviet Union, the designation of cybernetics did not disappear. On the contrary, the term was retained without reservation, and informatics or computer science, indeed the whole use of computer technology, was subsumed under it.[229] Only in the 1980s, the notion of *informatika* (informatics) gradually replaced the use of the word *kibernetika* (cybernetics) to describe the field of computing). In the GDR, automation, and rationalization, aided by technological advances, increasingly became a kind of panacea in the SED's political repertoire for responding to social, political, and economic challenges.[230] The SED leadership was convinced that the promised benefits of scientific and technological advances in computing and data processing could be much better exploited within the socialist system and

---

[226] Günter Laux and Fritz Scholz, *Der Kybernetik-Report* (ZKI-Informationen, special issue no. 2) (Berlin: Zentralinstitut für Kybernetik und Informationsprozesse der AdW der DDR, 1989).
[227] Pietro Morandi, "Die ordnungspolitische Gegenrevolution in der DDR der 60er Jahre. Die Absage an den Militarismus und die Verwirtschaftlichung der Nationalen Mission der DDR," in *Kommunikation und Revolution*, ed. Kurt Imhof and Peter Schulz (Zürich: Seismo, 1998), 263–83.
[228] Hubert Laitko, "Wissenschaftlich-technische Revolution: Akzente des Konzepts in Wissenschaft und Ideologie der DDR," *Utopie kreativ*, no. 73–74 (1996): 33–50.
[229] Fuchs-Kittowski "Zur Herausbildung von Sichtweisen der Informatik," 367n29.
[230] Donig, "Informatik im Systemkonflikt," 470.

its planned economy. Consequently, the SED's nascent sociotechnical imaginary of modern computer technology emphasized the paramount role of the working people in shaping the development and use of this new technology in the socialist society, as well as the claim that computer technology would serve the benefit of all people in the GDR.

3.2 Institutionalization of Computer Science and Informatics in Higher Education

The start of the 1960s saw the establishment of computer centers in the GDR within the context of a push to establish research and development in the field of cybernetics and information processing. Calculating machines and automatic control systems were developed for the use in industry, research, and administration. Therefore, a need arose to train personnel with a high degree of expertise in these new technologies. From an educational point of view, we can therefore start our search for the roots of computing as an academic discipline in the GDR with the following questions: Where were the first computers built, and in which academic disciplines were the people educated and trained to build and work on these machines, to maintain and repair them, and ultimately to work on the further development of computer technology?

The question of the disciplinary origins of computer science also alludes to the question of naming the new field of knowledge. The title of this chapter deliberately refers to two different terms: *Computer science* and *informatics*. While these two terms may be used interchangeably today, in the early days of computing they implicitly referred to two different emphases in the conception of the new field of knowledge. While *computer science* focused on the device of the computer itself, the hardware, and the design and construction of computer systems and programs, *informatics* emphasized the process of information processing and the fundamental role of information in a variety of systems and processes.[231] This fundamental distinction reflects the bivalent nature of the emerging discipline, which was situated somewhere between technology and engineering on the one hand, and mathematical computation and data processing on the other – although the boundaries between these two spheres were by no means clear-cut.

---

[231] Wolfgang Coy, "Was ist Informatik? Zur Entstehung des Faches an den deutschen Universitäten," in *Geschichten der Informatik: Visionen, Paradigmen, Leitmotive*, ed. Hans Dieter Hellige (Berlin/Heidelberg: Springer, 2004), 479–81.

In West Germany, the cybernetician and engineer Karl Steinbruch first introduced the term *Informatik* (eng.: Informatics) in 1957 as a portmanteau that combines the words "information" and "automatic" to name the new field of automated information processing.[232] In the West German tradition, informatics was thus technically oriented and focused on automated information processing machines and their theoretical-mathematical foundations. In France, the term *Informatique* was coined by Philippe Dreyfus in 1962 through the foundation of the "Société d'Informatique Appliquée," the Society of Applied Informatics. However, his conception encompassed a broader definition of the field that also included the social and economic effect of rationalization and automation, as well as aspects of communication. In 1966, the French Academy established the official use of *Informatique* to designate the "science of information processing."[233] In the Soviet Union, the term *Informátika* was introduced in 1966 by Mikhailov, Gilyarevskii, and Chernyi, to refer to a "scientific discipline that studies the structure and properties of scientific information."[234] In this way, *Informátika* brought the meaning of the term closer to library science rather than to a technically oriented computer science. But the meaning of the term subsequently changed, and towards the end of the 1970s and the beginning of the 1980s, informatics was interpreted as a "science focused on the theory of programming and the use of computing hardware."[235] However, the Russian *Kibernetika* (eng.: cybernetics) remained in use as a catch-all term to describe everything that had to do with automated information processing.[236] In the GDR, the name of scientific groups and institution concerned with computing changed from "cybernetics" and "data processing" in the 1960s to "information processing" in the 1970s and 1980s. The latter was a reference to the name of the field favored by the IFIP, of which the GDR became a member in 1970, with Nikolaus Joachim Lehmann as its delegate.[237] Only towards the very end of the GDR was the designation of

---

[232] Klaus Biener, "Karl Steinbruch – Informatiker der ersten Stunde," *RZ-Mitteilungen*, no. 15 (1997): 53–54, https://edoc.hu-berlin.de/handle/18452/6886 (accessed May 31, 2024).
[233] Philippe Breton, Alain-Marc Rieu, and Franck Tinland, *La Techno-Science en Question. Éléments pour une Archéologie du XXe Siècle* (Seyssel: Editions Champ Vallon, 1990), 186–92.
[234] Hans Wellisch, "From Information Science to Informatics: a terminological investigation," *Journal of Librarianship* 4, no. 3 (1972): 176.
[235] Viatcheslav A. Yatsko, "Informatics, Informations Science, and Computer Science," *Scientific and Technical Information Processing* 45, no. 4 (2018): 235.
[236] Klaus Fuchs-Kittowski, "Grundlinien des Einsatzes der modernen Informations- und Kommunikationstechnologien in der DDR. Wechsel der Sichtweisen zu einer am Menschen orientierten Informationssystemgestaltung," in *Informatik in der DDR – eine Bilanz*, ed. Friedrich Naumann and Gabriele Schade (Bonn: Gesellschaft für Informatik, 2006), 57.
[237] Fuchs-Kittowski, "Grundlinien des Einsatzes der modernen Informations- und Kommunikationstechnologien," 55.

3.2 Institutionalization of Computer Science and Informatics in Higher Education — 63

study courses and university institutes changed to "informatics" – once again in reference to the IFIP, which had decided in 1989 to rename the discipline from "information processing" to "informatics."[238]

The different cultures and traditions of naming the field of computing and information processing reflect different understandings of what the new science was supposed to be, what it was concerned with and what constituted it. In other words, the search for a name also mirrored the struggles to reach a common understanding regarding the newly emerging discipline's scope, paradigms, and methods, and from there, to deduce the implications for content and form of computer education and training in higher education.

As reflected in the name, the new field was broadly understood as the science of informational processes in the GDR, rather than putting a narrow focus on the computer itself. However, this understanding did not remain unchallenged, but formed the subject of academic disputes about the nature of the emerging new domain of knowledge, as exemplified by a fundamental disagreement between scientists at the Humboldt University in East Berlin and at the Technical University of Dresden in the 1960s and 1970s. The latter had a decidedly technical profile, and its Institute for Computing Machinery (ger.: Institut für Maschinelle Rechentechnik, IMR) was led by Nikolaus Joachim Lehmann, professor for applied mathematics since 1953, who had developed a series of pioneering programmable electronic computers, the "Dresden computers" "D1" (1956), "D2" (1959), as well as a smaller desktop computer, the "D4a" (1963).[239] At the Technical University of Dresden, the developing field of electronic computing followed a relatively narrow, technical conceptualization of information in data processing. The science of computing and informatics was considered a purely technical discipline. In contrast, researchers in information processing at the computing center of the Humboldt University of Berlin understood informatics as a broad and far-reaching fundamental discipline that was concerned with both human and machine information processes.[240] In 1972, Klaus Fuchs-Kittowski, a philosopher of science who had habilitated with a thesis on bio-cybernetics and had subsequently worked as a university lecturer on philosophical problems of cybernetics at the Humboldt

---

238 Fuchs-Kittowski, "Grundlinien des Einsatzes der modernen Informations- und Kommunikationstechnologien," 60.
239 Nikolaus J. Lehmann, "Zur Geschichte des 'Instituts für maschinelle Rechentechnik' der Technischen Hochschule/Technischen Universität Dresden," in *Zur Geschichte von Rechentechnik und Datenverarbeitung in der DDR 1946–1968*, ed. Erich Sobeslavsky and Nikolaus J. Lehmann (Dresden: Hannah-Arendt-Institut für Totalitarismusforschung, 1996), 123–57.
240 Fuchs-Kittowski, "Grundlinien des Einsatzes der modernen Informations- und Kommunikationstechnologien," 57.

University of Berlin, was appointed to the full professorship of information processing at the Humboldt University of Berlin.[241] His research on information processing focused on philosophical-scientific questions about the phenomenon of information, as well as the study of interactions between computer machines and social organizations in the form of sociotechnical systems.

Thus, despite identical or similar designations of the newly emerging discipline, the research profiles and the teaching offered at the various faculties and universities varied significantly. This was due to different university policy decisions regarding the institutional affiliation of newly established computing centers and information processing departments, but also due to the staffing of chairs and related research groups with people from different disciplinary backgrounds in mathematics, engineering, or philosophy. In other words, the science and teaching of computing in higher education were not formally defined for the whole of the GDR in these early days, but its defining paradigms and methods rather emerged from the everyday practices of groups of informaticians and computer scientists in research and teaching, with different conceptualizations of what the new field of knowledge should be. These processes of disciplinary formation, in turn, can be conceptualised through the emergence of a dominant sociotechnical imaginary of computer technology that came to define the academic discipline of information processing in the GDR.

As mentioned earlier in this chapter, the new field of computing and electronic data processing drew on core concepts of cybernetics and referenced a corresponding sociotechnical imaginary. Cybernetics, and later the new discipline of information processing, were conceived as new scientific fields that would provide the tools, methods, and theories for studying, controlling, and optimizing information processes in technical and social systems. In these early days of computing in the GDR, three facets of a nascent sociotechnical imaginary can be identified that played a crucial role in the early development of both computer technology and the academic discipline of information processing during the 1960s and early 1970s, shaping the goals and priorities of researchers and developers, as well as the ways in which this new knowledge found its way into university curricula.

The first is related to the material and technical effects of computing and focuses on the optimization of technical systems. It envisions techno-scientific progress to boost the socialist economy, and in turn, the computer as a tool for automating and streamlining processes of economic production. The emphasis here is on increasing efficiency and productivity, by replacing certain elements of

---

241 Frank Fuchs-Kittowski and Werner Kriesel, "Biografie von Klaus Fuchs-Kittowski," in *Informatik und Gesellschaft: Festschrift zum 80. Geburtstag von Klaus Fuchs-Kittowski*, ed. Frank Fuchs-Kittowski and Werner Kriesel (Frankfurt am Main: Peter Lang, 2016).

human labor with programmable machines, which were expected to be able to carry out tasks in a faster and more precise fashion. Accordingly, knowledge and skills related to information technology would need to be taught to students in engineering and electrotechnology, as well as in agricultural science, to modernize and boost productivity.

The second facet relates to a technocratic vision of social and economic planning and focuses on the optimization of social systems. This facet of the imaginary includes the idea that computers can be used to support and enhance the central planning and decision-making processes of the socialist government.[242] In the Soviet Union, a similar sociotechnical imaginary of the computer as a tool of technocratic control was prevalent among Party elites, which Eugen Loebl described as "Computer Socialism."[243] The running of simulations and the collection, analysis and interpretation of large data sets to inform economic planning and decision-making, as well as the use of new information and communication technologies, were imagined to greatly facilitate the flow of information in a centrally planned economy. This vision went hand in hand with the economic policy reforms under the NÖSPL introduced by Walter Ulbricht in 1963.[244] The envisioned potential of computer technology to optimize the planned economy is exemplified by the following statement, which was printed in the SED's official party newspaper *Neues Deutschland* in 1964:

> Only with the help of electronic data processing systems can the complicated, multifariously interwoven relationships in our economy be processed in such a way that flawless, forward-looking decision-making is made possible. In fact, it is really only with the electronic calculating machines that the planned economy obtains the necessary technical basis. Now, planning can be substantially optimized.[245]

To support the effective implementation of the NES, managers and economists would need to be familiarized with the tools of modern information processing and computer technology so as to make the best use of the new tools and methods in the organization of workplaces, production processes, planning and accounting.

Finally, the third facet is related to the social and ideological effects of computing, focused on the role of computer technology in the socialist struggle for a world in which work is no longer a source of exploitation, alienation, and oppression. Accordingly, this facet of the imaginary posited that only under socialism would the computer

---

242 Schmitt, "Socialist Life of a U.S. Army Computer," 141–43.
243 Loebl, "Computer socialism."
244 Jeffrey Kopstein, "Ulbricht Embattled: The Quest for Socialist Modernity in the Light of New Sources," *Europe-Asia Studies* 46, no. 4 (1994): 597–615.
245 "Rebellion der Tatsachen – gegen wen?," *Neues Deutschland*, March 27, 1964, 3 (translated from German).

serve as a means of empowering the people. The new computer technology is envisioned as a means of freeing people from tedious, repetitive work and allowing them to fully exploit their intellectual and creative potential for problem solving, as well as supporting and enhancing their efforts in scientific research innovation. This vision suggested the development of computers for scientific purposes in higher education and research institutions, as well as a broad effort to provide students in all disciplines with knowledge and skills in computing and information processing.

The three facets were part of a newly formed, still unpolished, sociotechnical imaginary shared by pioneering scientists, engineers and state authorities who drove the introduction of information processing technology and related methods and theories into university curricula. In particular, the focus on automation and rationalization, with its seductive promise of cybernetic control and planning of the socialist economy and society, quickly found common ground among engineers and politicians in the 1960s. This imaginary also resonated with proponents of the discourse of the "scientific-technological revolution,"[246] in which new technology was presented as a crucial means of achieving the goals of socialism and gaining the upper hand in the struggle between the socialist and capitalist systems.

The sociotechnical imaginary of computer technology as a tool for increasing productivity and scientific and technological innovation, as well as a means of making the planned economy more efficient, was reinforced by the successful domestic development and production of the first mainframe computers in the GDR. The "OPREMA" and the "ZRA1" were both developed at Carl Zeiss Jena, a manufacturer of optical systems and optoelectronics.[247] In 1955, the "OPREMA" was used in the GDR's first computer center at Carl Zeiss Jena to increase the speed and accuracy of optical calculations. Development of the "ZRA1" was completed in 1960 and around 30 "ZRA1" computers were subsequently used in industry, universities, and research for scientific and technical calculations. As mentioned above, Lehmann simultaneously developed a small digital vacuum tube computer called "Dresden 1" (D1) at the Institute of Applied Mathematics at the Technical University of Dresden.

The domestic production of computer technology in the GDR was a vital step towards realizing the anticipated gains in productivity and efficiency. The COCOM embargo[248] was imposed by the US and its allies on the Socialist Bloc to prevent a growing war potential of the socialist states during the Cold

---

[246] Andreas Malycha, "Im Zeichen von Reform und Modernisierung (1961 bis 1971)," *Informationen zur politischen Bildung*, no. 3 (2011): 41.
[247] Kerner, "Vorbereitung des Informatik-Unterrichts."
[248] The Co-ordinating Committee on Multilateral Export Controls (COCOM) was founded in 1950, initiated by the USA to prevent countries of the Socialist Bloc from access to strategically important technologies and goods such as weapons, nuclear technology, and microelectronics (Frank

War.[249] It prohibited the export of Western computer technology to the GDR, thus making sanctioned imports costly.[250] The small number of Western computers imported into the GDR in the 1960s, including many from the French manufacturer Bull, were primarily deployed in government departments such as the State Planning Commission, the Ministry of State Security, industry ministries or the Central State Administration.[251] Computers imported from the Socialist Bloc, on the other hand, often proved to be unreliable and technically inadequate for the tasks at hand. As a result, the serial production of domestic computers was essential for the introduction of computing courses in higher education. The use of GDR computers in computing centers in industry and universities made it possible to offer training courses in the programming, use, maintenance, and repair of electronic computers, and to gradually extend these courses to more students.[252]

Table 1: Higher education computer centers in the GDR (1964).

Higher Education Institution	Institute	Established
Technical University of Dresden	Institute for Computing Machinery	1956
University of Halle	Institute for Numerical Mathematics	1962
University of Leipzig	Institute for Computing Machinery	1962
Technical University of Ilmenau	Institute for Computing Machinery	1962
Technical University of Magdeburg	II. Institute of Mathematics	1962
College of Architecture and Civil Engineering, Weimar	Institute of Mathematics	1962
Technical University of Karl-Marx-Stadt	Institute of Mathematics	1963
University of Rostock	Computing Center	1964
College of Economics, Berlin	Institute for Economic Data Processing	1964
Humboldt University of Berlin	Institute of Mathematics	1964

Source: Adapted from Pieper, Hochschulinformatik, 185.

Cain, "Computers and the Cold War: United States Restrictions on the Export of Computers to the Soviet Union and Communist China." *Journal of Contemporary History* 40, no. 1 (2005): 131–47).
249 Cain, "Computers and the Cold War."
250 Schmitt, "Socialist Life of a U.S. Army Computer."
251 Schmitt, *Digitalisierung der Kreditwirtschaft*, Table 4: "Computeranschaffung in der DDR (auszugsweise), 1956–1990," https://zeitgeschichte-digital.de/doks/files/2130/schmitt_computer_tab4_DDR_allg_a.pdf (accessed May 31, 2024); Pieper, *Hochschulinformatik*, 180.
252 Laux and Scholz, *Kybernetik-Report*, 12.

All computing centers were involved in mathematical basic research or applied mathematics in economics and computer systems (see Table 1). Consequently, they initially focused on training of students in mathematics.[253] At the Universities of Berlin, Halle, Jena, Leipzig, and Rostock, as well as at the Technical University of Dresden, mathematics students received an additional computer-oriented education and training in one of three possible fields: "Mathematical Methods of Economics, Technology and Planning," which was offered at the Humboldt University of Berlin, "Mathematical Cybernetics," offered at the Universities of Leipzig, Jena, Berlin and the Technical University of Dresden, and "Numerical Mathematics and Machine Computing" which was taught at the Universities of Halle and Rostock, and the Technical University of Dresden.[254] The latter included a lecture on programming in ALGOL, as well as a practical introduction to electronic computers on the digital computer "ZRA1" and the analogue computer "Endim 2000."[255]

Against the background of the narrative of the "scientific-technological revolution" and the reforms of the New Economic System in the 1960s, the academic discipline of mathematics came under political pressure to increase its relevance to industrial production. The SED's narrative of "science as a primary productive force" made it imperative for academic disciplines to be as practice oriented as possible in their research and teaching.[256] For mathematics, as a "pure" science, this posed a serious challenge. In particular, the SED called for the "unity of scientific education and productive practice," that is, for university teaching to be more oriented towards the economy and the labor market, and for greater cooperation with industrial partners in order to meet their demands for research and education at universities. As a response, a new type of "industrial mathematician" was envisaged, whose economically oriented qualification would enable mathematicians to be employed in industry, to develop new technologies and take on tasks in the planning and management of production.[257] Expertise in the application of computer technology was seen as crucial for these mathematicians, who were expected to be able to help harness the potential of modern computer technology in order to increase economic productivity. The introduction of new fields of study in mathematics such as "Mathematical Cybernetics and Computer Technology," which would be reshaped into "Information Processing" in the 1980s, precisely served this purpose. This qualification was intended to open a broad

---

253 Pieper, *Hochschulinformatik*, 186.
254 Pieper, *Hochschulinformatik*, 187.
255 Pieper, *Hochschulinformatik*, 201.
256 Wolfgang Lambrecht, "Neuparzellierung einer gesamten Hochschullandschaft. Die III. Hochschulreform in der DDR (1965–1971)," *die hochschule*, no. 2 (2007): 181–84.
257 Pieper, *Hochschulinformatik*, 202–203.

field of employment for mathematicians in all areas of the national economy, in which automation projects involving electronic data processing systems played a central role.[258]

In 1964, the SED ratified the "Program for the Development, Implementation and Enforcement of Electronic Data Processing"[259] to kickstart organizational reforms and economic growth with the help of a domestic computer industry.[260] Electronic data processing had become an economic and political priority. In contrast to both the Soviet Union and the USA, where a substantial part of research and development in the field of computer technology was financed through military budgets,[261] the program stated clearly that the tasks of GDR electronics were not to be determined by military technology, but primarily by the civilian aspects of data processing and by operational control technology.[262] In line with the prevailing sociotechnical imaginary described above, the program listed a number of objectives.[263] These included the optimization of economic planning and management with the help of data processing machines, the automation and rationalization of research and development through the use of computer technology, which was expected to contribute decisively to increasing the speed of scientific and technological progress, and the automation of economic production by integrating computers into automated systems in industry.

The 1964 "Data Processing Program" involved the establishment of new or the expansion of existing computing centers and computing departments in universities, as well as the development of training programs for highly skilled experts in electronic data processing. It envisaged the training of 26,000 specialists for the production and operation of data processing systems by 1970.[264] This am-

---

258 Pieper, *Hochschulinformatik*, 204.
259 SAPMO DY 30 / J IV 2/2/936, minutes no. 21/64, meeting on June 23, 1964, agenda item no. 4: "Programm zur Entwicklung, Einführung und Durchsetzung der maschinellen Datenverarbeitung in der DDR in den Jahren 1964 bis 1970."
260 Sobeslavsky and Lehmann, *Geschichte von Rechentechnik und Datenverarbeitung*, 62–70.
261 Slava Gerovitch, "'Mathematical Machines' of the Cold War: Soviet Computing, American Cybernetics and Ideological Disputes in the Early 1950s," *Social Studies of Science 31*, no. 2 (2001): 253–87; Kenneth Flamm, *Creating the Computer: Government, Industry, and High Technology* (Washington, DC: Brookings Institution, 1988), 28–79.
262 Sobeslavsky and Lehmann, *Geschichte von Rechentechnik und Datenverarbeitung*, 57.
263 SAPMO DY 30 / J IV 2/2/936, minutes no. 21/64, meeting on June 23, 1964, agenda item no. 4: "Programm zur Entwicklung, Einführung und Durchsetzung der maschinellen Datenverarbeitung in der DDR in den Jahren 1964 bis 1970," 46–48.
264 Only two years later, in 1966, the Politburo of the Central Committee of the SED concluded that it would not be possible to train the required number of 26,000 IT specialists by 1970 and that the data processing program would have to be considered "not fulfilled." The main reason for this was that the target number of specialists to be trained had not been set in relation to the

bitious number included 19,000 specialists for computing stations, 3,000 specialists for the development and production as well as 2,000 for the organization and maintenance of data processing systems. It also set a target to train 1,000 to 2,000 planners, economists, and technologists, as well as 1,000 "mathematical-technical assistants" for the operation of computing machines (a job title that later changed to "programming engineer").[265]

The computer pioneer Nikolaus Joachim Lehmann and the engineer and informatician Gerhard Merkel, who at the time was deputy director of the Central Institute for Automation (ger.: Zentralinstitut für Automatisierung, ZIA), were key players in the development and implementation of the electronic data processing program. While Lehmann believed that computer experts should be trained as engineers, Merkel argued that they should be trained as mathematicians. This debate had had a direct impact on the institutional affiliation and design of the new degree programs to be established.[266] The rooting of higher education programs on electronic data processing and informatics in university departments of mathematics is therefore only one side of the story. During the same period, computer technology and electronic data processing also played an increasingly important role in engineering and electrotechnology, as exemplified by the research into computing machines at the Institute for Computing Machinery at the Technical University of Dresden under the direction of Nikolaus J. Lehmann.

As a technical discipline, primarily concerned with machine systems in industrial production, the need to introduce courses on information processing and computer technology in engineering and electrotechnology study programs seemed evident. According to the 1964 data processing program, the system of higher education was thus not only put in charge for the education of *mathematicians*, but also of *engineers* for the research and development of data processing technology, and the use of such technology in computing centers. In addition, economists and planners needed to be trained, responsible for the implementation, organization, and use of data processing technology in the industry and state administration of the GDR.

In consequence, institutions of higher education introduced a variety of study courses on computing throughout the 1960s.

---

number of school graduates and the capacity of the universities and technical schools (Pieper, *Hochschulinformatik*, 188).

265 Pieper, *Hochschulinformatik*, 173–74.
266 Birgit Demuth, Frank Rohde, and Uwe Aßmann, "50 Jahre universitärer Informatik-Studiengang an der TU Dresden aus der Sicht von Zeitzeugen in einem Zeitstrahl," *Informatik Spektrum* 45, no. 3 (2022): 185.

3.2 Institutionalization of Computer Science and Informatics in Higher Education

Table 2: Introduction of higher education study courses on computing and informatics in the GDR during the 1960s.

1962	Specialization study course "Theoretical Informatics"	Friedrich Schiller University of Jena, Faculty of Mathematics
1963	Engineering study course "Electronic Data Processing"	University of Rostock
1963	Study course "Computer Electronics"	Technical College of Ilmenau, Faculty of Light-current Technology
1965	Study course "Information Processing and Computer Electronics"	Technical University of Karl-Marx-Stadt, Faculty of Electrical Engineering
1965	Study courses "Electronic Data Processing Systems" and "Programming"	Engineering School for Mechanical and Electrical Engineering of Dresden
1965/66	Advanced study course "Mathematics and Data Processing"	Martin Luther University of Halle-Wittenberg, Faculty of Agriculture
1966	Specialization study course "Mathematical Cybernetics and Computer Technology"	University of Greifswald, Section of Mathematics
1967	Engineering study course "Data Processing"	Technical University of Magdeburg
1968	Mathematical study course "Informatics"	University of Rostock
1969	Specialization study course "Mathematical Cybernetics and Computer Technology"	Humboldt University of Berlin & University of Leipzig

Source: Data from Pieper, Hochschulinformatik, 362.

The examples listed in Table 2 illustrate the fact that starting in the 1960s, computer specialists in the GDR were in most cases either educated as mathematicians (Universities of Jena, Greifswald, Berlin, and Leipzig) or engineers (Technical Universities of Karl-Marx-Stadt, Magdeburg, Dresden, as well as at technical colleges such as in Ilmenau and Dresden). The University of Rostock trained computer specialists both as mathematicians in "Informatics" and as engineers in "Electronic Data Processing."

Starting in 1965, the Engineering School for Mechanical and Electrical Engineering in Dresden offered a postgraduate study program in electronic data processing for engineers, which covered technical subjects on machine computing technology and programming courses in ALGOL, as well as cybernetics, econometry, and the organization and planning of industrial production. In addition, "Electronic Data Processing Systems" was introduced as a new field of study.[267] In 1969, the Engineering School was re-established as the Dresden College for Engineering in Electronics and Data Processing (ger.: Ingenieurhochschule Dresden).[268] It was to serve as a lighthouse for the higher education and training of engineers in computer and automation technology as well as information processing with a strong orientation towards the qualification needs of industry. Dresden was as a hub for microelectronics and computer technology in the GDR. Key institutions for training, research, and development as well as production facilities in the field of data processing were located there, such as the Central Institute for Automation (ger.: Zentralinstitut für Automatisierung, ZIA), the German Academy of Science's Institutes for Data Processing (ger.: Institut für Datenverarbeitung, IDV) and Electronics (ger.: Institut für Elektronik Dresden, IED), as well as the IMR at the Technical University of Dresden and the Robotron combine.[269] Both the Technical University of Dresden and the Dresden College for Engineering in Electronics and Data Processing maintained a close cooperation with the Robotron combine, as well as the local microelectronics and automation research institutes. This was particularly important regarding their curricula in the field of computer technology, as a large part of their graduates – the engineers, mathematicians, and engineer-economists with qualifications in the field electronic data processing – would go on to work in the local data processing and automation industry, computing centers, or research facilities. Following the re-profiling of the Dresden College of Engineering, two new fields of study were introduced in response to the qualification needs of industry: "Systems Engineering of Data Processing" (renamed as "Information Processing" in 1972) and "Information Electronics," with subsequent specialization courses in programming, application preparation, production, measurement and test engineering, process control, and testing and maintenance.[270]

The development and introduction of new information technologies to automate economic production also went hand in hand with new theories and methods of data-driven management and organization in industry – fuelled by a sociotech-

---

267 Angela Buchwald, "Die Ausbildung von Informatikern in Dresden – frühe Anfänge," in *Informatik in der DDR – Grundlagen und Anwendungen*, ed. Birgit Demuth (Bonn: Gesellschaft für Informatik, 2008), 136.
268 Demuth, Rohde, and Aßmann, "50 Jahre universitärer Informatik-Studiengang," 186; Erwin Schmidt, "Zur Geschichte und Lehre der Informatik an der Technischen Universität Dresden bis zur Gründung des Informatik-Zentrum 1986," in *Informatik in der DDR – Grundlagen und Anwendungen*, ed. Birgit Demuth (Bonn: Gesellschaft für Informatik, 2008), 144.
269 Pieper, *Hochschulinformatik*, 227.
270 Buchwald, "Ausbildung von Informatikern," 138.

nical imaginary of computer technology as a powerful means of rationalization. This process contributed to the emergence of the field of business informatics. As a result, certain university departments and schools of economics such as the College of Economics in Berlin-Karlshorst trained specialists who were skilled in socialist business management and administration, as well as in data processing and computer technology. The successful use of new information technologies in industrial production was seen as dependent on new ways of planning and organizing work and production processes. Therefore, highly educated management and administrative staff with a thorough understanding of new technologies were considered essential to prepare and implement the introduction of computerized machines and systems in factories and plants.[271] The SED politburo candidate Günther Kleiber[272] argued in 1968 that managers of socialist enterprises did not need to be trained as electronic data processing specialists, programmers, or maintenance technicians. Rather, he elaborated, it was necessary to start with the basics of socialist management and economic cybernetics, and to teach "[. . .] such knowledge of data processing that the managers will be able to recognise the possibilities of its use, to determine the tasks to be worked on and to control them regularly; in short, they must be able to decide what is to be done where and in what period of time with the help of data processing."[273]

In 1968 the section "Economic Cybernetics and Operations Research" (ger.: Ökonomische Kybernetik und Operationsforschung) was established at the Humboldt University of Berlin.[274] The section included a department for "System De-

---

[271] "Computer fordern neue Berufe," *Neues Deutschland*, December 24, 1966, 12; "Elektronische Datenverarbeitung – Instrument moderner Leitung," *Neues Deutschland*, June 9, 1968, 4.
[272] Kleiber served as Deputy Minister for Electrical Engineering and Electronics, as well as State Secretary for the Coordination of the Deployment and Use of Electronic Data Processing at the Presidency of the Council of Ministers of the GDR between 1966 and 1971. He became a member of the SED's politburo in 1984. (Helmut Müller-Enbergs, "Kleiber, Günther," in *Wer war wer in der DDR? Ein Lexikon ostdeutscher Biographien*, ed. Helmut Müller-Enbergs et al. (Berlin: Christoph Links Verlag, 2010), https://www.bundesstiftung-aufarbeitung.de/de/recherche/kataloge-datenbanken/biographische-datenbanken (accessed May 31, 2024)
[273] "Elektronische Datenverarbeitung – Instrument moderner Leitung," *Neues Deutschland*, June 9, 1968, 4 (translated from German).
[274] When cybernetics lost political favor with the SED from 1969 onwards and efforts to reform the socialist economic system with the help of economic cybernetics and systems research were abandoned, the section was renamed 'Theory of Science and Organization of Science' (ger.: Wissenschaftstheorie und Wissenschaftsorganisation) and reprofiled accordingly. However, its department for "System Design and Automated Information Processing" was preserved (Klaus Fuchs-Kittowski et al., "Gründung, Entwicklung und Abwicklung der Sektion Ökonomische Kybernetik und Operationsforschung/Wissenschaftstheorie und Wissenschaftsorganisation an der Humboldt-Universität zu Berlin," in *Die Humboldt-Universität Unter den Linden 1945 bis 1990*.

sign and Automated Information Processing" for the training of electronic data processing organizers and systems analysts. The qualification of these highly skilled specialists was intended to contribute to the effective integration of automated data processing into social organizations and, focus on the sociotechnical aspects of creating and managing complex human-machine systems.[275]

The computing centers in institutions of higher education were also tasked with the education and training of field-specific specialists in other disciplines, usually in the form of specialization or advanced courses of study. The training agricultural students in "Mathematics and Data Processing" at the University of Halle can serve as an example of this. The use of computer technology to modernize farm accounting and administration, or of data-processing technology to monitor and increase the yield of crops or dairy cows, for example, were promoted early on in the GDR press as concrete applications of the new information technology.[276] Teaching the next generation of agronomists the basics of the new information technology was a means of promoting the use of computers and data processing in agriculture in the GDR. Advanced or specialization courses were aimed at providing students and experienced practitioners with the necessary computer skills and knowledge to find innovative ways of solving tasks and problems in their field of expertise, thus developing new computer-based solutions to improve productivity or product quality.

An important juncture in the disciplinary development of computing and information processing was the third reform of higher education in the GDR, which began in 1967 and was closely linked to the economic reforms of the New Economic System at the time.[277] The SED leadership's aim behind the major reform was threefold. On the one hand, it sought to pervade the universities both politically and ideologically, and to assert the party's claim to leadership. On the other hand, it also aimed to forge closer links between science and industry, particularly in the natural sciences and engineering, to increase the function of science as a "productive force" in the socialist economy. Finally, the reforms aimed to increase the efficiency of research and teaching through structural reorganization and the streamlining of study

---

Zeitzeugen – Einblicke – Analysen, ed. Wolfgang Girnus and Klaus Meier (Leipzig: Leipziger Universitätsverlag, 2010, 158–60).
275 Fuchs-Kittowski et al. "Gründung, Entwicklung und Abwicklung," 162–65 and 173–77.
276 "Computer für die Landwirtschaft," *Neue Zeit*, January 29, 1969, 6; "Stammkarte für jede Kuh: Rechentechnik und Kybernetik helfen der Viehzucht," *Neue Zeit*, October 20, 1968, 5; "Computer in der Landwirtschaft: Vollautomatisierte Leitungssysteme in der Erprobung," *Berliner Zeitung*, April 2, 1969, 4; "Futterbedarf mit dem Computer optimiert," *Neues Deutschland*, January 27, 1971, 6.
277 Lambrecht, "Neuparzellierung," 171.

programs. For the field of informatics and computing, the dismantling of the traditional structures of higher education institutions was a particularly significant development – namely, the dissolution of the existing institutes in favor of a much smaller number of new interdisciplinary units, the so-called "sections". In 1969/70, merely 170 sections had emerged from the formerly more than 900 institutes at the universities and technical colleges of the GDR.[278] For the field of information processing, this was an opportunity to step out of the shadows of the mathematics and engineering institutes which it had been a part of, and to form its own unit – an important step towards becoming its own discipline.

At the Karl-Marx-University in Leipzig, a new section for "Computer Technology and Data Processing" was created in 1969, in which the former Institute for Computing Machinery with the university computing center was integrated.[279] The section for mathematics had strongly, but unsuccessfully opposed the separation of its mathematical information processing branch. The new section for "Computer Technology and Data Processing" moved away from a strong focus on mathematical computing and broadened its scope in the teaching of electronic data processing specialists. In addition to the computer education and training of mathematicians in "Mathematical Cybernetics and Computer Technology," it also taught electronic data processing and computing to students in other disciplines, as well as the postgraduate course in electronic data processing for senior executives in the local economy.[280]

Similarly, new sections for mathematics and information processing were created in 1968 and 1969, respectively, at the Technical University of Dresden. The section for mathematics was responsible for the education of Diploma Mathematicians, which were taught in "Mathematical Cybernetics and Computing," probability theory and mathematical statistics, as well as numerical mathematics. Teaching in information processing focused on theoretical informatics and its mathematical foundations.[281] The section for information processing, on the other hand, offered a dedicated study program in information processing. After eight (later nine) semesters of studying, focusing more on the aspect of application and programming technology in electronic data processing systems, graduates were awarded the degree of Diploma Engineer.[282] In Dresden, the creation of

---

278 Lambrecht, "Neuparzellierung," 179.
279 Pieper, *Hochschulinformatik*, 207.
280 Pieper, *Hochschulinformatik*, 206.
281 Pieper, *Hochschulinformatik*, 242.
282 Silvia Kapplusch, "Geschichte der Informatik an der technischen Universität Dresden," Technische Universität Dresden, last modified September 6, 2016, https://tu-dresden.de/ing/informatik/

a dedicated information processing section in 1969 thus led to a shift away from mathematically oriented informatics towards a more engineering-oriented curriculum with a far greater emphasis on the technical aspects of computer systems. The establishment of the Dresden College of Engineering with a dedicated information processing section in the same year served to further expand the capacity for higher education and training of computer engineers.[283]

In contrast, at the Universities of Halle-Wittenberg, Rostock, and Berlin, no sections for information processing were established in the late 1960s, and university computing centers were instead integrated into the new sections for mathematics. This meant that the education and training of experts in computing and information processing was carried out under the auspices of the section for mathematics, leading to the degree of a Diploma Mathematician. The respective courses for both mathematicians and students of other disciplines focused on numerical mathematics and programming, including practical courses at the university computing centers.[284] In the case of these three universities, the field of computing remained strongly linked to the discipline of mathematics, both in terms of its institutional ties and the focus of its research and teaching. It was not until 15 to 20 years later, that a dedicated section of department for information processing or informatics was established. In Rostock, a section for information processing was created in 1984 which took over the training of informaticians from the section of mathematics.[285] In Halle, an institute for informatics was finally established in 1989,[286] and at the Humboldt University of Berlin, a department for informatics was created in 1990.[287]

Despite these differences between universities, the establishment of information processing departments at several universities and the creation of dedicated information processing study programs in Dresden – notably technically rather than mathematically oriented – was a breakthrough for the nascent discipline of informatics in the GDR. The Dresden Technical University's program for training engineers in information processing was subsequently developed by the Ministry

---

die-fakultaet/geschichte/geschichte-der-informatik-an-der-technischen-universitaet-dresden (accessed May 31, 2024).
2016; Pieper *Hochschulinformatik*, 242.
283 Schmidt, "Geschichte und Lehre der Informatik," 144.
284 Pieper, *Hochschulinformatik*, 207–15 and 226.
285 Benjamin Venske, *Das Rechenzentrum der Universität Rostock 1964–2010* (Rostocker Studien zur Universitätsgeschichte, Vol. 19) (Rostock: Universität Rostock, 2012), 39–40.
286 Pieper, *Hochschulinformatik*, 211.
287 Pieper, *Hochschulinformatik*, 215.

of Higher Education into a binding national academic curriculum in information processing, which was implemented in 1976.[288]

The study program consisted of four and a half years of studies and led to the award of a diploma in "Information Processing Engineering".[289] It had a decidedly technical orientation, and prepared students for the development and use of computer systems for automated information processing in industrial production and economic management and planning. Programming instruction and the theory, development, and implementation of automatic information processing systems accounted for the largest share of the study program in terms of teaching hours. Students were taught in the theory and methods of machine- and problem-oriented programming, "algorithmic thinking," and introduced to the hardware basics of automated information processing systems. In addition, a significant share of teaching was dedicated to fundamentals in physics and mathematics related to information technology (e.g., numerical mathematics, automation, and algorithm theory), as well as the basics of Marxism-Leninism and socialist business administration. The practical orientation of the course was supported by practical exercises on computers, a four-week internship in a socialist company, as well as a practical semester involving a work placement in industry.[290]

The national basic study program for information processing was an important step in the emergence of informatics as an independent discipline in the GDR, but clearly positioned it as a field of engineering science. At a colloquium in 1984 on the further development of the education and training system for engineers and economists in the GDR, the Minister for Higher Education, Hans-Joachim Böhme, spoke of the "emergence of informatics as an independent *engineering* discipline" in the context of the growing role of production automation and the accelerated development and application of microelectronics.[291]

In the field of mathematics and economics, on the other hand, information processing did not constitute a basic study program (ger.: Grundstudienrichtung) in its own right but was part of the mathematics and economics study programs in the form of a possible specialization amongst others (ger.: Fachstudienrichtung). While the basic study program for mathematics of 1976 listed "Mathematical Cyber-

---

288 Ministerium für Hoch- und Fachschulwesen, *Studienplan für die Grundstudienrichtung Informationsverarbeitung zur Ausbildung an Universitäten und Hochschulen der DDR* (Berlin: Ministerium für Hoch- und Fachschulwesen, 1976).
289 ger.: Diplomingenieur für Informationsverarbeitung.
290 Ministerium für Hoch- und Fachschulwesen, *Studienplan Grundstudienrichtung Informationsverarbeitung*, 5–15.
291 Hans-Joachim Böhme, "Aus- und Weiterbildung der Ingenieure und Ökonomen," *Das Hochschulwesen* 33, no. 3 (1985): 62 (translated from German, emphasis added).

netics and Computer Technology" as a specialization encompassing 256 hours of instruction,[292] the new study program of 1982 included 300 hours of instruction in "Information Processing" as part of the basic study in mathematics, followed by a possible specialization in the field encompassing another 360 hours of instruction.[293] At the Humboldt University of Berlin, however, a special variant of the basic study program in mathematics was offered, which expanded the subject area of information processing at the expense of instruction in mathematical analysis.[294] In addition to the basic education in mathematics, students of mathematics at the Humboldt University of Berlin received a more extensive basic training in informatics consisting of 740 hours, which included the programming languages PASCAL, PL/I, MODULA or CDL, as well as instruction in compiler technology and computing architectures, operating systems, and computer simulation.[295]

In economics, the national study programs of 1972 and 1982 both list the subject area "Mathematical Methods and Data Processing in the Economy" as a possible field of specialization. It was aimed at preparing students for the development and implementation of data processing systems in combines, government bodies, higher education, research institutions, as well as organization and computing centers.[296] In 1986, a dedicated section for "business informatics" was established at the College for Economics in Berlin. It included the scientific fields of data processing, operations research, accounting and statistics, and management of the socialist economy, as well as the computing center.[297] The creation of the section for business informatics in Berlin was an expression of a further branching out of the field of information processing into a new discipline. Positioned in the field of economics, the newly institutionalised discipline of "business informatics" was far less technically oriented than

---

292 Ministerium für Hoch- und Fachschulwesen, *Studienplan für die Grundstudienrichtung Mathematik zur Ausbildung an Universitäten und Hochschulen der DDR* (Berlin: Ministerium für Hoch- und Fachschulwesen, 1976), 17.
293 Ministerium für Hoch- und Fachschulwesen, *Studienplan für die Grundstudienrichtung Mathematik zur Ausbildung an Universitäten und Hochschulen der DDR* (Berlin: Ministerium für Hoch- und Fachschulwesen, 1982), 14–15.
294 Ministerium für Hoch- und Fachschulwesen, *Ergänzung zum Studienplan für die Grundstudienrichtung Mathematik. Mathematische Informatik* (Berlin: Ministerium für Hoch- und Fachschulwesen, 1982), 1.
295 Ministerium für Hoch- und Fachschulwesen, *Ergänzung Studienplan Grundstudienrichtung Mathematik*, 2–5.
296 Ministerium für Hoch- und Fachschulwesen, *Studienplan für die Grundstudienrichtung Wirtschaftswissenschaften* (Berlin: Ministerium für Hoch- und Fachschulwesen, 1972), 12.
297 Peter Zschockelt, "Wirtschaftsinformatik an der HfÖ Berlin," in *Informatik in der DDR – Tagung Berlin 2010. Tagungsband zum 4. Symposium 'Informatik in der DDR' am 16. und 17. September 2010 in Berlin*, ed. Wolfgang Coy and Peter Schirmbacher (Berlin: Humboldt-Universität zu Berlin, 2010), 148–49.

the conceptualization of information processing at the Technical University and College of Engineering in Dresden, and instead focused on aspects of computer-assisted organization, planning, and management of the socialist economy.

In the field of agriculture, students were trained either as agricultural engineers or agricultural economists at technical colleges, schools of engineering or schools of agricultural engineering.[298] Informatics was introduced as a subject in the degree programs following a decision by the SED leadership in 1986, which stipulated that a sound knowledge of modern information processing should become an objective of higher education for agricultural engineers and economists.[299] However, the amount of time dedicated to informatics, and the contents of the course varied greatly between agricultural and economic specializations. In the former, informatics was taught in merely 68 hours and included the development, organization and aims of data processing in agriculture, the basics of informatics, the use of standardized software, and the practical use of computers to model various aspects of an agricultural business. The students would, for example, use ready-made computer software to calculate and optimize the feeding of farm animals.[300] In contrast, 140 hours were dedicated to the teaching of informatics in economic specializations, focusing additionally on the basics of algorithms and programming, data collection and information processing project development, as well as the use of computers for the rationalization of management, planning, and administration.[301]

Informatics in the field of agriculture was thus divided into a more technically oriented "engineering" branch on the one hand, focusing on computerized agricultural machinery as well as computer hard- and software to automate tasks and control processes in agricultural production, and a more management-oriented "business economics" branch on the other, that focused on economic rationalization, socialist management, and planning on farms using computers as powerful information processing tools.

---

[298] Hartmut Koch, "Stand und Probleme der Informatikausbildung an landwirtschaftlichen Fachschulen der DDR," in *Agrarinformatik, Band 19: Referate der 11. GIL-Jahrestagung in Nürtingen, September 1990*, ed. Hans Geidel, Reiner Mohn, and Gerhard Schiefer (Stuttgart: Ulmer, 1990), 353.
[299] "Konzeption für die langfristige Entwicklung der Aus- und Weiterbildung der Agraringenieure und Agrarökonomen an den Hoch- und Fachschulen der DDR" [Decision of the Politburo of the Central Committee of the SED of 8 April 1986/Decision of the Council of Ministers of the GDR of 18 April 1986], *Das Hochschulwesen* 34, no. 7 (July 1986): 167–70.
[300] "Ausbildungsstätte für junge Agrar-Ingenieure in Zierow," *Berliner Zeitung*, January 5, 1988, 3.
[301] Koch, "Stand und Probleme der Informatikausbildung," 354.

Following the institutionalization of information processing and informatics in higher education in the form of dedicated sections and study programs, the next big step was the formation of a Society for Informatics in the GDR. In West Germany, such a Society had already been established in 1969.[302] The GDR equivalent, the "Gesellschaft für Informatik der DDR" (GIDDR) was founded in 1985 by computer scientists as a "scientific society" affiliated with the Academy of Sciences of the GDR.[303] Its members organized scientific meetings and conferences as well as continuing education events. The executive committee published the *GI-Mitteilungen*, a journal containing scientific contributions as well as information and reports on events and activities of the GIDDR. In particular, the Society used the term "informatics" rather than "information processing" in its name – an expression of its understanding of informatics as a scientific discipline, rather than a collection of automated methods of data processing.

At the same time, the basic study program "information processing" was further developed and renamed to "informatics." New curricula for informatics were implemented in 1986 at the Technical Universities of Karl-Marx-Stadt, Magdeburg, and Dresden, as well as at the Engineering College in Dresden, and the University of Rostock.[304] The new study program included specializations in "Theoretical Informatics," "Systems Software," "Applied Informatics," as well as "Computer Systems Design and Operation." In contrast to the previous study programming in information processing, the new basic study program in informatics provided an explicit definition of informatics as a scientific discipline with its own merits, scientific methods, and research interests, beyond its practical value for the socialist economy. In the context of the new study program for informatics, it was defined as follows:

> Informatics is a very young and rapidly developing science. It deals with the systematic processing of information, the basic procedures for obtaining, processing, storing, retrieving, and transmitting information and the general methods of application. It is based on abstraction and modelling of objects and processes of objective reality with the aim of effectively controlling informational processes. Informatics provides the scientific basis for

---

302 Fritz Krückeberg, *Die Geschichte der GI* (Bonn: Gesellschaft für Informatik, 2001), 14.
303 Gerhard Merkel, "Bildung und Wirken der Gesellschaft für Informatik der DDR," in *Informatik in der DDR – eine Bilanz*, ed. Friedrich Naumann and Gabriele Schade (Bonn: Gesellschaft für Informatik, 2006), 451–61.
304 Ministerium für Hoch- und Fachschulwesen, *Studienplan für die Grundstudienrichtung Informationsverarbeitung zur Erprobung der Ausbildung in Verwirklichung der Konzeption für die Gestaltung der Aus- und Weiterbildung der Ingenieure und Ökonomen in der Deutschen Demokratischen Republik an der Technischen Universität Dresden, der Technischen Hochschule Karl-Marx-Stadt, der Technischen Hochschule Magdeburg, der Wilhelm-Pieck-Universität Rostock und der Ingenieurhochschule Dresden* (Berlin: Ministerium für Hoch- und Fachschulwesen, 1986).

modern information and communication technologies. It uses computer and communication technology based on microelectronics.[305]

The 1976 curriculum, on the other hand, had viewed information processing under the primacy of economic development, which required intensive study of the new computer technology and the training of highly educated specialists to operate and develop it. While this earlier engineering course in information processing had been primarily a pragmatic educational response to the economic policy task of qualifying engineers for the development and use of computer systems in the economy, the new study course was an expression of the growing recognition of informatics as its own scientific discipline.

The new study program not only included goals for the professional qualification of the training, but also for the education of the future computer engineers to become "socialist personalities" with ideological and party-political steadfastness.[306] To prepare students for the job of a computer engineer in the socialist society, a higher education was required that would provide graduates with a high "sense of political responsibility and moral attitude" and a "firm socialist class stand."[307] For this purpose, the study plan included a basic education in Marxism-Leninism, which was mandatory for all students in higher education since the higher education reform of the 1950s, regardless of their field of study. Students in informatics were also expected to demonstrate an exemplary work ethic, creativity, perseverance, determination, and modesty.[308] In addition, the study program also listed four specific qualities that future computer engineers needed to have.[309] First, the work in collectives required good communication and cooperation skills. This was deemed as particularly important, as information processing technology was considered as a field where skilled workers from quite different fields and educational backgrounds would have to work together. Second, and related to the first point, was the readiness of computer engineers to familiarize themselves with new areas of expertise, as this would allow for a successful implementation and use of computer technology in new fields of applica-

---

[305] Ministerium für Hoch- und Fachschulwesen, *Studienplan Grundstudienrichtung Informationsverarbeitung*, 4 (translated from German).
[306] Ministerium für Hoch- und Fachschulwesen, *Studienplan Grundstudienrichtung Informationsverarbeitung*, 4–5.
[307] Ministerium für Hoch- und Fachschulwesen, *Studienplan Grundstudienrichtung Informationsverarbeitung*, 4.
[308] Ministerium für Hoch- und Fachschulwesen, *Studienplan Grundstudienrichtung Informationsverarbeitung*, 5.
[309] Ministerium für Hoch- und Fachschulwesen, *Studienplan Grundstudienrichtung Informationsverarbeitung*, 20.

tion. For example, the automation of fruit plucking on a farm would require for the computer engineer to get acquainted with the work processes on a fruit farm, to be able to model them and find an appropriate solution where computers could successfully take over certain tasks or even entire workflows. Third, the continued willingness to undergo further training was stated as a prerequisite, as both computer technology and the field of informatics were rapidly progressing, and thus required for experts in this area to continuously update their knowledge and skills. Finally, the study program listed the need for the ability to withstand intense physical and psychological stress to be able to successfully execute the demanding job of an informatics engineer in the GDR economy.

Until 1986, the Technical University of Dresden and the Dresden College of Engineering had been educating students in information processing in parallel. The same year that the new curricula in information processing were implemented, the two educational institutions were brought together in autumn of 1986 to form a unified university center for computer education and research in Dresden. The former College of Engineering was merged with the Technical University's section for information processing into the newly established "Informatics Center of Higher Education" at the Technical University of Dresden on 4 October 1986.[310] With an annual enrollment of 400 to 500 full-time and distance learning students, the "Informatics Center" quickly became the largest academic training center for informatics and computing in the GDR. The right to award doctorates, however, remained with the Faculty of Electrical Engineering and Electronics.[311]

During the 1970s and 1980s, the emerging field of informatics found fertile ground in several other disciplines, as its theories, methods and tools were seen as applicable to a wide range of sociotechnical systems. This is illustrated by the branching out of the field of computing and information processing in different directions within the spheres of engineering, mathematics, and economics, as described above, but also in agriculture and the natural sciences. Particularly in the 1980s, the development of smaller, more powerful, and cheaper microcomputers made it possible to expand computing education in higher education, as practical training no longer had to rely solely on mainframe computers in large computing facilities.

This expansion of computing in higher education is illustrated by the 1985 action plan devised by the Ministry of Higher Education, which envisaged a massive increase in computer equipment by 1990: In addition to more than 1,000 office

---

310 Ottomar Herrlich, "Gründung und Wirken der Sektion Informationsverarbeitung der Technischen Universität Dresden – ein Gedächtnisbericht," in *Informatik in der DDR – eine Bilanz*, ed. Friedrich Naumann and Gabriele Schade (Bonn: Gesellschaft für Informatik, 2006), 329.
311 Demuth, Rohde, and Aßmann "50 Jahre universitärer Informatik-Studiengang," 186.

3.2 Institutionalization of Computer Science and Informatics in Higher Education — 83

computers and data processing systems, the plan stipulated an annual need for 2,000 domestically produced small computers for the training of students in higher education and technical colleges.[312] This figure even exceeded the needs of the entire vocational training sector, estimated at 6,000 small computers in total,[313] as well as the needs of the elementary school system, estimated at 1,600 to 1,700 computers in the same period.[314]

The action plan, drawing on the SED's sociotechnical imaginary of computer technology, expressed the conviction that *all* students needed to receive an education and training in computing. Several facets of the SED's imaginary were invoked, specifically the political and economic importance of the new technology for the development of the GDR and in the struggle between socialism and capitalism, as well as its intrinsic value for the development of human forces and the socialist personality:

> In the unity of communist education and technical education, a deep understanding of the political and economic significance of informatics for the development of the productive forces in all spheres of society must be achieved among all students. Recognizing the value of informatics for the development of human creativity should instil in students the will to use their education in informatics [. . .] for the development of their own personality. Every student must become aware of the class character of the use of informatics in confrontation with the anti-human use of this key technology in the imperialist states.[315]

However, the aim was not to train all students in all disciplines to become computer *experts*. Rather, a four-tiered model was developed for a differentiated higher education of skilled personnel in computing and information technology.[316]

---

312 SAPMO DY 30/J IV 2/2/2138, "Maßnahmeplan des Ministeriums für Hoch- und Fachschulwesen zur Realisierung der Konzeption 'Standpunkte zu Konsequenzen aus der Entwicklung der Informatik und informationsverarbeitenden Technik für das Bildungswesen'," 5.
313 SAPMO DY 30/J IV 2/2/2138, "Maßnahmeplan des Staatssekretariats für Berufsbildung zur Realisierung der Konzeption 'Standpunkte zu Konsequenzen aus der Entwicklung der Informatik und informationsverarbeitenden Technik für das Bildungswesen'," 17.
314 SAPMO DY 30/J IV 2/2/2138, "Maßnahmeplan des Ministeriums für Volksbildung zur Realisierung der Konzeption 'Standpunkte zu Konsequenzen aus der Entwicklung der Informatik und informationsverarbeitenden Technik für das Bildungswesen'," 10.
315 SAPMO DY 30/J IV 2/2/2138, "Maßnahmeplan des Ministeriums für Volksbildung zur Realisierung der Konzeption 'Standpunkte zu Konsequenzen aus der Entwicklung der Informatik und informationsverarbeitenden Technik für das Bildungswesen'," 4 (translated from German).
316 BArch DR 3/26974, "Dienstbesprechung beim Minister," dated December 22, 1986: "Vorlage Nr. 189/86: Volkswirtschaftliche Gesamtkonzeption zur Qualifizierungsstrategie auf dem Gebiet CAD/CAM," 6–7.

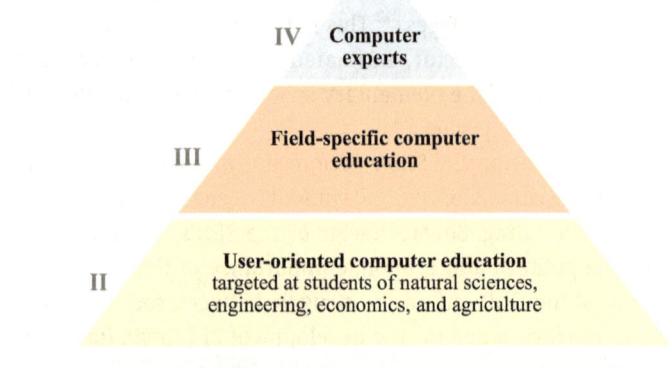

Figure 2: Four-tier model of computer technology education in higher education in the GDR. Source: Graphical representation by the author, based on SAPMO DY 30/J IV 2/2/2138, "Standpunkte zu Konsequenzen aus der Entwicklung der Informatik und informationsverarbeitenden Technik für das Bildungswesen," 5–8.

The lowest tier encompassed the teaching of basic computer science skills to large numbers of students in all disciplines, especially also in the humanities and social sciences.[317] It was aimed at teaching students the necessary knowledge and skills to use computer technology in their specific field. This strategy of introducing computer technology to all students across disciplines reflects the underlying sociotechnical imaginary, which posits computers as powerful tools that penetrate all areas of human thought and work. As a result, highly educated professionals in all branches of industry, business, and science needed to understand the potential of this new technology to advance their respective fields in line with the SED's vision of how computers could be used to modernize and grow the socialist economy of the GDR.

The second tier involved a more in-depth, user-oriented computer education and training in CAD/CAM, aimed at students in the natural sciences, engineering, economics and agriculture. Their instruction involved the knowledge of a higher programming language, as well as the necessary skills to use data management and

---

[317] SAPMO DY 30/J IV 2/2/2138, Attachment "Standpunkte zu Konsequenzen aus der Entwicklung der Informatik und informationsverarbeitenden Technik für das Bildungswesen," 5.

software systems to solve tasks in their respective field of expertise. The third level was aimed at providing an advanced education and training in informatics and computing for approximately 15% of the students in technical, economic, and agricultural disciplines, that would enable the sophisticated use of this technology for specialized purposes, such as the creation of application software specific to a particular field of application in research or industrial production. These students received in-depth programming training in a second programming language, instruction in computer and operating systems, as well as training in the implementation, further development, and maintenance of user software in their field. The training of subject-specific specialists at levels two and three corresponded to the SED's sociotechnical imaginary of computer technology as a means of increasing economic productivity by optimizing sociotechnical systems in technical fields, economics, and agricultural sciences. The introduction of comprehensive computer training for students in these disciplines was therefore prioritized.

The final and highest level of higher computer education was reserved for the training of highly qualified computer experts through a dedicated study program in informatics. These graduates in informatics needed to be enabled to work in close cooperation with users to develop hard- and software tools and systems according to specific user needs. Their task consisted in the further advancement of GDR computer hard- and software in both interdisciplinary and international cooperation within the Socialist Bloc.

In contrast to its historical roots in mathematics, informatics and computing had become a highly interdisciplinary field, both in terms of research and higher education. The availability of microcomputers, alongside a vigorous political push for modernizing the GDR's industry with the help of computer-aided manufacturing and design (CAD/CAM) in the 1980s[318] had further shifted the focus of higher education and training in informatics and computer technology closer to the technical sciences. By the second half of the 1980s, most graduates of universities and technical colleges that were educated in informatics or computer-aided manufacturing and design had a background in the technical sciences, followed by economic sciences, agricultural sciences and finally mathematics and natural sciences.[319]

---

318 Gerhard Merkel, "Computerentwicklungen in der DDR – Rahmenbedingungen und Ergebnisse," in *Informatik in der DDR – eine Bilanz*, ed. Friedrich Naumann and Gabriele (Bonn: Gesellschaft für Informatik, 2006), 52.

319 These three academic fields – engineering, economics, and agriculture – were considered "the principal fields of study for the social development of the GDR" (Horst Möhle, *Aus- und Weiterbildung Erwachsener auf Hochschulebene in der DDR* (Hagen: Zentrales Institut für Fernstudienforschung, 1986), 10).

Table 3: University graduates with a higher education in informatics and/or CAD/CAM, differentiated according to the level of computer technology education they received (see Figure 2).

Field of Study	Level	Number of Graduates in the year					
		1986	1987	1988	1989	1990	Total
Mathematics and Natural Sciences	I	1400	1400	1400	1400	1400	7000
	II	100	200	500	750	900	2450
	III	n/a	20	50	100	150	320
Technical Sciences	I	7500	7500	7500	7500	7500	37500
	II	3500	5000	7500	7500	7500	31000
	III	375	750	1125	1125	1125	4500
Economic Sciences	I	2750	2750	2750	2750	2750	13750
	II	100	1600	2750	2750	2750	9950
	III	n/a	n/a	130	275	410	815
Agricultural Sciences	I	1250	1250	1250	1250	1250	6250
	II	n/a	1250	1250	1250	1250	5000
	III	n/a	60	125	190	190	565
Information Processing (1990: Informatics)	IV	250	250	250	270	380	1400
Total	I	12,900	12,900	12,900	12,900	12,900	64500
	II	3,700	8,050	12,000	12,250	12,400	48400
	III	375	830	1,430	1,690	1,875	6200
	IV	250	250	250	270	380	1400

Source: BArch DR 3/26974, "Dienstbesprechung beim Minister," dated December 22, 1986, attachment no. 1, "Vorlage Nr. 189/86: Volkswirtschaftliche Gesamtkonzeption zur Qualifizierungsstrategie auf dem Gebiet CAD/CAM," 14.

Notable across all fields of study is the relatively low share of graduates trained in informatics on levels III (field-specific specialists) and IV (generalized computer specialists) compared to those who received a basic user training with no or very limited programming instruction. The SED's push for a computer education for all students must therefore be understood as an effort to provide students of all disciplines with a basic understanding of computing, but above all to inculcate in them the SED's sociotechnical imaginary. According to the SED's vision of a computerized future under socialism, highly qualified graduates were to be familiarized with the myriad possibilities offered by modern computer technology for achieving economic prosperity and social progress in the GDR. In contrast, only a limited number of students, especially in the technical sciences and in special courses of study on information processing, received in-depth specialized training in computer science.

3.3 Continuing Education and Training in Computing and Information Processing

Computer technology and its application in economic production developed rapidly during the 1970s and 1980s. The political program of the SED reflected these changes and made it imperative to respond quickly to new developments on an international scale and to the new skill requirements that resulted from them. The SED's economic strategy for the 1980s reaffirmed the paramount importance attached to microelectronics and automation technology in rationalization and boosting economic productivity, as well as to overcome work activities that were deemed detrimental to health, physically heavy, and monotonous.[320] The further development of microcomputers, industrial robots, and CAD/CAM systems for the use in industrial production was an essential component of this economic program. Consequently, the rapid adoption of the latest technological innovations in research and development and their transfer to the workplace was an essential aspect of the SED's plan to increase economic competitiveness of the GDR's socialist economy. This in turn meant that to realize the sociotechnical imaginary of computing as a driver of economic competitiveness and progress, it was essential that the highly skilled people working on and with computers constantly updated their skills and knowledge to keep up with this rapidly evolving technology. Against this background, lifelong learning became an increasingly important role for higher education institutions to facilitate the transfer of skills and knowledge between research and industrial practice. A 1977 report by the Training Center for Electronic Components at the Technical University of Karl-Marx-Stadt emphasized this close link between technological progress, economic competitiveness, and continuing education:

> Especially under the conditions of the scientific-technological revolution, the demands for continuous targeted further education as part of all-round personality development are particularly high, since scientific-technological progress is the main factor in the intensification of our national economy, and socialist intensification is the decisive issue in increasing the effectiveness of social production. [. . .] Once acquired, education and experience alone are no longer sufficient to survive in modern research, development, and production or to be able to positively influence and advance them.[321]

---

[320] Zentralkomitee der SED, *Bericht des Zentralkomitees der Sozialistischen Einheitspartei Deutschlands an den X. Parteitag der SED, Berichterstatter: Genosse Erich Honecker* (Berlin: Dietz Verlag, 1981), 56.
[321] BArch DR 3/26535, H. G. Schneider, H. Wiegand, and J. Frühauf: "Erfahrungen aus der Arbeit des Weiterbildungszentrums Elektronische Bauelemente an der Technischen Hochschule Karl-Marx-Stadt," 4 (translated from German).

The political and economic importance that the SED attached to modern computer technology and the characterization of informatics as a constantly and rapidly developing field of knowledge thus played an important part in promoting the notion of "lifelong learning" in the GDR and, at the same time, in assigning a key role to universities as providers of further education and training in this regard.[322] In anticipation of the rapid scientific and technological advances in the new information processing technologies, higher education institutions were tasked with providing various forms of continuing education and training. To meet the different needs of highly qualified personnel in research and industry, the following four forms of further education and training were offered:[323]

The first involved the transfer of knowledge between science and industry, in the form of so-called "problem seminars". These were regular organized exchanges of experience between academics in higher education and research, and practitioners in industry, focusing on specific new developments and findings in the field.

Secondly, institutions of higher education arranged further education and training for their own staff, namely university lecturers and junior researchers. This served both as a means of transferring knowledge and expertise between different scientific research and higher education institutions, as well as to ensure up-to-date education and training for students and emerging researchers.

Third, universities were tasked with the further development and expansion of postgraduate studies in informatics. These courses were aimed at highly skilled practitioners in industry to acquire an in-depth education in new information technology, the theories and methods of automated information processing, and their practical application.

Finally, institutions of higher education also offered a broad range of further education and training courses in the field of computing and information technology, in coordination with the Chamber of Technology (germ.: Kammer der Technik, KdT) and other training institutions. These courses were de-

---

[322] An increase in the importance attached to adult education in education policy in response to technological change and the associated upheavals in the labor market in the 1970s and 1980s can also be observed in the European Communities and Switzerland. See Michael Geiss, "Die Politik des lebenslangen Lernens in Europa nach dem Boom," *Zeitschrift für Weiterbildungsforschung* 40, no. 2 (2017): 211–28; Lucien Criblez, "Zur Entwicklung der Weiterbildung/Erwachsenenbildung während der Bildungsexpansionsphase der 1960er- und 1970er-Jahre," *Education Permanente*, no. 2 (2022): 8–19.

[323] SAPMO DY 30/J IV 2/2/2138, "Maßnahmeplan des Ministeriums für Hoch- und Fachschulwesen zur Realisierung der Konzeption 'Standpunkte zu Konsequenzen aus der Entwicklung der Informatik und informationsverarbeitenden Technik für das Bildungswesen'," 2–3.

signed as a short-term adult and continuing education measure, focusing on specific aspects of new computer technology and its applications, and tailored to the needs of local industry.

The new role of higher education institutions in providing continuing education and training for highly skilled scientific and industrial personnel was further consolidated with the institutionalization of centers on continuing education in key fields of the SED's science and technology policy. In 1978, a first center for continuing education in the field of microelectronics was set up,[324] which was to coordinate the content and organization of continuing education programmes on microelectronics and computer technology offered by the various educational institutions.[325] Fifteen universities and colleges formed part of the coordination center, as well as three engineering schools, all district and professional associations of the KdT, and a number of relevant institutes of the GDR's Academy of Sciences. A new center for coordinating continuing education in "informatics" was set up at the "Informatics Center for Higher Education" which had been established in 1986 in Dresden, replacing six territorially oriented "Leading Centers for Continuing Education in Information Processing" (ger.: Leitzentren der Weiterbildung für Informationsverarbeitung).[326]

However, continuing education not only played a role in keeping the academic and industrial expert's knowledge on computer technology up to date with scientific and technological advances, but also to strengthen and secure their ideological and political alignment with the SED's sociotechnical imaginary. In a report on the experiences gained at the center for continuing education on electronic components at the Technical University in Karl-Marx-Stadt, the authors explicitly pointed out that continuous further education was not only aimed at contributing to technical progress and increases in productivity, but always had to foster the "all-round development and education of socialist personalities" which "consciously shaped social life." The tasks for the GDR's system of further education were therefore "not be derived unilaterally from the development of

---

[324] These centers for coordinating continuing education in a certain field were called 'continuing education complex' (ger.: Weiterbildungskomplex).
[325] Wolfgang Lambrecht, *Wissenschaftspolitik zwischen Ideologie und Pragmatismus: Die III. Hochschulreform (1965–71) am Beispiel der TH Karl-Marx-Stadt* (Münster: Waxmann, 2007), 228.
[326] SAPMO DY 30/J IV 2/2/2138, "Maßnahmeplan des Ministeriums für Hoch- und Fachschulwesen zur Realisierung der Konzeption 'Standpunkte zu Konsequenzen aus der Entwicklung der Informatik und informationsverarbeitenden Technik für das Bildungswesen'," 4.

technology."[327] It was seen as essential that socialist workers, as the "owners of the means of production," possessed a sound understanding of the inner workings and regularities of society in order to be able to consciously shape the socialist society in a "planned and organized manner".[328] Further education therefore had to form a unit of both technical and political-ideological instruction to enable workers to assess and actively control the social impact of the development and use of new technologies according to socialist ideals and values, so as to achieve the desired results.[329]

Crucially, within the SED's sociotechnical imaginary, the computer was regarded as politically "neutral," allowing it to be adopted in the GDR despite its origins in the USA.[330] The SED's sociotechnical imaginary suggested that it was not the machine itself, but rather the way how it was *appropriated* and *used* within a certain political, social, and economic system, which would determine its effects on the people. In line with the socialist leadership's assertion that computer technology was used in a destructive way under capitalism,[331] it was argued that to avoid such detrimental social effects on the working class, its further development and use in the GDR needed to be closely aligned with the socialist ideology and, by extension, the political programmes of the SED. Students needed to be "made aware of the use of informatics in the imperialist states, which is against the interests of the working people,"[332] replacing people with machines. In the socialist GDR, by contrast, the SED promised that students would not end up unemployed, but able to use their skills and knowledge for the benefit of all: "Where people, not profits, are the measure of all things, even the most complex problems of the scientific and technological revolution are solved for the benefit

---

[327] BArch DR 3/26535, H. G. Schneider, H. Wiegand, and J. Frühauf: "Erfahrungen aus der Arbeit des Weiterbildungszentrums Elektronische Bauelemente an der Technischen Hochschule Karl-Marx-Stadt," 5.
[328] BArch DR 3/26535, H. G. Schneider, H. Wiegand, and J. Frühauf: "Erfahrungen aus der Arbeit des Weiterbildungszentrums Elektronische Bauelemente an der Technischen Hochschule Karl-Marx-Stadt," 6.
[329] BArch DR 3/26535, H. G. Schneider, H. Wiegand, and J. Frühauf: "Erfahrungen aus der Arbeit des Weiterbildungszentrums Elektronische Bauelemente an der Technischen Hochschule Karl-Marx-Stadt," 6.
[330] Schmitt, "Socialist Life of a U.S. Army Computer," 146.
[331] Modrow, "Schlüsseltechnologie," 734.
[332] SAPMO DY 30/J IV 2/2/2138, "Maßnahmeplan des Ministeriums für Hoch- und Fachschulwesen zur Realisierung der Konzeption 'Standpunkte zu Konsequenzen aus der Entwicklung der Informatik und informationsverarbeitenden Technik für das Bildungswesen'," 4 (translated from German).

of mankind. The spectre of unemployment does not grin at graduates here; elsewhere it is part of the wake of new technologies, and there it often devalues the efforts of years of training."[333]

Thus, only if the developers and users of computer technology in the GDR were firmly rooted in socialist values and beliefs, and their actions aligned with how the SED envisioned the further trajectory of computer technology in the GDR, could the new technology be considered "safe" from the party's ideological point of view. Particularly in higher education, where the computer experts who were to play a major role in the development and future applications of computer technology in the GDR were educated and trained, it was of paramount importance to the party leadership to ensure that the sociotechnical imaginary and the political priorities of the SED were firmly anchored in the study programs and regularly reaffirmed through continuing education. Accordingly, socialist higher education needed to instil a "sense of responsibility" in scientific and technological elites, to guide the future trajectory of modern computer technology into the "right" direction: "The higher the level of education and the greater the wealth of experience of a scientist, the better this sense of responsibility can develop. However, the sense of responsibility flattens with incomplete, superficial and out-of-date education, because then one's own possible actions and decisions can no longer be fully kept track of."[334] The role of continuing education was thus also to renew the "sense of responsibility" in computer experts by revitalizing the normative function of the SED's sociotechnical imaginary, thus preventing computer experts from "unaware wrong action and decision-making," which was "very often connected with a lack of willingness to rectify the situation, precisely because in such cases the lack of deeper insight, the lack of thorough knowledge and the ensuing impediment of the ability to communicate with the collective of co-workers, limit the rational, factually correct ability to think, act and decide."[335]

With regard to the social implications of the self-proclaimed "scientific-technological revolution," the SED leadership declared it imperative for institutions of higher education to undertake interdisciplinary research into how the application of microelectronics impact on planning and management of eco-

---

[333] "Training am Computer," *Neues Deutschland*, November 5, 1986, 2 (translated from German).
[334] BArch DR 3/26535, H. G. Schneider, H. Wiegand, and J. Frühauf: "Erfahrungen aus der Arbeit des Weiterbildungszentrums Elektronische Bauelemente an der Technischen Hochschule Karl-Marx-Stadt," 6 (translated from German).
[335] BArch DR 3/26535, H. G. Schneider, H. Wiegand, and J. Frühauf: "Erfahrungen aus der Arbeit des Weiterbildungszentrums Elektronische Bauelemente an der Technischen Hochschule Karl-Marx-Stadt," 6 (translated from German).

nomic production, the role of workers and the character of their work as well as the qualification and skill requirements of the workforce.[336] While the SED's plan for research on the broader impact of microelectronics and computer technology did not mention possible *undesirable* effects that could manifest in the GDR, it explicitly called for a "pro-active confrontation" with the social consequences of the application of microelectronics under capitalism, as well as the "demonstration of the superiority of the socialist social order" in this regard.[337] The results of such research endeavors were thus predetermined and merely served to confirm the SED's sociotechnical imaginary, in which the social effects of the widespread use of computers within a given society were determined by its dominant political ideology. In this context, the contrasting juxtaposition of the socialist and capitalist systems prevented a differentiated discussion of the desirable, but also the possibly undesirable negative social consequences of the use of technology in the GDR.

At the same time, the politically motivated confrontation between capitalist and socialist systems in the competition for new computer technology led to a focus on exclusive cooperation with countries of the Socialist Bloc in higher education and the training of computer specialists. The GDR's political leadership and educational policymakers were convinced that only together with the Soviet Union and the other countries of the socialist community could the potential of informatics for the GDR's performance growth be fully exploited.[338] It was argued that the rapid changes in production, the implementation of the scientific and technical revolution, the increase in the efficiency of socialist production, and the requirements of the "class struggle with imperialism" made it absolutely necessary to fully develop the potentials resulting from the socialist economic integration of the COMECON countries.[339] Socialist integration, in turn, was expected to

---

336 BArch DR 3/26535, "Orientierung der Zentralen Parteileitung für die Durchführung der Mitgliederversammlungen zur Auswertung der 6. Tagung des Zentralkomitees der SED auf dem Gebiet der Mikroelektronik," 5–6.
337 BArch DR 3/26535, "Orientierung der Zentralen Parteileitung für die Durchführung der Mitgliederversammlungen zur Auswertung der 6. Tagung des Zentralkomitees der SED auf dem Gebiet der Mikroelektronik," 6.
338 SAPMO DY 30/J IV 2/2/2138, "Maßnahmeplan des Ministeriums für Volksbildung zur Realisierung der Konzeption 'Standpunkte zu Konsequenzen aus der Entwicklung der Informatik und informationsverarbeitenden Technik für das Bildungswesen'," 4.
339 The 'Council for Mutual Economic Assistance' (COMECON) was an organization for economic, scientific, and technical cooperation among socialist countries that existed between 1949 and 1991. It was led by the Soviet Union and included the Eastern Bloc as well as other socialist countries, such as Cuba and Vietnam.

3.3 Continuing Education and Training in Computing and Information Processing — 93

place new demands on the content of education and to contribute to the deepening of cooperation in the education and training of cadres.[340]

Consequently, the SED leadership called for increased international cooperation in higher education with other socialist countries, especially regarding new technologies, through the exchange of students, emerging researchers, and guest lecturers, as well as the organization of international conferences and study visits. The knowledge gained from such international contacts were then to be transferred into practice in the GDR's research institutions and industry.[341] Moreover, it was suggested to follow the example of the Soviet Union and set up further education courses for foreign graduates from the socialist member states at GDR universities. As this was seen as a completely new task for the GDR higher education system, but one that would become increasingly important and necessary in the future, a limited number of continuing education courses were established for this purpose by 1980. The courses were designed to focus on scientific fields which were deemed attractive for socialist states, and which related to key fields of scientific and technical cooperation within COMECON, such as microelectronics.[342]

However, cooperation and the exchange of knowledge in the field of new computer technology between socialist countries was not as free and unrestricted as the rhetoric of systemic conflict between socialism and capitalism and the SED's repeated calls for collaboration between socialist "brother states" might suggest. In fact, the competition for economic and technological progress went beyond the level of ideology and manifested itself at the level of individual countries struggling to ensure the competitiveness of their national economies. This is illustrated by the fact that the same document, which had suggested the establishment of further education and training courses for foreign graduates which had studied at universities in the GDR, advised *against* the opening of already existing further education courses in the field of new technologies directed at GDR cadres to foreign specialists of socialist states, since these courses were "very specifically

---

[340] M. Laschke, "Probleme der Aus- und Weiterbildung als Bestandteil des einheitlichen sozialistischen Reproduktionsprozesses," dated 1972, cited in: BArch DR 3/26535, H. G. Schneider, H. Wiegand, and J. Frühauf: "Erfahrungen aus der Arbeit des Weiterbildungszentrums Elektronische Bauelemente an der Technischen Hochschule Karl-Marx-Stadt," 5.
[341] BArch DR 3/26535, "Orientierung der Zentralen Parteileitung für die Durchführung der Mitgliederversammlungen zur Auswertung der 6. Tagung des Zentralkomitees der SED auf dem Gebiet der Mikroelektronik," 2–3.
[342] BArch DR 3/26535, "Vorschläge zur Durchführung kurzfristiger Weiterbildungslehrgänge in den Jahren 1977 bis 1980 für Absolventen aus anderen ML/RGW, die DDR-Hochschulen absolviert haben," dated February 10, 1977, 2.

tailored to the needs of the GDR, and in many cases confidential in character."³⁴³ Instead, a number of courses on selected topics such as the production of integrated semiconductor components, theoretical electrotechnology, and microelectronics were set up specifically for the further education of *foreign* specialists from COMECON countries. In return, the GDR expected for its own graduates from foreign universities in socialist countries to be able to undergo short-term further education and training courses at these institutions of higher education abroad.³⁴⁴

Thus, even among the states of the Socialist Bloc, there was a certain degree of secrecy surrounding cutting-edge expert knowledge in microelectronics and computing, as it was shared within the scope of continuing education at university level. In contrast to the more broadly themed undergraduate programs, problem-oriented seminars and postgraduate studies were tailored to the immediate skill needs of GDR industry and focused on the latest research findings and technological innovations.

While the SED's imaginary posited computer technology as a matter of critical importance in the power struggle between socialist and capitalist systems, the decision to keep certain knowledge from its ideological partners, hints at the fact that in practice, the struggle was perhaps not so much about strengthening socialism in the ideological battle against capitalism, but rather to strengthen the GDR as a nation-state, its national economy and the legitimacy of its political leadership by showcasing its technological prowess and innovativeness in a global competition against both capitalist and other socialist states. Similarly, the Soviet Union assigned research and development in microelectronics and computer technology largely to the secretive military complex, thus keeping the latest findings in this field under wraps.³⁴⁵ Even the socialist partner countries were denied access to the Soviet Union's expertise in high technologies, severely hampering international cooperation in computer development within the Socialist Bloc, which the SED had counted on. The global economic and technological competition between industrialized nation-states was therefore not only fought out between the two blocs of the socialist East and the capitalist West, but also, to a

---

343 BArch DR 3/26535, "Vorschläge zur Durchführung kurzfristiger Weiterbildungslehrgänge in den Jahren 1977 bis 1980 für Absolventen aus anderen, ML/RGW, die DDR-Hochschulen absolviert haben," dated February 10, 1977, 3 (translated from German).
344 BArch DR 3/26535, "Vorschläge zur Durchführung kurzfristiger Weiterbildungslehrgänge in den Jahren 1977 bis 1980 für Absolventen aus anderen, ML/RGW, die DDR-Hochschulen absolviert haben," dated February 10, 1977, 6.
345 Barkleit, *Mikroelektronik in der DDR*, 32.

certain extent, between countries within the same bloc – each on its own against all the others, even among the socialist "brother states."

3.4 Chapter Conclusion

The emergence of modern computer technology in the GDR initially was concentrated in higher education institutions, such as the Institute for Computing Machinery at the Technical University of Dresden, where Nikolaus Lehmann developed pioneering programmable computing machines, or in large industrial combines such as Carl Zeiss in Jena, where the "ZRA1" and "OPREMA" were created. In its early days, it was a new technology that was developed and used by a small fraction of society. Few researchers and highly educated specialists had access to the large and expensive mainframe computers, which were stationed in computing centers associated with universities, research institutes, state ministries and large industrial complexes. In the 1960s, most computing centers in higher education were institutionally associated with mathematics departments. Their work was driven by an interest in numerical computing, as well as by efforts to make the "pure" science of mathematics more practice-oriented by further exploiting the potential of scientific computing for applications in industrial production. At the same time, computer technology also took root in more technically oriented disciplines such as engineering, where researchers were more interested in system design and architecture of hardware and software.

Concurrently with the development of the first modern computers, a sociotechnical imaginary of computer technology developed, shaped by pioneering researchers and engineers, that was soon readily adopted by the SED and imbued with the party's ideological and political stamp. Drawing upon the discourse on cybernetics, that both predated and accompanied the development of the first programmable digital computers in the GDR, the SED promoted a sociotechnical imaginary of computer technology as a potent instrument of information processing, that would form the basis of scientific, data-based political decision-making and control of the socialist planned economy. Moreover, computers were imagined as a technology that would revolutionizes industrial production and significantly boost the GDR's economic productivity. In the context of the Cold War, it was anticipated that these effects would catapult the GDR to the forefront of the competition between socialist and capitalist systems. An important element of the SED's imaginary was the claim that computer technology was inherently "neutral" and that its social effects were determined by the political and ideological system that guided its development and use. The sociotechnical imaginary of socialist computerization, which envisioned a desirable future of social and economic progress for the GDR, was thus contrasted with the dystopia of inhumane "capitalist" computerization, which was envisioned as unfolding to the detriment of the working class and prioritizing profits over the well-being of the people.

The sociotechnical imaginary's focus on increasing economic productivity and raising the effectiveness of centralised planning and governance of the socialist economy fostered the spread of computer technology and computational methods from mathematics and engineering into research and higher education in the field of economics, socialist business management and administration, and agriculture. With the advent of smaller and cheaper microcomputers in the 1980s, the dominant sociotechnical imaginary no longer conceived of the computer solely as a technology for increasing efficiency in management, planning and industrial production, but also as a tool that could be used universally in all kinds of disciplines to support and expand human mental abilities in research, teaching, and learning. As a result, the SED leadership decided to introduce computer education courses in all disciplines that would train students in the use of this new technology and show them the wide range of applications and benefits of computers for socialist society and the economy. However, more in-depth knowledge of computer technology and advanced programming skills remained reserved for a select number of students, the majority of whom were in technical disciplines.

The SED identified modern computer technology as an area of paramount importance for the future of the GDR, but one that would be subject to rapid change due to the rapid pace of scientific and technological progress – as epitomized by the rhetoric of the "race against time."[346] Consequently, an urgent need for continuing education in this field was identified, which became a task of growing importance for higher education institutions. Universities and technical colleges were tasked with providing a wide range of continuing education programs and short courses for staff in both academia and industry, in order to update the skills and knowledge of these highly skilled people with the latest advances in the field, and to ensure a seamless transfer of knowledge and exchange of experience between academics, researchers, and practitioners.

Despite the SED's rhetoric of fruitful socialist cooperation in higher technologies among the COMECON member states, and in particular the partnership with the Soviet Union, cutting-edge research results and the latest, highly industry-relevant knowledge and skills imparted in continuing education courses were kept to some extent confidential, even among the socialist partner states. This reluctance on the part of both the Soviet Union and the GDR to cooperate fully and without restrictions in the field of high-technology research and training suggests that com-

---

346 Sabrow, "Zukunftspathos als Legitimationsressource," 181.

puter technology was seen not only as a matter of competition between socialism and capitalism as to which *system* could better exploit its advantages and achieve a higher degree of social and economic progress, but also as a matter of *national* prestige and political legitimacy of a country's leadership. The sociotechnical imaginary promoted by the SED saw the computer as a technology of the future that would bring prosperity to the GDR and help to cure its economic and political ailments.[347] The SED publicly proclaimed and celebrated its "national" technological achievements, in what appeared to be an increasingly desperate effort to establish itself internationally as a serious economic competitor,[348] but also on a domestic scale, to secure the political support of the people in the GDR and increase their confidence in the government's ability to deliver on its ambitious goals and promises for a bright and prosperous future.[349]

---

[347] Schuhmann, "Zukunft der Arbeit," 174.
[348] Wolfgang Müller, "Hat die Mikroelektronik-Industrie in der DDR noch eine Zukunft?," *Wechselwirkung* 12, no. 44 (1990): 35.
[349] Barkleit, *Mikroelektronik in der DDR*, 42.

4 Training the Socialist Workforce for the Computer

The very first integrated circuit was created in 1959 by Jack Kilby and patented 1964 in the USA.[350] It formed the basis for microchip technology, which was at the core of the development of increasingly smaller and more affordable computers in the following decades. The low cost and high-speed information processing potential of microelectronics technology, its flexibility and wide range of applications promised considerable improvements in productivity in industrial production, office management and other areas of the economy and society. The production of early types of integrated circuits soon also started in Japan, Great Britain, France, and West Germany between 1965 and 1968.[351]

In the GDR, the first integrated circuits were manufactured in 1970.[352] In these early days, thus, the GDR was not all that far behind the international competition in the development of microelectronics. Unlike the socialist leadership of the Soviet Union, the GDR's SED leadership had already officially recognized and promoted since the mid-1950s also the civilian use of electronic computer technology outside the military complex as an important engine for social and economic progress. However, when Walter Ulbricht had to resign from his post as the SED's First Secretary and Erich Honecker took his place in 1971, Honecker's new political program called for "unity of economic and social policy" and set new priorities. The socialist leadership would continue its efforts to promote technological innovation but shifted priority to producing more consumer goods and housing to improve living standards and ensure the political loyalty of the people. As an immediate consequence, the SED leadership decided to drastically reduce the investment in high technologies.[353] Thus, during the early 1970s, as microelectronics and electronic computing became key technologies in many advanced industrial states, the political and economic conditions in the GDR for building up a potent microelectronics industry were rather unfavorable.

---

350 Haigh and Ceruzzi, *New History of Modern Computing*, 98–99.
351 Barkleit, *Mikroelektronik in der DDR*, 20.
352 Olaf Klenke, *Kampfauftrag Mikrochip: Rationalisierung und sozialer Konflikt in der DDR* (Hamburg: VSA-Verlag, 2008), 51.
353 Barkleit, *Mikroelektronik in der DDR*, 21.

4.1 Building a Computer Industry in the GDR

Over the 1970s, the GDR's technological backlog compared to Western competitors and Japan increased significantly.[354] The impact of Honecker's restrained technology policy was exacerbated by the effect of the COCOM embargo, which had been imposed by the USA on socialist countries. It successfully prevented the legal import of specialized technological equipment and state-of-the-art computing technology that GDR scientists and engineers could have taken advantage of for reverse-engineering and imitation. The undermining of these regulations with the help of the GDR's secret service consumed large amounts of the already scarce foreign currency and was only successful in isolated cases.[355] From a strategic point of view, the GDR's approach of imitating Western technology instead of developing their own innovations was unsuitable to eliminate the technological backlog. For the most part, it was not possible to acquire the very latest devices to serve as models for their own copies. And until the efforts of reverse-engineering bore fruit, more years passed, so that the domestically produced components and devices were already far behind the state-of-the-art. In the best case, the copycat strategy could help the GDR not to fall back too far behind the leading competitors in the field of computer technology – but it could not create the kind of innovation that would have been necessary to put the GDR at the global forefront of technological development.[356]

In addition, a serious willingness to cooperate in the development of microelectronics among the Eastern Bloc was largely absent, especially in the case of the Soviet Union. There, the development of high technologies took place primarily in the top-secret military complex, to which the GDR was not given access to information that could have helped its efforts in microelectronics research and development. The latest technological developments from these "closed-off" areas of military research could therefore not be transferred to the field of civilian production.[357] Especially as the US-Soviet relations worsened in 1983, the Soviet Union tried to involve the states of the Eastern Bloc more closely in their armament programs. As a result, the Soviet Union increased pressure on the political

---

354 Barkleit, *Mikroelektronik in der DDR*, 21.
355 Olaf Klenke,"Globalisierung, Mikroelektronik und das Scheitern der DDR-Wirtschaft," *Deutschland Archiv* 35, no. 3 (2002): 424.
356 On the import of US computer technology (in particular of the companies IBM and Control Data Corporation) by the GDR in the 1960s, see Donig, "Appropriating American Technology."
357 Dittmann, "Microelectronics under Socialism," 49.

leadership of the GDR to prioritize military uses of microelectronics in their research and development efforts.[358]

But by the mid-1970s, the civil use of microelectronics to build state-of-the-art industrial machinery and boost the productivity of industrial production became a key concern of GDR economists and political leaders. As export numbers for GDR machine tools had plummeted, the accelerated and intensified development of microelectronics was seen as vital to secure the exportability of important economic products, especially in mechanical engineering.[359] In addition, microelectronics had internationally become a key field to prove the competitiveness of national economies and ability of countries to keep up with the rapid progress in high-technology research and development. As such, it played an important role – ideologically, economically, and militarily – in the political competition between socialist and capitalist systems.[360] At the domestic level, the development of microelectronics and computer technology also offered an opportunity for socialist propaganda and in fulfilling the SED's political pledge regarding the betterment of working and living conditions in the GDR. Computers and computerized machines promised to relieve the working people of monotonous and heavy labor, and to boost productivity to increase economic prosperity and make desired consumer products available more quickly and in greater numbers. Especially also the supply of modern electrotechnical and electronic consumer goods for the GDR population was crucial in the comparison with West Germany, as GDR citizens were well aware of the shortcomings of the socialist economy in fulfilling their consumption needs.[361]

Thus, in line with Honecker's paradigm of "unity of economic and social policy," the SED's sociotechnical imaginary in the second half of the 1970s emphasized in particular the aspects of computer technology as a powerful means of increasing economic productivity and competitiveness of the GDR industry on the world market, as well as the positive social effects of the use of computer technology in economic production, namely better working and living conditions for the people of the GDR.

In response to the increasingly pressing economic and political challenges, and the promises of the dominant sociotechnical imaginary which posited computer technology as a potent remedy, the SED leadership decided in 1977 to launch an initiative to build and expand its own capacity for the development

---

358 Barkleit, *Mikroelektronik in der DDR*, 29.
359 Klenke, "Globalisierung, Mikroelektronik und Scheitern," 422.
360 Donig, "Appropriating American Technology," 32.
361 Barkleit, *Mikroelektronik in der DDR*, 42.

and use of microelectronics and computer technology in GDR industry.³⁶² While the Soviet Union continued to perceive microelectronics technology primarily as a means in the attempt to gain the upper hand in the arms race with NATO, the GDR primarily (but not exclusively) invested in the development and use of microelectronics and computer technology to overcome its national economic crises in the late 1970s and over the course of the 1980s.³⁶³ In 1978, the SED initiated the creation of the Microelectronics combine in Erfurt as a socialist conglomerate of different factories and enterprises related to the production of semiconductors, microprocessors, and other microchip technology.³⁶⁴ The VEB combine Robotron became the focal point of electronic data processing in the GDR, and a leader in the field of computer technology among the states of the Socialist Bloc. The Robotron combine was responsible for the development, production, and distribution of electronic data processing machines, small and microcomputers amongst other electronic equipment and consumer goods, as well as operating systems and software.³⁶⁵ In addition, Robotron also provided its clients with training and services for electronic computer technology and office machines.³⁶⁶

The GDR's political leadership anticipated that the massive investment in building up a domestic microelectronics industry would pay off, on the one hand, by triggering modernization effects for the entire economy, including enormous increases in productivity. On the other hand, it was hoped that exports of microelectronic components and devices, computer technology, and computerized machines would bring much-needed foreign exchange into the country.³⁶⁷ Against this background, the "microelectronics program" launched in 1977 was first and foremost an extensive political program for economic modernization and rationalization, driven by the SED's sociotechnical imaginary of computer technology as a driver of economic progress. But to get the working people of the GDR on

---

362 Barkleit, *Mikroelektronik in der DDR*, 34–35.
363 Dittmann, "Microelectronics under Socialism," 53.
364 Combines (ger.: Kombinate) can be considered the basic form of social organization of socialist large-scale production with the aim of rationalizing and centralizing control of economic production processes. Combines in the GDR were large economic units that combined several enterprises, R&D, and sales departments of one industrial branch linked together by a common technological process or the production of a common range of goods under one roof. Within the GDR, most of the former "Associations of Publicly Owned Enterprises" (ger.: Vereinigungen Volkseigener Betriebe, VVBs) were turned into combines between 1968 and 1979. See: Sabine Spangenberg, *The Institutionalised Transformation of the East German Economy* (Heidelberg/New York: Physica-Verlag, 1998), 114.
365 Krakat, *Schlussbilanz*, 17–27.
366 Merkel, *VEB Kombinat Robotron*, 36.
367 Barkleit, *Mikroelektronik in der DDR*, 111.

board with this ambitious strategy of widespread computerization, the SED's propaganda also had to include promises of the expected positive effects on the people, by envisioning the new technology as a harbinger of "social progress" – in the form of better conditions of working and living in the GDR.

4.2 A Workforce for the Era of Mainframe Computers

In 1967, the Robotron 300 (R 300) mainframe computer went into serial production, which was to become one of the most widely used mainframes in the GDR for both scientific and commercial applications (see Figure 3). The R 300 was a fully transistorized computer of the second generation, and more powerful than any of the previously developed computers in the GDR.[368] It was praised as an important working tool for organizing the complete and continuous processing of all the masses of data in the reproduction process and heralded as a means of modernizing the GDR's economy, but also as a potentially lucrative export product.[369] The operation of the R 300 required a high number of skilled workers: It was estimated that in a two-shift operation, as many as 75 workers were necessary for operating a single machine.[370] While less than a third of these workers would require a higher education, roughly two thirds were to be trained through vocational and adult education. The latter included 25 workers fully skilled in electronic data processing, and 25 with a partial training in this field, as well as two mechanics and a steno typist.

In addition to preparing workers to operate and maintain the machine, the introduction of the R 300 into companies and other institutions also required extensive and careful planning on the part of top management. To coincide with the start of serial production, a two-day seminar was held for the directors of the select number of companies and institutions that were due to receive an R 300 in the same year or the following year.[371] The seminar's purpose was to prepare the directors for the deployment of the powerful and expensive computer they were

---

[368] The 'Robotron 300' (R 300) was a medium-sized electronic data processing system by the 'VEB Rafena Werke Radeberg', formerly a manufacturer of television devices. The R 300 was presented to the international public for the first time in September 1966 at the "Interorgtechnika 66" trade fair in Moscow ("'Robotron 300' auf der 'Interorgtechnika'," *Neues Deutschland*, August 31, 1966, 2).

[369] "Robotron 300 in Moskau," *Neue Zeit*, September 2, 1966, 5.

[370] BArch DQ 400/2410, "Referat des Genossen Hofmann vor der Kommission Berufsbildung am 14.4.1967 zum Problem der Aus- und Weiterbildung von Facharbeitern für die Bedienung der EDV," 1–2.

[371] The first five units were delivered in 1967 to the Institute for Data Processing in Dresden, the 'VEB Bürotechnik' in Leipzig, the 'VEB Maschinelles Rechnen' in Suhl, and the Social Security Ad-

Figure 3: Magnetic tape units (top) and control panel (bottom) of the R 300.
Sources: "Facharbeiterin für Datenverarbeitung am Magnetbandgerät, 1973" Photograph by Eugen Nosko, © Deutsche Fotothek / Eugen Nosko; "Robotron R300," 2009, Procolotor, CC BY-SA 3.0, via Wikimedia Commons.

ministration of the National Board of the Free German Trade Union Federation ("Robotron 300 in Serie," *Neue Zeit*, March 9, 1967, 2).

about to receive, focusing in particular on the tasks that the SED leadership had set with regard to the computerization of the GDR's economy, and the envisioned economic benefits that were to be achieved with the help of the R 300.[372] The fact that the seminar was chaired by none other than Günther Kleiber, State Secretary for Data Processing, underlines the economic and political importance attached to the operational preparations for the deployment of the R 300.

Since there were not yet any official, GDR-wide vocational curricula and job profiles for the future operators and maintenance staff of the computer, local training centers set up specialised short-term courses in the use and maintenance of the R 300. The training center of the VEB Bürotechnik in Leipzig, for example, was equipped with an R 300 for the practical training of engineers and technicians. Within just three years, more than 15,000 workers were to be trained there in short courses lasting between three and six weeks.[373]

Within a short time, the limited but growing availability of mainframe computers created an urgent need for skilled operators and maintenance staff, especially if the potential of the expensive mainframes was to be exploited to its full potential through shift operation.

Consequently, and in face of the envisioned scale of deploying electronic data processing machines in the GDR, their operation could no longer be limited to professionals with a higher education degree. Instead, new job profiles for electronic data processing specialists were developed for vocational training and adult education, to meet the increasing demand for qualified personnel that was able to operate and program the new electronic data processing machines.[374] These early computers of the second generation used magnetic tapes and drums for data storage and had to be programmed in machine and assembly language. Paper tape and punched cards were used for the input and output of information.

In the 1960s, the operation sites of new data processing technology – research facilities, combines and companies – were responsible for planning their own need of qualified workers for data processing, and arranging adequate training measures.[375] Some companies, combines and research facilities trained "technical computers" since the early 1960s and "skilled workers for data processing" since

---

[372] "Seminar über Robotron 300," *Neue Zeit*, May 17, 1967, 2.
[373] "15000 lernen Robotron 300 kennen," *Neues Deutschland*, July 22, 1967, 2.
[374] BArch DQ 400/2410, "Rahmenausbildungsunterlage für die sozialistische Berufsausbildung. Berufsbezeichnung: 'Facharbeiter für Datenverarbeitung'," dated July 20, 1967, 1.
[375] BArch DQ 400/2410, "Referat des Genossen Hofmann vor der Kommission Berufsbildung am 14.4.1967 zum Problem der Aus- und Weiterbildung von Facharbeitern für die Bedienung der EDV," 4.

1965,[376] but only to cover their own demand for qualified personnel, and tailored to their specific technical skill needs.[377] These workers underwent a short training in the practical operation of specific computer machines available in their workplace, without being taught an adequate level of fundamental theoretical knowledge.[378] However, in face of the fast-paced advancement of computer technology, vocational policymakers saw a need for a broader education and training in the use of electronic computers that went beyond a practical introduction to and working knowledge of one specific machine. Instead, the education and training of skilled data processing workers needed to provide a solid theoretical foundation to enable them in the future to familiarize themselves quickly and easily with newly introduced machines and devices and use them for their work.[379] It was expected that over the next few years the proportion of data processing workers employed on punch-card machines would gradually decrease, while the share of workers employed on electronic computers and as programmers would increase significantly.[380] It was argued that a more theoretically sound and broader practical education and training in electronic data processing would allow a smoother transition of workers originally trained on punch card machines to newer electronic computers.

At the time, there was no centralized oversight over the number of qualified workers in these jobs, as these occupations as well as ad-hoc training measures had been created somewhat spontaneously on a local level as the need arose. Therefore, these qualifications had not been planned for, and thus, also did not feature in the reports on plan fulfillment to central state authorities.[381] By and large, there was a wild proliferation of special training and further education measures to qualify

---

[376] BArch DQ 400/2410, "Zuarbeit zum Bericht für den Beirat für Datenverarbeitung über 'Sicherung und Ausbildung der Kader'," dated April 10, 1967, 1.
[377] BArch DQ 400/2410, "Zuarbeit zum Bericht für den Beirat für Datenverarbeitung über 'Sicherung und Ausbildung der Kader'," dated April 10, 1967, 4–5.
[378] BArch DQ 400/2410, "Anlage 1: Besonderheiten der Ausbildung des 'Facharbeiters für Datenverarbeitung'," undated, 2.
[379] BArch DQ 400/2410, "Anlage 1: Besonderheiten der Ausbildung des 'Facharbeiters für Datenverarbeitung'," undated, 2.
[380] It was estimated that the proportion of workers on punch card machines would fall from 69.2% in 1967 to 54% in 1970, while the proportion of workers in technology/programming and operators of electronic data processing machines would rise from 13% to 40% over the same period (BArch DQ 400/2410, "Anlage 1: Besonderheiten der Ausbildung des 'Facharbeiters für Datenverarbeitung'," undated, 2).
[381] BArch DQ 400/2410, "Zuarbeit zum Bericht für den Beirat für Datenverarbeitung über 'Sicherung und Ausbildung der Kader'," dated April 10, 1967, 1–2; BArch DQ 400/2410, "Zum Problem der Koordinierung der Ausbildungskapazitäten und des Einsatzes der Lehrkräfte bei der Aus- und Weiterbildung in den Berufen der Bedienung von EDVA," 2–3.

workers for the use of electronic data processing, such as short and weekend courses, training cycles of vocational academies and evening schools, as well as courses in adult education centers.[382] This hotchpotch of various short courses and unsystematic training efforts was the result of a short-term response to urgent skills demands, as computer technology began to take hold in industry.

The lack of territorial coordination in the further training of skilled workers for the use of electronic data processing technology had led to fragmentation of training contents, as well as insufficient planning and overview of existing training capacities and teaching staff.[383] Individual in-company training academies developed their own continuing education courses for electronic data processing personnel without planning and implementing training beyond their own skill needs. To counter such fragmentation, vocational education experts suggested the centralized development of training materials and curricula for the qualification of skilled workers in the use of electronic data processing machines and systems. As a result, in 1967, the two occupations of "technical computer" and "skilled worker for data processing" were merged to form a new basic occupation under the name of "skilled worker for electronic data processing."[384]

The "VVB Maschinelles Rechnen,"[385] which was among the first to receive an R 300 mainframe computer, was tasked with developing an official job profile and a framework document for the vocational education and training of "skilled workers for electronic data processing" in the GDR.[386] According to the official job description contained in this document, skilled workers for data processing needed to be able to stand for long periods of time, as the large mainframe computers had to be operated whilst standing, and capable of logical thinking, as op-

---

382 BArch DQ 400/2410, "Referat des Genossen Hofmann vor der Kommission Berufsbildung am 14.4.1967 zum Problem der Aus- und Weiterbildung von Facharbeitern für die Bedienung der EDV," 4.
383 BArch DQ 400/2410, "Referat des Genossen Hofmann vor der Kommission Berufsbildung am 14.4.1967 zum Problem der Aus- und Weiterbildung von Facharbeitern für die Bedienung der EDV," 6–7.
384 BArch DQ 400/2410, "Zum Problem der Koordinierung der Ausbildungskapazitäten und des Einsatzes der Lehrkräfte bei der Aus- und Weiterbildung in den Berufen der Bedienung von EDVA," 6; BArch DQ 400/2410, "Referat des Genossen Hofmann vor der Kommission Berufsbildung am 14.4.1967 zum Problem der Aus- und Weiterbildung von Facharbeitern für die Bedienung der EDV," 13.
385 ger.: Association of Publicly Owned Enterprises in Machine Computing.
386 BArch DQ 400/2410, "Zum Problem der Koordinierung der Ausbildungskapazitäten und des Einsatzes der Lehrkräfte bei der Aus- und Weiterbildung in den Berufen der Bedienung von EDVA," 6; BArch DQ 400/2410, "Berufsbild für den Ausbildungsberuf Facharbeiter für lochkartenmaschinelle und elektronische Datenverarbeitung (Datenverarbeiter)."

erators had to be able to judge whether the computer's output was reasonable or whether an error had occurred. The main duty of data processing workers was to translate the tasks, that had previously been structurally prepared and conceptually resolved by mathematicians, economists or engineers, into a machine program and then process it on the computer.[387] Several newspaper articles picked up on the development of the new job, and presented the skilled occupation of data processor as particularly suitable for girls and women.[388] The strategy of vocational policymakers to advertise it as such seems to have been fruitful: By the early 1970s, around 80% of trainees in data processing were female.[389] It was also emphasized that the career prospects of skilled workers for data processing were bright, as computer technology was deemed "future-proof": "The job is well paid, its prospects are enormous, because in many cities – especially in industry – computing machines are currently being set up; in Berlin they are already operating in large companies."[390]

Within the framework of the so-called scientific-technological revolution, which Walter Ulbricht had proclaimed in 1967 as a key mission for the GDR, the political salience of technical education and training increased significantly.[391] In the eyes of the SED leadership, the technological foundation for this "revolution" as well as the planned modernization and automation of the socialist economy was formed by electronics, operational technology, and data processing.[392] According to their sociotechnical imaginary of a sweeping computerization of the GDR's socialist economy, the use of new computer technology was expected to soon spread to workplaces beyond commercial and scientific data-processing centers, and that all working people in the GDR would be confronted with this new technology directly or indirectly, to a greater or lesser extent.[393] It was therefore considered necessary to train not just a few specialists in the new computer tech-

---

[387] BArch DQ 400/2410, "Berufsbild für den Ausbildungsberuf Facharbeiter für lochkartenmaschinelle und elektronische Datenverarbeitung (Datenverarbeiter)," 1.
[388] See for example: "Wer füttert Robotron 100?," *Berliner Zeitung*, January 27, 1965, 7; "Neuer Beruf," *Neues Deutschland*, July 7, 1965, 3; "Neue Berufe für Frauen und Mädchen," *Neues Deutschland*, February 11, 1967, 2.
[389] "Vertrag der Jugend mit der Zukunft – Neue Rahmenausbildungen für Lehrlinge," *Neues Deutschland*, April 23, 1971, 3.
[390] "Wer füttert Robotron 100?" *Berliner Zeitung*, January 27, 1965, 7 (translated from German).
[391] For a discussion of the concept of the "scientific-technological revolution' and its relevance to the SED's science and technology policy, see Laitko, "Wissenschaftlich-technische Revolution."
[392] BArch DQ 4/450, "Grundlagen der Elektronik – Grundlagen der BMSR-Technik – Grundlagen der Datenverarbeitung. Lehrpläne für die sozialistische Berufsausbildung," 6.
[393] BArch DQ 4/450, "Grundlagen der Elektronik – Grundlagen der BMSR-Technik – Grundlagen der Datenverarbeitung. Lehrpläne für die sozialistische Berufsausbildung," 6.

nologies, but to involve *all* working people in preparing for the planned all-embracing computerization of socialist society in the GDR. Vocational education and training policymakers concluded that a modern vocational education and training had to provide all workers with the necessary theoretical and practical tools in the field of automation technology and data processing so that they could become "conscious and active masters and co-creators of modern production" and contribute to the realization of the scientific and technical revolution in the socialist society of the GDR.[394]

Consequently, it was decided on 11 June 1968 to introduce three new basic subjects into initial VET to strengthen technical education and training, covering the foundations of electronics, data processing, and "BMSR" – the latter referring to the common abbreviation used in the GDR for operational measurement, control, and regulation technology which was considered essential for the automation of industrial production.[395] Together, the three new subjects added up to a total of 190 hours of instruction dedicated to data processing and automation technology. All three technical subjects were declared mandatory for all apprentices who had finished the 10th grade of the polytechnic secondary school, regardless of the occupation they were trained for. The new subjects served as a form of general computer technology education in basic vocational education, and as a foundation for further specialized vocational training to build upon. The theoretical and practical instruction in data processing and automation technology was meant to empower learners to master and control the new technology, and "mentally open up to them the wealth and abundance of tasks for the realisation of the scientific-technological revolution in socialist society."[396] The subject "Basics of Electronics" sought to provide learners with an understanding of the electrical engineering fundamentals and functionality of electrical components that make automation and electronic data processing possible in the first place.[397] The goal of "Basics of BSMR-Technology" was to familiarize students with the control and regulation of technical systems, and some of the core con-

---

[394] BArch DQ 4/450, "Grundlagen der Elektronik – Grundlagen der BMSR-Technik – Grundlagen der Datenverarbeitung. Lehrpläne für die sozialistische Berufsausbildung," 6.

[395] This decision was part of the resolution of the People's Chamber of the GDR of June 11, 1968 on the "Principles for the further development of vocational training as a component of the unified socialist education system" (ger.: Beschluss der Volkskammer über die "Grundsätze für die Weiterentwicklung der Berufsausbildung als Bestandteil des einheitlichen sozialistischen Bildungssystems"); BArch DQ 4/450, "Grundlagen der Elektronik – Grundlagen der BMSR-Technik – Grundlagen der Datenverarbeitung. Lehrpläne für die sozialistische Berufsausbildung," 6.

[396] BArch DQ 4/450, "Grundlagen der Elektronik – Grundlagen der BMSR-Technik – Grundlagen der Datenverarbeitung. Lehrpläne für die sozialistische Berufsausbildung," 6 (translated from German).

[397] BArch DQ 4/450, "Grundlagen der Elektronik – Grundlagen der BMSR-Technik – Grundlagen der Datenverarbeitung. Lehrpläne für die sozialistische Berufsausbildung," 15–16.

cepts of cybernetics. In addition, the subject covered the principles and functioning of automation technology in industrial production, for example, the automated measurement and control of pressure, temperature, and flow rate.[398] Finally, the subject "Basics of Data Processing" was meant to introduce students to the use of data processing machines, and in particular, digital computer technology in their future workplace.[399] But the goals of instruction, as stated in the syllabus, did not stop at the technical details on the modes of operation of various machines and computers. A higher priority was given to the objective that learners understood the aims, requirements, possibilities, and organization of data collection, as well as the tasks of the socialist skilled worker in the introduction and application of data processing technology.[400] The latter referred to an understanding of the "revolutionary" impact of data processing technology on the management and division of labor in socialist companies and combines, and demanded an active participation of workers in overcoming outdated organizational structures. This included the acceptance of changing working tasks and conditions, as well as a willingness to continue to acquire the necessary qualifications in response to new and changing skill demands.[401]

Together, the three basic vocational subjects were intended to form a sound foundation of technical skills and knowledge in face of an uncertain technological future. Vocational education experts hoped that this education would allow skilled electronic data processing workers to familiarize themselves with new devices and equipment in the future, as new computer technology would become available in their workplace. In the short term, a solid technical education was intended to help workers adapt flexibly to new skill requirements by building on their initial education with short courses of continuing and adult education. In the medium term, however, training materials would need to be regularly adapted to reflect the transition to new generations of computer technology, which would potentially lead to fundamental changes in the vocational skills required of workers operating these new computers and computerized machines.

However, in contrast to what the sociotechnical imaginaries propagated by the SED seemed to imply, technological progress did not map directly and linearly onto the use of technologies in the real world of work. This was especially true in the

---

398 BArch DQ 4/450, "Grundlagen der Elektronik – Grundlagen der BMSR-Technik – Grundlagen der Datenverarbeitung. Lehrpläne für die sozialistische Berufsausbildung," 25–27.
399 BArch DQ 4/450, "Grundlagen der Elektronik – Grundlagen der BMSR-Technik – Grundlagen der Datenverarbeitung. Lehrpläne für die sozialistische Berufsausbildung," 39–41.
400 BArch DQ 4/450, "Grundlagen der Elektronik – Grundlagen der BMSR-Technik – Grundlagen der Datenverarbeitung. Lehrpläne für die sozialistische Berufsausbildung," 39.
401 BArch DQ 4/450, "Grundlagen der Elektronik – Grundlagen der BMSR-Technik – Grundlagen der Datenverarbeitung. Lehrpläne für die sozialistische Berufsausbildung," 40–53.

case of the GDR, where there was a significant backlog of five to seven years compared to the international state-of-the-art in computer technology,[402] and new technological developments only slowly and gradually found their way into industry. Initially, skilled electronic data processing workers were still mainly working on punch-card machines, but from the late 1960s and early 1970s they were gradually required to switch to new electronic computers such as the Robotron 300,[403] depending on how quickly the latest technology was made available to their particular industry and workplace, in accordance with the economic plans of the SED authorities. A concept for the deployment of computer technology laid down in the mid-1970s provided for all second-generation computers (such as the Robotron 300) to be phased out by 1980 and replaced by more potent mainframes.[404] But many of these old computers remained in use for longer, especially in small and medium-sized enterprises. The chronic shortage of computers and microelectronics components necessitated not only a relatively slow change of generations, but also an economic use of scarce computer capacities that were controlled and managed by the state. Thus, the use of old mainframes and large second-generation computers continued, even while new microcomputers were introduced piece by piece starting from the mid-1980s, spreading the use of computer technology to new areas of the economy.[405]

In the second half of the 1960s, the international state-of-the-art in computer technology shifted to integrated circuits, where a variety of transistors were placed on silicon chips. These third-generation, microelectronics-based computers were more powerful and efficient than the previous transistor computers. The technology of integrated circuits also allowed for a significant reduction in the size of the machines. Finally, the large-scale integration of circuits on a single microchip allowed from the 1970s onwards even more powerful and smaller computers. These microcomputers no longer relied on magnetic tape or punch cards, but on semiconductor memory (such as RAM and ROM), keyboards, monitors, and printers for data input and output. Not only were they smaller and much more affordable than their predecessors, but also more approachable for a wider user base. Over the course of the 1980s, microcomputers became a mass consumer product in many industrialized countries, heralding the new era of personal and home computing.

---

[402] Barkleit, *Mikroelektronik in der DDR*, 27.
[403] BArch DQ 400/2410, "Anlage 1: Besonderheiten der Ausbildung des 'Facharbeiters für Datenverarbeitung'," undated, 3.
[404] "Rechnerdefizit zwingt zum Einsatz alter Technik," Computerwoche, June 6, 1980, http://www.cowo.de/a/1189741 (accessed June 30, 2023).
[405] "Rechnerdefizit."

Table 4: Generational shifts in the GDR's domestically produced computer technology in comparison to international developments.

	International State of the Art	GDR Computers
1st Generation (vacuum tubes)	1945–1959	ZRA 1 (1960)
2nd Generation (transistors)	1960–1964	R 300 (1968)
3rd Generation (integrated circuits)	1964–1970	R 21 (1970) EC 1040 (1974)
4th Generation (microprocessors)	Since 1970	Z 9001 (= KC 85/1) (1984) HC 900 (= KC 85/2) (1984)

Source: Adapted from Eberhard Prager and Evelyn Richter, Software – Was ist das? (Berlin: Verlag die Wirtschaft, 1986), 25–27.

In the GDR, these technological shifts took place with a delay of several years compared to its western competitors (see Table 4). Until the year 1980, the GDR microelectronics industry had produced over 36 million integrated circuits of 210 different types. But only a fraction of these qualified as large-scale integrated circuits.[406] Moreover, it was not until 1980 that the first 8-bit microprocessor was manufactured by the VEB Mikroelektronik Karl Marx in Erfurt, laying the foundation for the development of microcomputers made in the GDR. This lagging and uneven transition to newly available computer technology, whilst the use of older technology persisted, made it difficult if not futile to attempt and establish solid and standardized curricula and training materials that would be applicable to the skill demands in a wide variety of computerized workplaces.

Consequently, a pragmatic approach to preparing workers for the use of new technology dominated, that focused primarily on the specific type of computer technology and software currently available and used in the trainees' workplace. The specialized preparation of workers for the introduction of new computer technology was usually organized in the form of on-the-job training on newly acquired machines and equipment, building on the basic computer training they had received as part of initial vocational training. In this sense, continuing training in the 1960s and 1970s provided a flexible approach for rapidly updating the existing skills and knowledge of workers who had operated older computers and computerized machinery, in line with the changing skill requirements of their individual workplaces.

---

[406] "Unsere Mikroelektronik und ihre Schlüsselrolle," *Neues Deutschland*, April 15, 1981, 6.

4.3 Envisioning the Computerization of Industrial Production and Offices

In the wake of the SED's campaign for industrial modernization during the 1980s, industrial robots, computer-aided design and manufacturing (CAD/CAM) as well as microcomputers became powerful key words that were circulated in the GDR's magazines and daily newspapers, loaded with sociotechnical imaginaries of a more prosperous future for the GDR's people and economy. Numerous reports were published on how workplaces in industry and offices had been and would continue to be transformed by the introduction of new technology, highlighting the increased efficiency of work processes and the new skills that workers on these computerized machines would need to acquire.

In particular, CAD/CAM in industrial production was frequently mentioned in the popular media from the mid-1980s onwards, as it promised to enable manufacturers to produce products with much greater precision and efficiency.[407] It allowed for developers and design engineers to simulate and compare a variety of draft variants in order to select the version with the most promising parameters for production.[408] Computer-aided industrial production was thus expected to increase the quality of products while also reducing the consumption of materials and energy. It also promised a faster and more flexible response to changing consumer demand, both domestically and in the global marketplace.[409] At the same time, CAD/CAM was seen as a technology that would enrich the content of work, freeing workers from physical and monotonous tasks and allowing them to engage in more creative, mentally stimulating activities as part of the production process.[410] In this sense, computer-aided design and manufacturing was perceived as a tangible, concrete example of how computer technology could serve to realize the desired future of economic prosperity and social progress as envisioned within the framework of the SED's sociotechnical imaginary.[411] Similar imaginaries were also propagated in capitalist countries, for example in the discourse on computer-integrated manufacturing (CIM) in West Germany.[412]

The use of such technology in the textile industry was a frequent topic of news articles, as the production of fashionable clothing was a fitting example to

---

[407] "CAD/CAM wirkt bis an den Arbeitsplatz," *Berliner Zeitung*, February, 22.2.1986, 9.
[408] "Arbeitsplatzcomputer und die Freude des Entdeckens," *Neues Deutschland*, July 26, 1986, 10.
[409] "Arbeitsplatzcomputer und die Freude des Entdeckens," *Neues Deutschland*, July 26, 1986, 10.
[410] Hummel and Rosenkranz, "CAD/CAM," 1106.
[411] Enders 1986; Salecker 1986; Hummel and Rosenkranz, "CAD/CAM."
[412] Julia G. Erdogan, "Wie die Fertigung (zunächst) nicht in den Computer kam. Der schwierige Prozess der Umsetzung von CIM," *Technikgeschichte* 90, no. 1 (2023): 3–24.

illustrate the exciting possibilities of new technology to satisfy rapidly changing consumer needs – a change that would be felt in the everyday life of people in the GDR (see, for example, Figure 4).[413]

Figure 4: Article in the newspaper Berliner Zeitung dated July 26, 1988, entitled "Guided by a computer, the blouses float through the room – Berlin ladies' fashion relies on new technology / Over 7200 more products per year" (translated from German).
Source: "Computergelenkt schweben die Blusen durch den Raum," Berliner Zeitung, July 26, 1988, 3.

The newspaper articles provided concrete examples of how the sociotechnical imaginary of a computerized future was already on the way of being realised in the GDR, for example by deploying computerized machines that would mark and sort laundry in the combine Rewatex in Berlin, thereby relieving the often female workers from strenuous physical labor and reducing the frequency of errors.[414]

---

[413] "Grosse Wäsche mit weniger Handarbeit," *Berliner Zeitung*, January 12, 1985, 11; "Schneiderinnen nutzen Computer für schnellen und massgerechten Zuschnitt," *Neues Deutschland*, December 11, 1987, 8; "Computergelenkt schweben die Blusen durch den Raum," *Berliner Zeitung*, July 26, 1988, 3; "Programme aus dem Computer für Chic und Charme der Berlinerinnen," *Neues Deutschland*, January 20, 1989, 8; "Märchenhafte Muster machen Maschen-Mode," *Neue Zeit*, February 7, 1989, 8.
[414] "Grosse Wäsche mit weniger Handarbeit," *Berliner Zeitung*, January 12, 1985, 11.

In the VEB *Berliner Strickmoden*, the computer-aided design of knitting patterns and programmable knitting machines allowed for a faster adaptation of production to new fashion trends, resulting in a reduction of monotonous work and an increase in work satisfaction for the predominantly female workforce.[415] In this sense, by focusing on the female-dominated textile industry, the newspaper articles also emphasized the "feminine" side of the new computer technology, stressing that women were just as capable and suited to operate computerized machines as men, and encouraging them to acquire the skills and knowledge necessary to use the new technologies in their workplaces.[416]

The SED propagated a sociotechnical imaginary of socialist computerization, according to which a highly skilled workforce with a positive attitude towards learning, further training and technological progress would enthusiastically deploy the computer as a new work tool to automate routine tasks and create new spaces for human creativity. The SED's sociotechnical imaginary of computer and automation technology, thus, was decidedly techno-utopian in nature: Computer technology would bring "unencumbered joy of discovery" to a growing number of workers, which would, in turn, result in "positive economic effects."[417] In other words, the SED claimed that exploiting the potential of new computer technology for the economic growth of the national economy would go hand in hand with social prosperity in the GDR.

Importantly, the computer was imagined merely as a tool in the hands of humans, and it was humans that would decide on the purposes, values and beliefs guiding its use: "All such automata offer the opportunity to increasingly relieve humans of intellectual routine work, which they also perform faster and more precisely. Nevertheless, they remain auxiliary means for humans, tools. [. . .] As ever, only human beings can consciously change and shape society, although they will make use of increasingly powerful and sophisticated tools – such as the 'intelligent' computer."[418] The techno-utopian imaginary, thus, was closely linked with socialist ideology. Capitalism as a social and political system was deemed incapable of harnessing the potential of new technology for social progress and the benefit of all.[419]

---

415 "Programme aus dem Computer für Chic und Charme der Berlinerinnen," *Neues Deutschland*, January 20, 1989, 8; "Märchenhafte Muster machen Maschen-Mode," *Neue Zeit*, February 7, 1989, 8.
416 "Aktuell auch für Frauen," *Berliner Zeitung*, March 25, 1985, 3.
417 "Arbeitsplatzcomputer und die Freude des Entdeckens," *Neues Deutschland*, July 26, 1986, 10.
418 "Computer mit Verstand – Fiktion oder Realität?," *Berliner Zeitung*, September 13, 1975, 13 (translated from German).
419 Wolfgang Fleischer, "Die Zukunft meistern," *technikus*, no. 10 (1985): 1–2; Wolfgang Fleischer, "Vertrauen ins Morgen," *technikus*, no. 9 (1986): 1–2.

Instead, the SED posited that computerization under capitalist conditions would lead to growing unemployment, an anti-human and discontented society. Workers would have to submit to the compulsion of technology to maximize profits and maintain the political, military, and economic influence of the capitalist powers in the world.[420] Negative sociotechnical imaginaries regarding the social effects of computer technology were thus dismissed as being exclusively problems of capitalist systems. Under socialism, on the other hand, a techno-utopian future would be attainable, as computer technology would be used exclusively for the benefit of the people, to better their conditions of working and living.[421]

Closely related to this was another key concern that the SED's sociotechnical imaginary for technological change in the GDR economy highlighted: the improvement of working conditions. Robots and computerized machines, the SED leadership proclaimed, would take over physically demanding tasks form many workers, thus reducing occupational health hazards. Computers and robots would be deployed to automate monotonous and repetitive work tasks, and as a result, create more room for satisfying, qualified, creative activities.[422] In November 1981, the Politburo reported that 9,260 industrial robots were in use across GDR industry.[423] The new five-year plan foresaw that until 1985, an additional 40,000 to 45,000 industrial robots were to be produced and deployed in the domestic industry, and as a result, as many as 112,500 workers would be freed up to take on new tasks elsewhere in the socialist economy.[424] Considering the GDR's constant labor shortage,[425] the anticipated streamlining and labor-saving effects of new technology were an exciting prospect for economic planners.

---

[420] Viktor Böse and Friederike Krump, *Computer in der DDR-Presse: eine ZEFYS gestützte Mikrostudie* (Berlin, 2016), 4, https://nbn-resolving.org/urn:nbn:de:0168-ssoar-47244-8 (accessed May 31, 2024).
[421] Fleischer, "Die Zukunft meistern;" Fleischer, "Vertrauen ins Morgen."
[422] SAPMO DY 30/2101, vol. 43: 19.–20.11.1981, "Bericht des Politbüros an die 3. Tagung des Zentralkomitees der SED," 64.
[423] SAPMO DY 30/2101, vol. 43: 19. –20.11.1981, "Bericht des Politbüros an die 3. Tagung des Zentralkomitees der SED," 63.
[424] SED, *Direktive des X. Parteitages der SED zum Fünfjahrplan für die Entwicklung der Volkswirtschaft der DDR in den Jahren 1981–1985* (Berlin: Dietz Verlag, 1981), 41.
[425] In response to the labor shortage, the GDR also invited large numbers of immigrant contract workers from Poland, Hungary, Vietnam, and other countries. See, for example, Rita Röhr, "Die Beschäftigung polnischer Arbeitskräfte in der DDR 1966–1990," *Archiv für Sozialgeschichte*, no. 42 (2002): 211–36; Mike Dennis, "Working under Hammer and Sickle: Vietnamese Workers in the German Democratic Republic, 1980–89," *German Politics* 16, no. 3 (2007): 339–57; Christina Schwenkel, "Rethinking Asian Mobilities," *Critical Asian Studies* 46, no. 2 (2014): 235–58.

In 1986, at the occasion of the 11th Party congress of the SED, the 5-year-plan for the years until 1990 was announced, which continued along the same lines as the previous plan. Namely, the politically enforced computerization of the economy was to be pushed even further and faster. The economic planners of the GDR proclaimed that 160,000 to 170,000 personal and office computers were to be produced, and the number of computer-aided workstations (CAD/CAM) in the GDR was to be increased from 16,000 in 1986 to anywhere between 85,000 and 90,000 in 1990. The number of skilled workers using this technology, they announced, would increase at least fivefold to half a million or more over the same time span.[426] However, very modest progress had been achieved with regard to the modernization and automation of industry. In 1984, official reports mentioned the deployment of 32,000 industry robots in the GDR. But these were largely very basic automated machine systems only capable of carrying out a single set of tasks following a fixed program without deviation. More sophisticated systems which allowed for flexible automation, on the other hand, were scarce in the GDR industry.[427] Moreover, the SED's proclamation that 100,000 would be using CAD- and CAM-stations by the end of 1986 was not as ambitious as it first might appear. Related to the total number of people working in the industry of the GDR at the time (3.2 million), this envisioned figure of computerized workplaces amounted to merely 3% of the workforce.[428] In comparison with western capitalist countries, the GDR was far behind in terms of the computerization. In 1986, the GDR reached merely 14 CAD/CAM stations per 100,000 employees, compared to 111 in West Germany and 215 in the USA. In terms of microcomputers, the GDR was able to provide 393 microcomputers per 100,000 employees, compared to 3,472 in West Germany – almost ten times as many.[429] In view of the adverse circumstances that left the GDR little choice but to build up its own, largely self-sufficient microelectronics industry, it is nevertheless remarkable how much was achieved in just a decade.

Guided by the SED's sociotechnical imaginary of modern computer technology spreading into all areas of social and economic life over the following decades, the question increasingly arose as to what consequences this development would have for the basic and specialized vocational training of the various skilled trades. In quantitative terms, significantly more skilled workers would be needed for both the development and use of hardware and software, as computers and

---

426 "Arbeitsplatzcomputer und die Freude des Entdeckens," *Neues Deutschland*, July 26, 1986, 10.
427 Wolfgang Hörner, "Technisch-ökonomische Entwicklung und Reformen im Bildungswesen der DDR," *Bildung und Erziehung* 40, no. 1 (1987): 21.
428 Hörner, "Technisch-ökonomische Entwicklung," 21.
429 Klenke, *Kampfauftrag Mikrochip*, 46.

4.3 Envisioning the Computerization of Industrial Production and Offices — 117

computerized machines gained in importance in the eyes of the GDR's economic planners.

This issue had become particularly pertinent since the Robotron combine was commissioned to develop the GDR's first microcomputer in the 1980s.[430] The number of prospective users was expected to increase significantly, as soon as microcomputers could be deployed in a wide range of occupations, including the sphere outside of large industrial plants. While only one or two decades before, computers had been large, centralised data processing machines operated by a whole group of workers, the small microcomputers of the 1980s could be used by a single person in a clerks' office, at the counters of post and savings bank,[431] in design and construction offices, and control centers of factories and combines. In addition, microcomputers allowed the use of more accessible problem-oriented programming languages such as BASIC,[432] and the development of user-friendly software further lowered the skill barrier to computer use. While the use of bulky and cost-intensive computer technology had previously been concentrated in scientific research and industrial production, the advent of office computers, databases, computer networks, and data communications now also provided a sound technical basis for the widespread use of modern computer technology in office administration, management, and planning.

In contrast to the economic strategy of the first half of the 1980s, which had predominantly focused on the production of industrial automation hardware,[433] the SED's five-year-plan for the years 1986 to 1990 now also mentioned software development as an area of considerable importance to the technological progress and economic prosperity in the GDR.[434] With the introduction of microcomputers in more and more areas of the economy and the extension of the range of possible applications, software played an increasingly significant role as a flexible means to adapt computer technology to different user needs, and to solve the specific technical problems at hand. While the provision of hardware remained a major concern,[435] software development came to the fore as the essential means of transferring functions of human intelligence onto machines.[436] The growing user base of computer

---

[430] Klaus-Dieter Weise, *Erzeugnislinie Heimcomputer, Kleincomputer und Bildungscomputer des VEB Kombinat Robotron* (Dresden: Förderverein für die Technischen Sammlungen Dresden, 2005), 7, http://robotron.foerderverein-tsd.de/322/robotron322a.pdf (accessed May 31, 2024).
[431] On the computerization of savings banks in the GDR and West Germany, see Schmitt, *Digitalisierung der Kreditwirtschaft*.
[432] The acronym 'BASIC' stands for 'Beginner's All-purpose Symbolic Instruction Code'.
[433] Zentralkomitee der SED, *Bericht an den X. Parteitag*, 56 and 65.
[434] Zentralkomitee der SED, *Bericht an den XI. Parteitag*, 30.
[435] Zentralkomitee der SED, *Bericht an den XI. Parteitag*, 28–29.
[436] Prager and Richter, *Software – Was ist das?*, 8.

technology also placed new demands on the user-friendliness[437] and quality of software, requiring a greater commitment of human and financial resources to software development. In 1984, the VEB Robotron-Projekt Dresden was established as part of the Robotron combine, focusing on the development of operating systems and user software.[438] In order to meet the new demand for a wide variety of software applications faster and despite very limited capacity, many Western software products were copied and distributed under a new name in the GDR, bypassing Western licensing rights and costs: The database program "Redabas" was based on "dBASE" from the US software corporation Ashton-Tate, the word processor "Text30" was a clone of the popular "WordStar" developed by the US software company MicroPro.[439]

As a result of these developments, preparing workers to use the new computer technology of the 1980s had to take a very different form from training data-processing machine operators in the 1960s and 1970s. A deep knowledge of the hardware and the use of punched cards became less and less important. Instead, it was expected that an increasing number of skilled workers would work with what was called "decentralized information processing technology," such as terminals and microcomputers.[440] With the availability of ready-made software, it was no longer necessary for every computer user to be able to develop their own programs. In 1984, a report devised by the Central Institute for VET stated the following: "Overall, [. . .] there was unanimity in the view that skilled workers, no matter in which direction they are trained, do not develop software. That is the domain of experts. Rather, skilled workers should use ready-made software packages for their tasks in the best possible way. The programming of, for example, CNC machines or office computers, is reserved for technologists or economists. The typist at an automatic typewriter must also mainly have secure skills in operating it. This includes saving texts, the procedure for changing texts, etc.

---

437 Winfried Hacker, "Software-Ergonomie; Gestalten rechnergestützter geistiger Arbeit?!," in *Software-Ergonomie '87: Nützen Informationssysteme dem Benutzer?*, ed. Wolfgang Schönpflug and Marion Wittstock (Stuttgart: B. G. Teubner), 34.
438 Hans-Jürgen Lodahl, "Das Softwarehaus des VEB Kombinat Robotron," in *Informatik in der DDR – Grundlagen und Anwendungen*, ed. Birgit Demuth (Bonn: Gesellschaft für Informatik, 2008), 231–42; Detlev Fritsche, "Mit Prototyprekonstruktion zum Welthöchststand? PC-Software in den letzten Jahren der DDR," *Dresdener Beiträge zur Geschichte der Technikwissenschaften*, no. 30 (2005): 110.
439 Fritsche, "Mit Prototyprekonstruktion zum Welthöchststand?," 110; Francis Hunger, "Sozialistische Co-Innovation – Wie in der DDR die relationale Datenbank DABA-1600 entwickelt wurde," *Zeitschrift für Medienwissenschaft* 14, no. 2 (2022): 67.
440 BArch DQ 4/5271, "Lehrplaneinheiten für die sozialistische Berufsbildung: Grundlagen zur Bedienung dialogorientierter dezentraler Informationsverarbeitungstechnik (Beispiel Bürocomputer)," 3.

How this is technically realized and programmed internally is beyond their knowledge and ability."[441] Consequently, a very basic knowledge of programming in problem-oriented programming languages such as BASIC, and some rudimentary knowledge about hardware and the general principles of how new information technology functioned were deemed sufficient for most.

With the expected long-term increase in demand for large numbers of workers with a variety of skills related to new computer technology, the GDR's vocational training system could no longer rely primarily on workplace training and short-term continuing education measures as it had in the 1960s and 1970s. So how would the system of *initial* vocational training have to be adapted to prepare trainees in all areas of economic life for technological change?

4.4 Introducing Computer Technology in Vocational Education and Training

In the 1980s, the GDR's occupational system was divided into over 300 basic occupations and around 630 fields of specializations. A vast majority of around 85% of school leavers in the GDR entered a vocational apprenticeship in one of these skilled trades.[442] However, their career choice[443] was heavily guided by career counsellors who were mandated to steer young people onto appropriate career paths in accordance with the SED's economic strategy and workforce planning.

In most occupations, the duration of initial VET was two or two and a half years. While for many training professions the completion of the 10th grade was a prerequisite, some professions required prospective apprentices to have only completed the 8th grade of the Polytechnic Secondary School (ger.: Polytechnische Oberschule, POS). A specialty of VET in the GDR was the double qualification of Vocational Training with Abitur (ger.: Berufsbildung mit Abitur, BmA). Some, but not all professions allowed a selection of well performing apprentices to attain a

---

441 BArch DQ 400/1161, "Konzeption zu Untersuchungen auf dem Gebiet 'Informatik in der Berufsbildung'," dated November 1984, 9 (translated from German).
442 Oskar Anweiler, "Berufsbildung in der Deutschen Demokratischen Republik unter vergleichenden Aspekten," in *Wissenschaftliches Interesse und politische Verantwortung: Dimensionen vergleichender Bildungsforschung*, ed. Jürgen Henze, Wolfgang Hörner, and Gerhard Schreier (Wiesbaden: VS Verlag für Sozialwissenschaften, 1990), 87.
443 Restrictive measures to steer the signing of new apprenticeship contracts towards priority occupations according to plan had been lifted in the 1950s (Uwe Vollmer, "Vollbeschäftigungspolitik, Arbeitsplatzplanung und Entlohnung," in *Die Endzeit der DDR-Wirtschaft – Analysen zur Wirtschafts-, Sozial- und Umweltpolitik*, ed. Eberhard Kuhrt (Opladen: Leske+Budrich, 1999), 323–73.

general higher education entrance qualification as part of their vocational education. Within just three years, these apprentices acquired the skills and knowledge required for completing their vocational training, as well as for entering higher education.[444] While the practical part of vocational instruction was reduced for these apprentices, they attended special classes for their vocational and general education with a higher requirement level, and finally had to pass the same final exam as pupils who attended the Extended Secondary School (ger.: Erweiterte Oberschule, EOS). The BmA, in conjunction with appropriate entry quotas, represented a significant instrument for adapting the training and education of young people to the needs of the national economy and for raising the level of qualification, especially in new technologies.

The principle of a dual system of vocational training can be considered as a common historical root of VET in both West Germany and the GDR. However, the different political and economic systems led to a diverging development of the structure and organization of VET in the two countries. In West Germany the dual-system referred to the two mutually complementary pillars of VET: On the one hand, practice-oriented in-company training of apprentices within the responsibility of the private sector, and on the other hand, school-based vocational education under the auspices of the state, focused on theoretical aspects of and general education. Within the socialist command economy of the GDR, cooperation, and division of responsibilities between state and the private sector were absent, as companies were publicly owned and subject to centralized planning by government authorities. Essentially, they had extremely limited influence over their own production level, allocation of human and material resources, as well as contents and organization of vocational training. The "dual system" of VET in the GDR thus primarily referred to the conceptual divide between predominantly practical and theoretical vocational training and instruction. Practical vocational training was provided by experienced skilled workers who served as instructors part-time or full-time, as well as engineering pedagogues. Theoretical instruction, on the other hand, was delivered by graduate teachers with university degrees for the teaching of the various vocational subjects.[445]

However, there was no meaningful division of vocational training into two sites of learning – the company and the vocational school, with their separate administrative and didactical systems. In the GDR, vocational schools for theoretical instruction

---

444 Anweiler, "Berufsbildung," 88.
445 Anweiler, "Berufsbildung," 85.

were either under company or communal sponsorship.[446] By the second half of the 1980s, around 80% of all apprentices attended vocational schools under company sponsorship, the so-called "industrial vocational colleges" (ger.: Betriebsberufsschule).[447] These training facilities covered both practical and theoretical VET and formed an integral part of the larger socialist combines of the GDR, consisting of training workshops, classrooms for theoretical instruction and often also apprentice dormitories. Industrial vocational colleges served to train apprentices to fulfill the combine's own need for young talent, but also took on the task of educating apprentices from other companies that did not have their own in-company vocational college for theoretical instruction. Directors and teachers at industrial vocational colleges were appointed by the management of the combine but needed to be approved by local state authorities. Communal vocational schools, which provided for the more theoretical parts of VET for the trainees of smaller companies, were directly subordinate to the local authorities for vocational education, training, and counselling.[448]

The responsibilities for the planning, organization, and carrying out of VET in the GDR were laid out in the decree on skilled worker professions of December 21, 1984.[449] Accordingly, the State Secretariat for Vocational Education determined the basic requirements for the planned continued development of the contents and profiles of the skilled worker professions. Its main concern was the management and coordination of the overall process of continued development of the skilled worker professions. The State Secretariat for VET worked closely together with the line ministries and other central state organs to determine the consequences for vocational education contents and occupational profiles in the various fields of the economy according to the SED's five-year plans (see Figure 5). In the first half of the 1980s, the State Secretariat for VET reviewed and revised the training content and documentation of skilled trades to address the growing influence of microelectronics and flexible automation, focusing on the most affected professional fields. In the second half of the 1980s, with the rise of microcomputers and spreading of computer technology across all spheres of the economy, curricular revisions focused on the use of per-

---

[446] The number of communal vocational schools decreased from 744 in 1955 to 244 in 1985. In turn, vocational schools under company sponsorship (ger.: Betriebliche Berufsschulen) increased from 610 in 1955 to 719 in 1985 (Staatliche Zentralverwaltung für Statistik, *Statistisches Jahrbuch 1985 der Deutschen Demokratischen Republik* (Berlin: Staatsverlag der Deutschen Demokratischen Republik, 1985), 298).
[447] Anweiler, "Berufsbildung," 85.
[448] Anweiler, "Berufsbildung," 85.
[449] SAPMO ZB 20049, "Verordnung über die Facharbeiterberufe vom 21. Dezember 1984," Gesetzblatt der Deutschen Demokratischen Republik 1985, part 1, no. 4 (February 20, 1985): 25–28.

sonal computers and CAD/CAM. In line with the SED leadership's imaginary of a virtually all-encompassing computerization of the GDR economy and society, efforts were concentrated on curricular revisions in basic technical training for all skilled trades, as well as the introduction of occupation-specific computer training for skilled trades in which computer technology would be playing a decisive role. This applied to jobs in the production of microelectronic components and devices, in industrial manufacturing involving microelectronics technology such as the production of computerized machine tools, and in the broad and diverse field of workplaces involving the use of personal computers and computerized machines.

Figure 5: Simplified governance structure for the determination of curricular content in VET. Source: Adapted from Helge Körner, "Berufsfachkommissionen," in Aspekte der beruflichen Bildung in der ehemaligen DDR, ed. Arbeitsgemeinschaft Qualifikations-Entwicklungs-Management Berlin (Münster: Waxmann, 1996), 21.

For each of the skilled trades, the State Secretariat for VET together with the competent line ministry in each case designated the "responsible authorities." These were usually companies or combines that were accorded special priority in the respective industrial sector, had a high internal demand for skilled workers in the respective occupation, and were equipped with modern technical equipment, as well as the necessary professional competence regarding the further development of the skilled occupation concerned.[450] For example, the VEB Kombinat Robotron Dresden was appointed as the responsible authority for the creation of the skilled trade "maintenance mechanic for data processing and office machines"

---

[450] Körner, "Berufsfachkommissionen," 15.

in the second half of the 1960s,[451] as well as its further development over the following decades, as new electronic data processing machines and computers were introduced.

The responsible authorities for each skilled occupation then appointed a "Vocational Expert Committee" (ger.: Berufsfachkommission) as a permanent, part-time working body in the professional field concerned. For the activities in these committees, scientific-technical and economic experts, proven practitioners in the relevant field, as well as occupational pedagogical and occupational hygienic experts were recruited.[452] It was up to the Vocational Expert Committee to propose updates to the vocational profile of the skilled occupation they were responsible for, reforms of the curricular materials for VET, and make propositions for new teaching and learning materials, as well as the further training of vocational trainers and teachers. As such, the Vocational Expert Committees played a vital role in adapting job profiles and vocational training curricula to the present and anticipated technological change. However, curricula for certain vocational subjects, in particular theoretical instruction, were centrally predefined by the State Secretaries for VET and General Education, amongst them the three basic technical subjects, and their successor "Basics of Automation."[453] However, in the special case of the latter, the Vocational Expert Committees were at least granted some influence in formulating the contents and goals of a subject intended to introduce all apprentices to new computer technology.

A close cooperation of the Vocational Expert Committees with technical experts of the respective responsible authorities, Central Offices for VET, and scientific institutions was intended to ensure that the Vocational Expert Committees were well informed not only about the current status, but also about emerging trends in the technological and framework conditions of production, and the latest scientific findings in the respective field.[454] In the late 1970s, the State Secretariat for VET mandated the "Central Institute for VET" (ger.: Zentralinstitut für Berufsbildung, ZIB) to support the various Vocational Expert Committees with consultations, scientific insights, and studies regarding technological change and new skill demands. The role of the ZIB was to prepare a "scientific" foundation for curricular reforms

---

451 BArch DQ 400/2410, "Einführung der EDV 1964–1968."
452 Körner, "Berufsfachkommissionen," 15.
453 Business Economics, Socialist Law, Basics of Automation, Civic Education, Sports (Körner, "Berufsfachkommissionen," 30).
454 Dieter Grottker, "Ungleiche Partner – Das Dogma der sozialistischen Allgemeinbildung und die berufliche Bildung in der DDR," *Syllabus. Gesammelte Aufsätze zur Berufs- und Bildungswissenschaft*, no. 1 (January 2020): 24, https://syllabus-dresden.de/2020/01/18/223/ (accessed May 31, 2024).

in VET, that is, to analyze the vocational implications of introducing new computer technology into workplaces in all spheres of the economy, give recommendations regarding the adaptation of VET to cater for new skill requirements, and make sure that vocational curricula were reformed accordingly.

In May 1979, the ZIB published an extensive study on the prospective effects of microelectronics on the further developments of the contents of skilled trades.[455] The aim of the study was to identify, which occupations would be affected over the course of the 1980s, and how the profiles and work contents of these jobs would change. Regarding the application of computerized machines and systems, there were still only few practical experiences in the GDR in the second half of the 1970s. To predict general trends for the 1980s, the researchers of the ZIB drew upon the examples of a few pioneering domestic companies to discuss possible applications in industrial production. But they also relied on international literature on the digital transformation of finance, insurance, banking, and office technology, which had not yet manifested itself in the GDR because it had not yet been a political priority for the SED.

However, the study's authors also asserted that the specific form that technological change would take was not simply a direct result of or inherent to microelectronics but was mediated by a variety of contextual factors and depended on certain conditions that had to be primed for the desired technological effects to manifest. They expressed their conviction, that only such a "complex view" of sociotechnical dynamics and taking into account the specific context of the GDR economy would allow for an accurate assessment of the digital transformation under way.[456] Thus, although the GDR's technical backlog offered the opportunity to learn from the more advanced capitalist countries and to anticipate coming developments more easily, there was still a great deal of uncertainty as to how these developments would translate to the political and economic context of the socialist GDR.

The study predicted that until 1985, industrial automation would significantly increase, but primarily in the form of partial automation, that is, the automation of individual devices and workplaces. Complex automation was not anticipated to be implemented in the GDR on a substantial scale until the second half of the 1980s. Until the mid-1980s, computerized controls would continue to be predominantly used in the production of machine tools and only marginally extend to other areas of industrial production, for example in the form of welding robots,

---

455 BArch 400/2022, "Studie zu voraussichtlichen Wirkungen der Mikroelektronik auf die Weiterentwicklung des Inhalts der Facharbeiterberufe," dated May 7, 1979.
456 BArch 400/2022, "Studie zu voraussichtlichen Wirkungen der Mikroelektronik auf die Weiterentwicklung des Inhalts der Facharbeiterberufe," dated May 7, 1979, 12.

or modern computerized controls in textile machines, plastic processing machines, and a variety of industrial robots.[457] However, with regard to the second half of the 1980s and the 1990s, the study authors predicted that the application of computer technology would largely permeate all areas of economic and social life: "This is when the effects will be most intensive in quality and extent, but also most differentiated."[458] Quantitatively, the focus of computer education and training requirements would shift from workers in hardware manufacturing to a large and diverse crowd of computer users, software developers, as well as service and maintenance personnel for computer technology.

Most notably perhaps, the study's authors pointed out that, regarding the use of new technology as a working tool, computerized machines and systems would not substitute conventional technology, but rather complement it. Thus, workplaces in the GDR would be characterized throughout the 1980s by a coexistence of both old and new technologies, which would necessarily lead to a differentiation of work contents and skill demands, depending on what kind of technology was available to workers in performing certain tasks.[459] Fully integrated computerized systems, as envisioned in the imaginaries of SED leaders, would only be built up gradually over years or decades in the industry of the GDR, due to the lack of technological resources. The resulting patchwork of partially automated and computer-controlled systems meant that worker had to assume the function of human "links" where computerized solutions were missing.[460] Whereas political leaders had promised that computers would relieve workers from tedious work, such linking positions required for workers to do stressful work in tune the rhythm of computerized machines. Often, this involved doing repetitive and monotonous tasks that left little room for human agency and variation. While these "linking positions" filled by humans were intended to be substituted by computerized machines, the slow and patchy progress meant that incomplete computerization had, paradoxically, created such unpleasant work tasks in the first place.

The ZIB's early study on the effects of technological change on workplaces and skill requirements, thus, clearly pointed out that the introduction of com-

---

457 BArch 400/2022, "Studie zu voraussichtlichen Wirkungen der Mikroelektronik auf die Weiterentwicklung des Inhalts der Facharbeiterberufe," dated May 7, 1979, 28.
458 BArch 400/2022, "Studie zu voraussichtlichen Wirkungen der Mikroelektronik auf die Weiterentwicklung des Inhalts der Facharbeiterberufe," dated May 7, 1979, 30 (translated from German)
459 BArch 400/2022, "Studie zu voraussichtlichen Wirkungen der Mikroelektronik auf die Weiterentwicklung des Inhalts der Facharbeiterberufe," dated May 7, 1979, 31.
460 BArch 400/2022, "Studie zu voraussichtlichen Wirkungen der Mikroelektronik auf die Weiterentwicklung des Inhalts der Facharbeiterberufe," dated May 7, 1979, 34.

puter technology would lead to both an increase in highly skilled, but also low skilled jobs. With regards to the political initiatives of the 1980s to significantly raise the qualification level of the whole workforce as a response to technological change, certain tensions were therefore anticipated. The GDR's political leadership seems to have been aware of the paradox, but insisted, that the tension of requiring well qualified workers to do low qualified work would only further promote efforts to computerize these jobs so workers would be freed up to do more skill adequate, and mentally stimulating work. Essentially, the creation of new low skilled jobs involving repetitive and mentally unstimulating tasks was considered a temporary phenomenon, and in turn, the overqualification of workers in such positions was a compromise that the SED was willing to accept. Overqualification was legitimized with the argument that, in face of the rapid progress in science and technology, workers needed to be prepared well in advance for new tasks involving computer technology in the future. As skill demands were expected to change quickly and multiple times over a working life, workers had to be broadly skilled so they could be deployed in several different jobs and functions in the future. After all, in face of changing economic demands, economic planners expected for workers to be required to move between job positions and functions more often and quickly, as new technologies made some job positions redundant and created new manpower needs in other areas.[461]

In addition, for ideological reasons, it was deemed unacceptable within the socialist system to deny a part of the labor force a sound education in order to place them in unskilled jobs. After all, such a divide between a technically literate elite and unskilled labor was exactly what GDR officials had repeatedly criticized with regard to how computerization in capitalist countries had exacerbated inequalities. Instead, to uphold the sociotechnical imaginary that computer technology under socialism would free workers form such undesired jobs to engage in more creative and mentally stimulating tasks, vocational education and training in the GDR had to provide all people with a high level of technical skills and knowledge. Disappointed overqualified workers in low-skilled positions were fobbed off with the promise that technological change would soon free them, too, from their unsatisfactory work tasks and provide them with more rewarding work where they could finally make full use of their skills and knowledge. However, until the end of the GDR, for many workers this promise was never fulfilled.

In its reports of 1980 and 1981, the ZIB criticized that the Vocational Expert Committees rarely made use of the study results prepared by ZIB, and that many

---

[461] Michael Guder, "Die Einstellung der beruflichen Bildung in der DDR auf neue Technologien," *Berufsbildung in Wissenschaft und Praxis* 16, special issue 'Berufsbildung in der DDR' (1987), 13.

did not follow up on its offers for consultations.[462] A number of Committees had voiced that they did not see a need for far-fetching revisions of curricula in initial VET, would rather expand further and adult vocational training to respond to new skill requirements once they were needed, depending on the introduction of new technology in individual workplaces. Thus, it seems to have been obvious to vocational training experts and pedagogues on the level of individual companies and combines, that the real conditions of industrial production and workplace structures in the GDR had not yet been transformed by the new technologies to nearly the extent that might be assumed from the SED's reports on progress in the relevant areas of innovation. While they had to prepare apprentices for an envisioned future world of work with computer technology, they also had to account for the actual current situation in workplaces. It was concluded that, on the one hand, the vocational training system should systematically provide the anticipated skills needed to deal with the new information technologies "in preparation". On the other hand, it was considered equally important not to ignore the persistence of traditional work processes and the coexistence of "old" and "new" technologies.[463]

However, in view of the accelerated use of microelectronics technology and the breakthrough of the first domestically developed and produced microcomputers in the GDR, towards the mid-1980s the question arose as to whether vocational education and training would have to respond with faster and more comprehensive curricular reforms. In order to prepare skilled workers in a wide variety of occupations for the demands of using computer technology, it seemed questionable in the whether the current conception of VET in this field, relying mostly on further education and training measures, could withstand "the onslaught of the 1990s."[464] The researchers of the ZIB were well aware of the GDR's technological backlog of six to twelve years in international comparison, which is why they saw an urgent need to prepare a skilled workforce well in advance to catch up with technological progress over the coming two decades. The constantly expanding circle of users of computer technology was cited as justification for the introduction of a broad-based vocational

---

462 BArch DQ 400/2021, "Bericht über die Unterstützung der Berufsfachkommissionen solcher Facharbeiterberufe, bei denen der Inhalt der Arbeit sofort und unmittelbar von der Entwicklung der Mikroelektronik beeinflusst wird."
463 Oskar Anweiler, "Berufsbildung in der Deutschen Demokratischen Republik unter vergleichenden Aspekten," *Berufsbildung in Wissenschaft und Praxis* 16, special issue 'Berufsbildung in der DDR' (1987), 6.
464 BArch DQ 400/1161, "Konzeption zu Untersuchungen auf dem Gebiet 'Informatik in der Berufsbildung'," dated November 1984.

computer training. In addition, it was expected that in the near future, the industry of the GDR would be able to provide microcomputers at reasonable prices for educational institutions. Microcomputers were no longer imagined just as a subject matter in education, but also as a means of instruction – an educational technology.[465]

In December 1983, the Politburo of the SED's Central Committee and the Ministers' Council decided on "Measures for the further improvement of Vocational Education and Training." According to this plan, several measures were taken to reform VET: The reorganization of the "system of skilled worker occupations" (ger.: Systematik der Facharbeiterberufe), a curricular reform of the three basic technical subjects in initial VET, as well as the development of new curricula for all skilled occupations in response to technological change.[466]

The revision of vocational curricula for all occupations were placed in the hands of the vocational expert committees. The most progressive, standard-setting combines were involved as key stakeholders in defining the form and content of the modernized vocational profiles. The new vocational curricula were gradually introduced over the following years. Starting in September 1987, apprentices in 78 basic occupations were already trained and educated according to new curricula, covering the vocational training of over 80% of all new apprentices.[467] The curriculum reforms were aimed primarily at taking into account new technological developments in the field of microelectronics, automation and computer technology.

A new occupational profile of "skilled worker for electronic components" was created to meet the increased demands on skills and knowledge in the manufacture of microelectronic components. In the training of skilled worker for operational measurement, control and regulation technology, a new field of specialization for "microcomputer-controlled systems" was introduced. The training of technical draughtsmen was fundamentally reformed to include learning and working at CAD-stations. As a last example, skilled workers for "writing technology" (ger.: Facharbeiter für Schreibtechnik) – a clerical occupation involving office communications, shorthand and typing, word processing, filing, and general office organization – according to the new curriculum also were to be gradually introduced to working with office computers.

---

[465] BArch DQ 400/1161, "Konzeption zu Untersuchungen auf dem Gebiet 'Informatik in der Berufsbildung'," dated November 1984.
[466] Hörner, "Technisch-ökonomische Entwicklung," 25.
[467] "Die richtige Qualifikation für die Aufgaben von morgen," *Neues Deutschland*, August 9, 1986, 10.

4.4 Introducing Computer Technology in Vocational Education and Training — 129

The reform of basic vocational subjects targeted specifically the three subjects that had been introduced 1968 – Data processing, Electronics, and "BMSR"[468] – which were now to be replaced by the single subject "Basics of Automation." The previous subjects had covered a wide range of topics related to automation and early computer technology, focusing on hardware as well as skills and knowledge in electrical engineering. At the heart of the new subject was microelectronics-based computerization. The aim was to teach some theoretical foundations of computer and automation technology and to introduce digital information acquisition, storage, transmission, and processing with the help of computers. However, as the curriculum for the new subject was drafted in early 1984, the practical use of a microcomputer was not yet envisaged and learning to program in BASIC was merely intended as a fully analogue pen-and-paper exercise. Vocational education policy-makers deplored this fact, as in their opinion, having an actual microcomputer in the classroom would allow teachers to make their teaching about information technology more vivid, being able to demonstrate learnings on an actual computer. This would both help to make teaching and learning more effective and motivate learners to study the subject matter in greater depth.[469] Thus, it was not pedagogical considerations that held back the introduction of computers into classrooms. Rather, it was due to economic reasons: In the first half of the 1980s, computers in the GDR were still too expensive and not available in sufficient numbers to equip vocational training facilities. However, as indicated at the end of the previous sub-chapter, this changed within the matter of a very short time, as the first microcomputers entered mass production in the GDR. The decision of the Politburo of the SED's Central Committee and the Ministers' Council in November 1985 emphasized the need for applied and practical computer training to prepare prospective workers for the use of computer technology in the workplace. Furthermore, the decision provided for the gradual equipping of vocational training institutions with computer rooms starting in 1986.[470] Each facility was to be equipped with a set of 10 microcomputers, preferably of the type "KC 85/2," providing for nine shared student workplaces and one for the teacher.[471] In addition, every computer required a

---

468 'BMSR' was the commonly used abbreviation in the GDR for 'Betriebsmess-, Steuerungs- und Regelungstechnik' (eng.: Operational measurement, control, and regulation technology).
469 BArch DQ 400/2346, "Konzeptionelle Überlegungen und Vorschläge zur Einbeziehung der Informationsverarbeitung und -technik in die Berufsbildung," dated May 1984, 4–5.
470 SAPMO DY 30/J IV 2/2/2138, "Standpunkte zu Konsequenzen aus der Entwicklung der Informatik und informationsverarbeitenden Technik für das Bildungswesen."
471 Starting from 1986, an annual quota of 1.200 KC 85/2 microcomputers was set for the equipment of VET institutions to achieve a somewhat uniform technical standard. From 1988 onwards, the gradual introduction of a new type of computer in VET was planned, especially with regard to the proposed development of a purpose-built educational computer (BArch DQ 4/5294, "Abstim-

cassette recorder model Geracord GC 6020, manufactured domestically by the VEB Elektronik Gera, and a Junost 402 B television made in the Soviet Union.[472]

When the new course "Basics of Automation" was finally introduced into initial vocational training in 1986, it included the practical use of microcomputers for programming and software applications. On the other hand, education and training regarding industrial computerized machines remained fairly limited. For example, the topic of CNC machines[473] made up 11% of theoretical instruction in basic VET (16 out of 144 hours), and merely 3% of practical instruction (10 out of 322 workdays).[474] Generally, hardware as a topic of instruction had been pushed back in favor of the use of software. This was in line with the ZIB researchers' belief that for most students the use of ready-made software applications was the most essential computer skill they would need in their future workplace: "The group of people who work with computer technology as users is clearly in the majority, compared to the manufacturers and the service or repair personnel. But also, the way skilled workers use computer technology [. . .] supports the thesis that the majority of skilled workers should be trained to work with software rather than being equipped with detailed knowledge of the hardware of information processing technology."[475]

In contrast to articles in the media, which popularized the notion that computers required users to be capable of programming, vocational education research in the mid-1980s highlighted that higher programming languages such as BASIC only played a minor role in the everyday dealings of skilled workers with computers.[476] For the majority of production workers it seemed pointless to acquire more than a bare minimum of programming skills, as their job involved – if requiring any computing skills at all – predominantly the use of ready-made user software. This kind of software for industrial and office applications was

---

mungsprotokoll zwischen dem Ministerium für Elektrotechnik und Elektronik und des Staatssekretariat für Berufsbildung zur Bereitstellung von Kleincomputern für die Berufsbildung.")

472 BArch DQ 400/2793, "Hinweise zur Gestaltung von Computerkabinetten in der Berufsbildung," dated January 1986, 5.

473 CNC stands for Computerized Numerical Control. CNC machines are machine tools that can be controlled by a computer program to automate the production of high-precision workpieces. Common CNC machining processes include milling, turning, drilling, grinding, and sawing.

474 Hörner, "Technisch-ökonomische Entwicklung," 26.

475 BArch DQ 400/2346, "Konzeptionelle Überlegungen und Vorschläge zur Einbeziehung der Informationsverarbeitung und -technik in die Berufsbildung," dated May 1984, 5 (translated from German).

476 BArch DQ 400/1161, "Konzeption zu Untersuchungen auf dem Gebiet 'Informatik in der Berufsbildung'," dated November 1984; BArch DQ 400/2346, "Konzeptionelle Überlegungen und Vorschläge zur Einbeziehung der Informationsverarbeitung und -technik in die Berufsbildung," dated May 1984.

largely developed by specialists with a degree in computer science. In industrial production, a working understanding of microelectronic hardware was required, as workers also needed to be able to troubleshoot computer-controlled machines. Outside of the producing sector, however, skills and knowledge related to computer hardware were considered non-essential. Instead, good communication skills, mental agility as well as logic and algorithmic thinking were deemed more important.[477] Moreover, a profound understanding and mastery of traditional technologies of their job was considered an essential prerequisite for the use and understanding of new technologies.[478]

In this sense, the design of a subject curriculum for "Basics of Automation" represented a compromise, the lowest common denominator, so to speak, between different demands and needs of a wide range of professional domains. The ambition behind the introduction of the new subject was simply to teach the basics of the new computer and information processing technology to build upon this foundation within the scope of workplace-specific education and training. Thus, in addition to the basic education and training in computing and automation technology, apprentices in selected occupations were taught occupation-specific computer knowledge and skills, in each case geared to the concrete requirements of the target occupations: Machine tool operators were trained in CNC technology,[479] technical draftsmen were prepared for CAD workstations,[480] and clerical workers introduced to word processing, file management and spreadsheet software.[481]

The State Secretariat for VET centrally administered the curriculum for the vocational subject "Basics of Automation" and issued a binding syllabus for the vocational expert committees of various trades to follow. But unlike other centrally administered syllabi, they were granted some degree of influence over the development and content design of the subject "Basics of Automation" in coopera-

---

[477] 'Algorithmic thinking' or 'algorithmic problem-solving' was mentioned by education policy-makers in the GDR to refer to the specific way of thinking required to translate a task into a formalized solution strategy, breaking it down into a series of systematic logical steps that could be processed by a computer. However, what exactly this meant and how it could be taught remained largely unsettled questions in the 1980s.
[478] BArch DQ 400/1161, "Konzeption zu Untersuchungen auf dem Gebiet 'Informatik in der Berufsbildung'," dated November 1984, 10.
[479] BArch DQ 400/800, "Ausbildung der künftigen Facharbeiter für Werkzeugmaschinen für CNC-Technik an der BBS 'Ernst Thälmann' des VEB Schwermaschinenbaukombinat 'Ernst Thälmann' Magdeburg – Ausbildungsdokumentation," dated October 1988.
[480] BArch DQ 4/5295, "Erprobungsbericht über die Erprobung zur Einführung des Lehrplanes für den Lehrgang 'CAD-Grundlagen' für die Zeichnerberufe," dated August/September 1987.
[481] BArch DQ 4/5271, "Bericht über die Erprobung der Informatikausbildung – 1. Lehrjahr – bei Facharbeitern für Schreibtechnik," dated July 1987; Hörner, "Informationstechnische Bildung," 632.

tion with the ZIB.[482] Moreover, the State Secretariat for VET provided the vocational expert committees with multiple versions of the syllabus to account for diverging interpretations of the subject and setting of priorities in line with different job-specific requirements. The 1988 edition of the "Basics of Automation" syllabus consisted of two parts, one dedicated to the "Introduction to Automation," the other to "Informatics."

The course "Introduction to Automation" was offered in two variants, both of which consisted of 68 total hours of instruction each but differed slightly in the distribution of time to individual topics of learning. While version one was the "generic" syllabus in "Introduction to Automation" for all apprentices, version two was adapted for skilled occupations involving management, planning, accounting, and administrative tasks to support trainees in this fields for the mastery of modern office technology.[483]

The second part of the syllabus on "Informatics" consisted of three different options: Informatics "basic requirements," "extended requirements," and "BmA." The basic version was taught in 33 hours and encompassed a practical introduction to working with microcomputers, an introduction to programming in BASIC, and the use of simple, ready-made user applications. In the other two versions of the syllabus, the informatics part was increased to 66 hours of teaching. The version with extended requirements was geared towards skilled occupations where microcomputers, currently or prospectively, played a significant role. More teaching hours were dedicated to programming instruction, and instead of training the use of simple software applications, the syllabus required to use of "complex user programs" relevant to the respective occupational field. In addition, the syllabus included a section on possible applications of computer technology in the workplace, discussing how a microcomputer could be used to solve various professional tasks. Finally, the third version was geared towards teaching informatics to trainees in a BmA program. In addition to the introduction to working with microcomputers and a more extensive programming course in BASIC, this version also included the development of their own computer programs by trainees as well as the use of existing software to solve scientific, technical, and economic tasks. The version for BmA students was meant to prepare them for their later studies at a technical college. It emphasized algorithmic problem-solving skills in particular, as microcomputers and especially software had gained in importance in applied research as well as applications in economics and engineering.

---

482 Körner, "Berufsfachkommissionen," 30–31.
483 BArch DQ 400/2793, "Lehrplan für die Facharbeiterausbildung. Grundlagenfach: Grundlagen der Automatisierung (einschliesslich Informatik)," edition of 1988, 14.

Despite the great importance attached to the modernization of the production process by deploying new computer technology, in the eyes of the SED, the "socialist work ethic" of the workforce remained the most significant driving force in economic production, it even gained in significance. Accordingly, the inculcation of a "communist attitude to work" remained a central learning objective of polytechnic and vocational education in the wake of the introduction of new technologies in production.[484] The curricular goals of the subject "Basics of Automation" stated the need for trainees to recognize that the automation and computerization of the GDR were a necessity, which under socialism (and only under socialism) would go hand in hand with social progress.[485] Unlike numerous features in newspapers and popular magazines which foregrounded almost exclusively the civil uses of new technology, the syllabus explicitly mentioned also the significance of computer technology in armed conflict in the context of the system conflict between socialism and capitalism: "[. . .] automation is an essential means for increasing labor productivity, the effective use of energy and materials and the constant perfection of weapons technology for the reliable military protection of socialism."[486] The stated aim of the subject was to inspire the readiness in all trainees to apply the acquired knowledge and skills in the field of automation technology in their daily life, that is, "in their professional work and during their honorary service in the armed forces." In addition, the syllabus required the trainees to be willing to constantly further their own education in accordance with the new demands of scientific and technical progress. It was made their responsibility as members of the socialist society to constantly update their skills as new computer technology was developed and became available in the GDR.

4.5 The Computerization of the Socialist Workplace

In 1986, the GDR's popular science and technology magazine for young readers *technikus* published a short fictional story under the title "The New Millennium."[487] The story was about a diligent engineer who one day at work was informed about a substantial restructuring of his workplace. His tasks would be taken over by a computer, the facility closed, and workers were relegated to new jobs elsewhere. Studying the restructuring plans, the engineer grew increasingly

---

484 Hörner, "Technisch-ökonomische Entwicklung," 30.
485 BArch DQ 400/2793, "Lehrplan für die Facharbeiterausbildung. Grundlagenfach: Grundlagen der Automatisierung (einschliesslich Informatik)," edition of 1988, 8.
486 BArch DQ 400/2793, "Lehrplan für die Facharbeiterausbildung. Grundlagenfach: Grundlagen der Automatisierung (einschliesslich Informatik)," edition of 1988, 8 (translated from German).
487 Wolfram Kober, "Das neue Jahrtausend," *technikus*, no. 1 (1986): 10–11.

angry and distraught: "I cannot discover my workplace anywhere. Instead, a computer with its manipulation system is squatting in my place! Twelve years I have worked there, often more than six hours, voluntarily into the night . . . and for what? To warm the seat for this soul-dead microtronic *(sic!)* creature?"[488] But instead of becoming unemployed, he was offered to go to Ghana with his entire work collective to work in development aid for the next five to ten years. However, the engineer was unwilling to relearn, rethink, and start all over again. Disgruntled, he went home to tell his wife how unfairly he had been treated, only to be scolded by her for his selfishness. The next day, he found that everyone had signed up for the new mission in Ghana, except for him.[489]

The moral of the story – at least from the standpoint of the SED ideologues – is easy to discern: The engineer had acted selfishly by not accepting change as an inherent consequence of technological progress. Unwilling to further his education and retrain for a new job assignment, where his work was needed to further the socialist cause, he had isolated himself from the socialist collective. He had behaved in an "unsocialist" manner, putting his own desires first instead of subordinating himself to a higher, collective purpose as part of the SED's modernization and computerization strategy for economy of the GDR.

The "right attitude" of workers in face of the computerization of workplaces was an important concern for the SED, in line with the centrality of humans in their sociotechnical imaginary of socialist computerization. The SED leadership was convinced that the direction, pace, and success of technological change in the economy of the GDR was strongly influenced by the attitude of the people towards computer technology, and their readiness to set in motion and support the necessary changes in their workplace. This included not only the willingness of young school leavers to enter vocational training in technical fields particularly relevant to the development and use of computer technology, but also of existing workers to undergo further education and retraining, if necessary.

The introduction of a basic vocational education in automation technology and informatics for all apprentices was perhaps the most emphatic materialization of the SED leadership's sociotechnical imaginary, according to which the entire working class of the GDR was to be tied into the project of comprehensive computerization and made jointly responsible for the realization of the expected economic and social progress. However, the vocational curricula in computer technology education sought not only to teach trainees the technical ins and outs of using new technology. Rather, the new curricular contents and goals were also

---

488 Kober, "Das neue Jahrtausend," 10 (translated from German).
489 Kober, "Das neue Jahrtausend," 11.

aimed at getting prospective workers to understand that they were expected to make a significant personal contribution – or perhaps even sacrifice – by taking responsibility for the process of modernizing and rationalizing their workplace, even if it meant making their skills redundant and automating their job, continuously updating their skillset to work on new computerized machines, and further increasing their productivity "for the common good." Thus, the socialist work ethic required them to put aside their own needs, desires, and inclinations when the SED's economic plans demanded it.

In capitalist countries, the demand for vocational education and training in new computer technology was driven by individual incentives: On the one hand, people qualified to fully exploit the potential of computers in their field of work were promised a bright career, excellent job prospects and salaries. Workers who had not kept up with the pace of the introduction of computer technology in their field and who had not undergone further training in time, on the other hand, were put at risk of being replaced by computerized machines and ending up in unemployment.[490] In contrast, in the socialist GDR, the state guaranteed the constitutional right to work for all citizens, combined with the duty to work for everyone who was able.[491] In addition, taking on professional tasks that required in-depth qualifications in computer use was not necessarily accompanied by a significant wage increase.[492] On the contrary, in industrial production, the relief of physically demanding work following the introduction of computerized machines could even mean a financial loss if, as a consequence, financial compensation payments for health burdens were withdrawn.[493]

Within the SED's depiction of socialist computerization of workplaces as represented in its sociotechnical imaginary, the rationalization of industrial production and increase of productivity with the help of computer technology seemed to go hand in hand with a betterment of working conditions conflict-free. CAD/CAM

---

[490] Doris Blechinger et al., *The impact of innovation on employment in Europe: An analysis using CIS data* (Mannheim: Zentrum für Europäische Wirtschaftsforschung, 1998).
[491] SAPMO ZB 20049, "Verfassung der Deutschen Demokratischen Republik vom 6. April 1968," Gesetzblatt der Deutschen Demokratischen Republik 1968, part 1, no. 8 (April 9, 1968): 199–222, Art. 24.
[492] The wage ratios between the groups of employees in the GDR were strongly levelled. University and technical college graduates earned only about 15% more than production workers, in West Germany up to 70% more. More important than, for example, the industrial sector in terms of wage levels were supplementary payments for shift workers in continuous shift systems, amounting to up to 30% of gross wages (Helga Stephan and Eberhard Wiedemann, "Lohnstruktur und Lohndifferenzierung in der DDR," offprint from *Mitteilungen aus der Arbeitsmarkt- und Berufsforschung* 23, no. 4 (1990): 550–62.
[493] For a detailed discussion of wage policy in the GDR in the microelectronics industry, see Klenke, *Kampfauftrag Mikrochip*, 149–201.

technology was heralded as a new, highly productive tool for rationalizing production processes. But it was also claimed that it would free workers from many routine tasks, while at the same time extending the scope for intellectually stimulating and creative work.[494] The SED's political propaganda thus foregrounded the anticipated positive effects of computerization for workers, such as the substitution of hard physical labor by machines, and the reduction of harmful emissions such as dust, chemical pollutants, and loud noise. In some male-dominated sectors of industrial production, the resulting alleviation of physical labor even promised to make it possible for the first time to employ women for certain jobs that now required the operation of computer-controlled machines rather than hard physical labor.[495]

At the same time, however, the SED leadership and economic planners were well aware that the introduction of computer technology into the economy paradoxically also entailed the creation of low skilled jobs that offered little mental stimulation, and a de-qualification of skilled workers in certain areas as computers and robots took over their work tasks.[496] In many cases, the monotony of working at automated systems could only be reduced by assigning workers to the simultaneous supervision and control of multiple machines, or adding preparatory, auxiliary and maintenance tasks to their job specification.[497] In general, the negative effects of the rationalization and restructuring associated with the introduction of new technology affected disproportionately vulnerable groups of less flexible workers. These included women and mothers, older workers, and those with health problems, and low-skilled or unskilled workers.

The negative aspects of computerization and automation did not feature in the SED's propaganda regarding the introduction of new technologies, of course. Rather, the conveyance of an entirely positive sociotechnical imaginary, presented as an attainable and desirable future, served to reduce reservations and doubts of workers with regards to learning about and using computer technology.[498] Within

---

494 Hummel and Rosenkranz, "CAD/CAM."
495 SAPMO DY 34/14639, "Vorlage für die Sekretariatssitzung am 16.8.1982," dated August 23, 1982, 4.
496 SAPMO DY 34/14639, "Information – Welchen Einfluss nehmen die gewerkschaftlichen Leitungen auf die Aus- und Weiterbildung von Frauen zur Lösung der Aufgaben von Mikroelektronik und Robotertechnik?," dated August 2, 1982, 6.
497 SAPMO DY 34/14639, "Zuarbeit zur Konzeption über die gewerkschaftliche Einflussnahme auf die Aus- und Weiterbildung der Frauen," dated July 7, 1982, 1; SAPMO DY 34/14639, "Information – Welchen Einfluss nehmen die gewerkschaftlichen Leitungen auf die Aus- und Weiterbildung von Frauen zur Lösung der Aufgaben von Mikroelektronik und Robotertechnik?," dated August 2, 1982, 6.
498 SAPMO DY 34/14639, "Vorlage für die Sekretariatssitzung am 16.8.1982," dated August 23, 1982, 7.

socialist state propaganda, the SED's sociotechnical imaginary served as an emphatic motivator and reminder to the working people, that the struggles of the "scientific- technological revolution" and the computerization of the economy could and would be solved under the ideological leadership of the SED. It promised that the resulting improvement in living and working conditions and the economic prosperity of the future socialist society would be a collective gain for the GDR, that was more than worth the personal and professional sacrifices of individual workers, as well as their efforts to undergo further education and training to face new skill needs. Thus, the invocation of the sociotechnical imaginary of a desirable computerized future was of special propagandistic significance, where the short-term effects of the introduction of new technologies consisted in the deterioration of working conditions for some and the devaluation of professional qualifications for others. These struggles were largely dismissed as temporary side effects of an incomplete "scientific-technological revolution" that would be made to disappear in the future with further technological progress.[499]

A more comprehensive or even public debate on the possible negative social effects of automation and computerization in the GDR could not gain a foothold for two reasons. First, the politically emboldened excessive optimism about science and technology relied on a deterministic understanding of technology, whereby the development and use of a new technology such as computers would lead quasi-automatically to social and economic progress. Such an understanding side-lined more nuanced debates about sociotechnical interdependencies and the possibility of undesired social repercussions. Second, the SED based its hold on political power on the ideological presumption of an already completed construction of socialism, and thus, an ultimately faultless society. Negative social effects of automation and rationalization were reflected as a phenomenon occurring only in the capitalist world, accompanied by the claim that this would not happen under socialist production conditions.[500]

Nevertheless, research and conferences of experts on social issues of automation and computerization of work still took place in the GDR. The reality of computerized workplaces, where negative effects on workers manifested, called for corresponding investigations and interventions in the field of occupational health and psychology. As a logical consequence, it followed from the claim that monotonous work or unemployment must not occur in socialism or must be avoided or

---

[499] On the paradigm of progress in the SED's political rhetoric, see Sabrow, "Zukunftspathos als Legitimationsressource."
[500] Fuchs-Kittowski, "Grundlinien des Einsatzes der modernen Informations- und Kommunikationstechnologien," 65.

compensated for, that these very real problems needed to be taken care of, both in theory and practice.[501]

When dealing with the discrepancies between the positive sociotechnical imaginary publicly propagated by the SED, and the more ambiguous reality in workplaces, the "Free German Trade Union Federation" (ger.: Freier Deutscher Gewerkschaftsbund, FDGB) assumed a key role in easing the resulting tensions. The FDGB was the sole national trade union center of the GDR, nominally representing all workers. However, it was not an independent representative body of the labor force, but rather a basic element and tool within the SED's power structure. The hierarchical structure of the FDGB served to extend the party's ideological control all the way down into companies and individual workers' collectives. In preparing the introduction of computer technology into workplaces, the FDGB was tasked with familiarizing workers with the restructuring plans and far-reaching changes in the job demands and tasks during socialist rationalization. The affected workers had to be made aware of the new qualification requirements and motivated to undergo the necessary further training.

In 1982, the FDBG noted that the mass political campaign of the executive committees and leaders was to explain to workers the profound economic significance, as well as the social and labor implications of the introduction of new computer technology. In addition, it was deemed necessary that workers were actively involved in the preparation and introduction phase of new technology in their workplaces.[502] For this purpose, the FDGB organized informational events in companies and combines on the strategic importance of microelectronics and robotics technology as part of socialist modernization and rationalization, and informed workers about concrete planned measures in their workplace. These events were focused on highlighting the positive effects of new technology in raising the effectiveness of work and improving working conditions. In addition, the FDGB took on a central role in reconciling personal interests and needs of workers on the one hand, with societal and economic requirements on the other.[503] This included, that workers would readily accept challenges and inconveniences, such as taking on shift work or undergo further training and education in their spare time, as part of their duty as individual workers within the socialist society.

---

501 Fuchs-Kittowski, "Grundlinien des Einsatzes der modernen Informations- und Kommunikationstechnologien," 66.
502 SAPMO DY 34/14639, "Vorlage für die Sekretariatssitzung am 16.8.1982," dated August 23, 1982, 1.
503 SAPMO DY 34/14639, "Information über die gewerkschaftliche Einflussnahme auf die Aus- und Weiterbildung der Frauen entsprechend den Anforderungen der Mikroelektronik und Robotertechnik" (draft), dated October 12, 1982, 2.

Female workers were particularly affected by rationalization and negative effects of the introduction of new technology. Several skilled occupations in the textile and light industry, in which women formed the large majority of workers, were partially automated. Consequently, the remaining tasks, which were often simple and repetitive, took up a larger share of the workday. The resulting monotony was countered in some cases by qualifying these women for multiple work assignments so that they could alternate workplaces and activities several times a day. In addition, women were encouraged to undergo further training to be able to operate, monitor, and troubleshoot industrial robots in addition to their usual duties, thus upgrading the content of their job.[504] In other cases, workplaces typically occupied by women were eliminated altogether. For example, the use of 17 industrial robots at the VEB Büromaschinenwerk Sömmerda rendered obsolete the manual winding of VHF chokes (a type of passive electronic component), which previously had been the job of 47 women employed as home workers. Unable to work on site at the plant due to a lack of day-care places or family problems, these women ultimately lost their jobs.[505] In most other cases, the workers whose jobs had been replaced by computerized machines were usually transferred to a new position within the company.

At the beginning of the 1980s, women accounted for 49.4% of all workers in the GDR and were employed on average at a level of 85.7%.[506] Accordingly, they played a vital role in the GDR's economy. At the same time, however, 17.1% of all women working in material production did not have a skilled worker's certificate or were employed in a low-skilled occupation not related to their qualification.[507] Moreover, only a small percentage of women acquired qualifications for the emerging skilled trades of electronics and operational technology, which in many areas were prerequisites for working on new technology or for accessing higher education in automation and computer technology.[508]

---

[504] SAPMO DY 34/14639, "Information – Welchen Einfluss nehmen die gewerkschaftlichen Leitungen auf die Aus- und Weiterbildung von Frauen zur Lösung der Aufgaben von Mikroelektronik und Robotertechnik?," 6.

[505] SAPMO DY 34/14639, "Information – Welchen Einfluss nehmen die gewerkschaftlichen Leitungen auf die Aus- und Weiterbildung von Frauen zur Lösung der Aufgaben von Mikroelektronik und Robotertechnik?," 6.

[506] SAPMO DY 34/14639, "Vorlage für die Sekretariatssitzung am 16.8.1982," dated August 23, 1982, 2.

[507] SAPMO DY 34/14639, "Vorlage für die Sekretariatssitzung am 16.8.1982," dated August 23, 1982, 2.

[508] SAPMO DY 34/14639, "Bericht über die Ergebnisse der Aus- und Weiterbildung der Frauen entsprechend den Anforderungen der Mikroelektronik im VEB Mikroelektronik 'Bruno Baum', Zehdenick," dated July 14, 1982, 3.

For the SED, the active involvement of women in introducing new computer technology into the GDR's economy was vitally important for several reasons.[509] First, the equal participation of both male and female workers in the process of the SED's "scientific-technological revolution" was a basic principle to be heeded within the framework of socialist ideology. Secondly, from a more pragmatic point of view, the economy of the GDR could simply not do without the major contribution of female workers. This was true in terms of the constant general labor shortage, but especially with regard to the urgent need for skilled workers to be deployed on new computerized machines. Third, the focus on female workers in the process of computerizing the GDR's economy was also due to sociopolitical reasons. Women whose jobs had been replaced by computerized machines needed to be retrained for new occupations. Job losses mainly affected workers with low or no qualifications, among whom the share of women was disproportionately high.[510] Solid retraining, thus, was essential, especially for emerging technical occupations such as those in the setup and maintenance of computerized machinery where the demand for skilled workers was particularly high.

In 1982, the national executive committee of the FDGB investigated the effects of technology on the work of women in companies that had been impacted early on by applications of microelectronics and computers technology.[511] The report of the investigation pointed out the need to address the specific issues of working women and mothers in the process of introducing new computer technology into the workplace – the worsening working conditions in the form of increasing shift work and health hazards, and the cutting of women's jobs as their work was replaced by computerized machines.[512] Furthermore, it highlighted the need to find

---

[509] Christiane Lemke, "Frauen, Technik und Fortschritt. Zur Bedeutung neuer Technologien für die Berufssituation von Frauen in der DDR," in *Die DDR in der Ära Honecker*, ed. Gert-Joachim Glaeßner (Opladen: Westdeutscher Verlag, 1988), 481–98; Susanne Kranz, "Women's Role in the German Democratic Republic and the State's Policy Toward Women," *Journal of International Women's Studies* 7, no. 1 (2005): 69–83.
[510] SAPMO DY 34/14639, "Bericht über die Ergebnisse der Aus- und Weiterbildung der Frauen entsprechend den Anforderungen der Mikroelektronik im VEB Mikroelektronik 'Bruno Baum', Zehdenick," dated July 14, 1982, 2.
[511] This included, for example, the machine tool manufacturer VEB Wema Union, the VEB Elektronik in Gera, the VEB Mikroelektronik 'Bruno Baum' in Zehdenick and the VEB Mikroelektronik Mühlhausen, the VEB Funkwerk Erfurt, and the office machinery manufacturer Büromaschinenwerk Sömmerda.
[512] SAPMO DY 34/14639, "Konzeption für die Information über die gewerkschaftliche Einflussnahme auf die Aus- und Weiterbildung der Frauen entsprechend den Anforderungen der Mikroelektronik und Robotertechnik und die Lösung damit verbundener sozialer Probleme in ausgewählten Betrieben und Kombinaten," dated April 27, 1982.

and offer adequate conditions and forms of further training and education of women, and especially mothers. It was declared as the tasks of local trade union leaderships[513] and their women's commissions to resolve these issues, as well as to prepare women in time for the new skill requirements, as laid down in the in-company training concepts and plans for the advancement of women.

A major issue for female workers was the increase of shift work in industrial production. To fully exploit the potential of the few available computerized machines, companies were instructed to keep them running around the clock. This often required the expansion of shift work from a 3-shift rhythm to a continuous 4-shift system. As a result, it became more difficult to assign women who could not or would not commit to shifting work schedules to regular day shifts.[514] This concerned, for example, mothers who had the legal right to refuse night shift work according to §243 of the Labor Code of the GDR, and older women with health impairments.[515] Although women and men participated almost equally in the labor force, the traditional social role of women as caregivers and housemakers was maintained – resulting in a double burden for working mothers.[516] In most cases, it was considered the mother's responsibility to provide childcare during her working hours and to find alternatives or stay home when the children were sick.[517]

One notable example that the FDGB investigated in 1986 was the preparation of female workers for the use of computer systems at the general post office in Berlin. A few years previously, computer terminals had been introduced at the coun-

---

[513] On the local level, company trade union leaderships functioned as the FDGB's anchorage in the individual companies. Unlike a workers' council, they were not an independent body representing the interests of the entire workforce of a company. Rather, the company trade union leadership was, together with its sub-organizations, the bottom-level recipient of orders within the system of democratic centralism, which had to implement the directives of the FDGB and mobilize the workers accordingly (Ulrich Gill, "Betriebsgewerkschaftsleitung (BGL)," in *FDGB-Lexikon: Funktion, Struktur, Kader und Entwicklung einer Massenorganization der SED (1945–1990)*, ed. Dieter Dowe, Karlheinz Kuba, and Manfred Wilke (Berlin 2009), http://library.fes.de/FDGB-Lexikon (accessed May 31, 2024).
[514] SAPMO DY 34/14639, "Bericht über die Ergebnisse der Aus- und Weiterbildung der Frauen entsprechend den Anforderungen der Mikroelektronik im VEB Mikroelektronik 'Bruno Baum', Zehdenick," dated July 14, 1982, 4.
[515] SAPMO DY 34/14639, "Information – Welchen Einfluss nehmen die gewerkschaftlichen Leitungen auf die Aus- und Weiterbildung von Frauen zur Lösung der Aufgaben von Mikroelektronik und Robotertechnik?," 7.
[516] Grit Büchler, *Mythos Gleichberechtigung in der DDR: Politische Partizipation von Frauen am Beispiel des Demokratischen Frauenbunds* (Frankfurt am Main/New York: Campus, 1997); Kranz, "Women's Role," 77.
[517] Kranz, "Women's Role," 80.

ters to raise the efficiency of postal services. The predominantly female employees of the general post office were placed in a higher pay grade after receiving computer training to work on the new terminals.[518] However, the new computer technology did not necessarily make the jobs of female terminal workers easier or more pleasant. The computer terminals were very prone to disruptions, and a lack of spare parts and appropriate skilled workers on site to make the necessary repairs resulted in significant downtime. This, in turn, often led to a high stress load and overtime for the female workers at the postal counter.[519]

Just as in industry, computer technology in the service sector was to be exploited as fully as possible and therefore used in multi-shift operation. However, many women who had been qualified for working on the computers were no longer willing to work shifts after they had given birth to a child and returned to their workplaces. The only employment option without shift work at the post office was in many cases in the mail delivery service, where mothers were employed far below their skill level and their computer skills were not used.[520]

The financial bonus that was paid in addition to the salary as an incentive for taking on shift work was apparently not a strong enough stimulus for women to bother with the disadvantages of shift work – especially when they had children and a household to care for at home.[521] Moreover, trade union representatives noted that not all women were ready to undergo further training for the work on new technology, even if this would mean a pay increase. Part-time working women, in particular, considered their monthly income as sufficient and did not see a need to acquire additional skills to advance in their career.[522]

The FDGB therefore saw a need for "constant political and ideological intervention with workers to achieve positive results"[523] – that is, to remind workers of their socialist duty and to persuade them to undergo further training or work shifts,

---

518 SAPMO DY 34/14639, "Information über die Ergebnisse der Untersuchung zur gewerkschaftlichen Einflussnahme zur Vorbereitung der Produktionsarbeiterinnen für den Einsatz an Schlüsseltechnologien im Hauptpostamt Berlin-Oberschöneweide," dated July 24, 1986, 4.
519 SAPMO DY 34/14639, "Information über die Ergebnisse der Untersuchung zur gewerkschaftlichen Einflussnahme zur Vorbereitung der Produktionsarbeiterinnen für den Einsatz an Schlüsseltechnologien im Hauptpostamt Berlin-Oberschöneweide," dated July 24, 1986, 7.
520 SAPMO DY 34/14639, "Information über die Ergebnisse der Untersuchung zur gewerkschaftlichen Einflussnahme zur Vorbereitung der Produktionsarbeiterinnen für den Einsatz an Schlüsseltechnologien im Hauptpostamt Berlin-Oberschöneweide," dated July 24, 1986, 6.
521 SAPMO DY 34/14639, "Bericht über die Ergebnisse der Aus- und Weiterbildung der Frauen entsprechend den Anforderungen der Mikroelektronik im VEB Mikroelektronik 'Bruno Baum', Zehdenick," dated July 14, 1982, 5.
522 SAPMO DY 34/14639, "FDGB Bundesvorstand Informationsbericht," dated July 30, 1982, 2.
523 SAPMO DY 34/14639, "FDGB Bundesvorstand Informationsbericht," dated July 30, 1982, 2.

if deemed necessary for economic reasons. For this purpose, special meetings for female workers were held in some companies, with commented slide shows on the subject of new technologies. Local trade union representatives organized these events, to demonstrate to the women the importance of introducing microelectronics and computer technology into the workplace as an "objective requirement to successfully master the tasks of the 1980s and thus securing peace."[524] These information events were intended to win over female workers to acquire the necessary skills and actively contribute to the successful integration of new computer technology into their workplaces.[525]

In addition, local representatives of the FDGB sought out personal conversations with workers regarding new qualification demands, to reconcile the personal needs of workers with those of the company and the socialist society. If the job required additional qualifications due to the introduction of new technologies, a so-called "qualification contract" was drawn up with the planned training and development measures, which the employee had to sign and commit to.[526] Young mothers were singled out for individual meetings with union representatives and urged to take on shift work.[527]

Furthermore, the local FDGB units were responsible for organizing the necessary qualification measures for workers, whose jobs were replaced by computerized machines or altered in such a way that new knowledge and skills were required for them to continue to be working in their position within the company. Further training for existing staff usually took place outside working hours, in the evenings or at weekends, and often in the form of self-study courses so as not to interfere with the required daily workload.[528] Participation in regional training programs for managers and executives, e.g. at industry academies, uni-

---

524 SAPMO DY 34/14639, "Welchen Einfluss nehmen die gewerkschaftlichen Leistungen auf die Aus- und Weiterbildung der Frauen zur Lösung der Aufgaben von Mikroelektronik und Robotertechnik?," dated August 12, 1982, 2 (translated from German).
525 SAPMO DY 34/14639, "Bericht über die gewerkschaftliche Einflussnahme auf die Aus- und Weiterbildung der Frauen entsprechend den Anforderungen der Mikroelektronik und Robotertechnik und die Lösung der damit verbundener sozialer Probleme in ausgewählten Betrieben und Kombinaten," dated August 5, 1982, 3.
526 SAPMO DY 34/14639, "FDGB Bundesvorstand Informationsbericht," dated July 30, 1982, 2.
527 SAPMO DY 34/14639, "Bericht über die gewerkschaftliche Einflussnahme auf die Aus- und Weiterbildung der Frauen entsprechend den Anforderungen der Mikroelektronik und Robotertechnik und die Lösung der damit verbundener sozialer Probleme in ausgewählten Betrieben und Kombinaten," dated August 5, 1982, 6.
528 SAPMO DY 34/14639, "Information über die Ergebnisse der Untersuchung zur gewerkschaftlichen Einflussnahme zur Vorbereitung der Produktionsarbeiterinnen für den Einsatz an Schlüsseltechnologien im Hauptpostamt Berlin-Oberschöneweide," dated July 24, 1986, 3.

versities and technical colleges, or in courses offered by the Chamber of Technology, equally posed a problem for women with family obligations, insofar as they were held outside their place of residence and work.[529] To alleviate these burdens, some larger companies set up special women's courses (ger.: Frauensonderklassen) during working hours for the relevant training and education in the new technologies of mothers or women with care responsibilities in the family.[530] In addition, older workers and young mothers returning to work after a longer absence due to maternity leave or a sick child were provided with self-study materials to acquire the necessary skills and knowledge in their free time at their own speed.[531]

In contrast to the SED's call for a necessary higher qualification of workers in response to the introduction of computer technology into economic production in the GDR, many workers needed no additional qualifications to continue to do their job. This was especially true for female workers in industrial production. In many cases, the content of their work merely changed insofar as they switched from operating a single machine to operating multiple machines with the aid of computerized controls.[532] These machine operators only required basic user knowledge and practical skills in operating the computer control panel, which could be acquired on the job. Maintenance and service personnel, on the other hand, required more thorough further training – but these were occupations predominantly held by men.[533]

The call for a broad qualification of all workers in the use of new computer technology in industry, thus, went hand in hand with the problem of an overqualification of parts of the workforce, and in particular, female workers. As stated

---

[529] SAPMO DY 34/14639, "Information – Welchen Einfluss nehmen die gewerkschaftlichen Leitungen auf die Aus- und Weiterbildung von Frauen zur Lösung der Aufgaben von Mikroelektronik und Robotertechnik?," 5.

[530] SAPMO DY 34/14639, "Bericht über die gewerkschaftliche Einflussnahme auf die Aus- und Weiterbildung der Frauen entsprechend den Anforderungen der Mikroelektronik und Robotertechnik und die Lösung der damit verbundener sozialer Probleme in ausgewählten Betrieben und Kombinaten," dated August 5, 1982, 5.

[531] SAPMO DY 34/14639, "Information über die gewerkschaftliche Einflussnahme auf die Aus- und Weiterbildung der Frauen entsprechend den Anforderungen der Makroelektronik und Robotertechnik" (draft), dated October 12, 1982, 6.

[532] SAPMO DY 34/14639, "Information – Welchen Einfluss nehmen die gewerkschaftlichen Leitungen auf die Aus- und Weiterbildung von Frauen zur Lösung der Aufgaben von Mikroelektronik und Robotertechnik?," 5.

[533] SAPMO DY 34/14639, "Information – Welchen Einfluss nehmen die gewerkschaftlichen Leitungen auf die Aus- und Weiterbildung von Frauen zur Lösung der Aufgaben von Mikroelektronik und Robotertechnik?," 5.

already in the early studies on the educational consequences of introducing new computer technology into workplaces, there was not only an increase in highly skilled experts, but also a substantial need for workers performing monotonous, low-skilled work – especially in the young industry of microelectronics, where many women were employed:

> In various production sections where microelectronic components are manufactured, a conspicuously large number of highly qualified female electronics workers perform monotonous tasks that are disproportionate to their extensive knowledge. Often, their deployment in such jobs is justified on the grounds that women are well suited for such work because of their dexterity and sensitivity. Some factory directors even claim that many women like monotonous work because they can do it routinely and thus keep their minds free for domestic concerns.[534]

The reality of low-skilled, mentally unstimulating work created alongside highly skilled jobs in the process of introducing of new technologies into economic production was only one aspect of how the lived experiences of some workers conflicted with the SED's dominant sociotechnical imaginary that promised an overall betterment of working conditions. This imaginary of a computerized workplace and all the perks it would offer also created tensions in many cases where young people completed their training and entered a workplace that had been left completely untouched by the new technology. During their vocational training, learners had been taught about the revolutionary ways in which computer technology would transform their work and allow for better quality and higher efficiency. They had been trained in using such technology and were eager to harness the potential of such modern equipment and demonstrate their newly acquired skills when they entered their job. However, one can imagine their frustration when they encountered old, traditional machinery and manually operated devices that relegated them to old work techniques that had been presented to them during training as inferior to the computerized workflow.

A notable example of this was secretarial work. In the GDR, such jobs were filled by so-called "skilled workers for writing technology" (ger.: Facharbeiter für Schreibtechnik), who did clerical work and were responsible for office communication in companies, combines, or government institutions.[535] This occupation was one of the most popular among female school graduates in the GDR when choosing their future career, and over 99% of new trainees entering VET for

---

[534] SAPMO DY 34/14639, "Information über die gewerkschaftliche Einflussnahme auf die Aus- und Weiterbildung der Frauen entsprechend den Anforderungen der Makroelektronik und Robotertechnik" (draft), dated October 12, 1982, 8 (translated from German).
[535] Zentralinstitut für Berufsbildung der DDR, *Facharbeiter für Schreibtechnik. Berufsbild für die Berufsberatung* (Berlin: Staatsverlag der DDR, 1985).

skilled workers for writing technology in the 1980s were women.[536] They were traditionally trained in shorthand and type writing, filing of documents, and general office organization.[537] In September 1987, new curricula for initial VET of skilled workers for writing technology were implemented to prepare them for the use of computer technology, and specifically text processing software.[538] Trained in both traditional typewriting technology and text processing on microcomputers, newly trained skilled workers for writing technology were aware of the efficiency gains resulting from the use of new computer technology. Sections of texts could be copied, saved, deleted, or rearranged without having to retype the whole page. In addition, text processing software allowed for the use of text modules and templates for letters or forms that could be flexibly adapted.

But microcomputers were only slowly introduced into offices in the GDR in the second half of the 1980s.[539] As a result, some apprentices were trained to use computers, but ended up in a workplace where no computer technology was available. A young female apprentice wrote in her final assignment for the completion of her vocational education as a skilled worker for writing technology in spring 1990 regarding her practical work experience: "During my work in the 5th Criminal Division of the Supreme Court, I have noticed that the work with pre-printed forms is not yet very efficient, and the visual image of the completed forms is not very aesthetic. Due to the many deletions for what is not applicable, the form looks unattractive. But because of inadequate technology, it has not been possible to design these forms differently."[540]

However, the young woman's completion of her vocational training coincided with the turbulent period of change following the fall of the Berlin Wall, which raised her expectations that the most modern and powerful computers would soon be made available to her for use in her future career as a typist or secretary. Her statement is an example of how, for many young people like her who com-

---

536 Gunnar Winkler, eds., *Frauenreport '90* (Berlin: Verlag Die Wirtschaft, 1990), 44.
537 Zentralinstitut für Berufsbildung der DDR, *Facharbeiter für Schreibtechnik*.
538 "Die Ausbildung künftiger Facharbeiter befähigt zur effektiven Nutzung der Schlüsseltechnologien," *Neues Deutschland*, October 9, 1987, 5.
539 In the state-controlled allocation of new computer technology, priority was given to the sectors and branches considered strategically most important within the SED's economic policy, such as the chemical industry, electrical engineering, electronics, and appliance manufacturing, as well as mechanical and vehicle engineering. The service sector was considered to be of secondary importance. ("Zur EDV-Anwendung in der DDR: Rechnerdefizit zwingt zum Einsatz alter Technik," *Computerwoche*, June 6, 1980, http://www.cowo.de/a/1189741 (accessed May 31, 2024)).
540 Regine Schuster, *Welche Anforderungen werden an einen Facharbeiter für Schreibtechnik gestellt?* [Hausarbeit zur Prüfung als Facharbeiter für Schreibtechnik] (Berlin, 1990), 12 (translated from German). Available at the library of the federal archive in Berlin-Lichterfelde, shelfmark 05 C 1307.

pleted their vocational training in the GDR in the late 1980s and had acquired some skills and knowledge in the use of computers according to the new curricula, the belief in their usefulness remained intact beyond the socialist system of the GDR. The belief in the sociotechnical imaginary of a computerized future, in which skilled computer users would thrive in their workplaces and find ways to use the new technology to make their work more satisfying and rewarding, seems to have lived on – even after the end of the GDR and the SED's authoritarian regime. However, it was dissociated from the SED leadership's claim that only under a socialist order would workers benefit from the computerization of the economy and society. Instead, computer knowledge and skills, which in many cases could not be fully utilized in the workplace in the GDR due to a lack of or inadequate technical equipment, now became potential facilitators for working people in the transition to Western computer technology and the capitalist system after the reunification of the two Germanys. In 1992, the newspaper *Neue Zeit* reported a rapidly increasing demand for vocational further education and retraining measures in East Germany, in particular for computer courses, stating that anything called "electronic data processing" or mentioning "personal computer" triggered a run.[541] In this regard, the sociotechnical imaginary of a computerized economy helped to make computer skills and knowledge a sought-after asset for individual to compete in the fundamentally transformed labor market, as the state no longer guaranteed secure jobs for everyone. The "Bundesagentur für Arbeit" (eng.: Federal Employment Agency) implemented and funded retraining, qualification, and further education measures for people at risk of unemployment or who were currently unemployed.[542] This, in turn, created a burgeoning market for private companies to provide adult education and training in computing for professional development. In the early 1990s, numerous advertisements for further education and training in computing started popping up in East German newspapers, promising "a secure office job,"[543] a "new career start,"[544] or a "lad-

---

541 "Ausbildung am Computer besonders gefragt," *Neue Zeit*, March 5, 1992, 4.
542 The basis for this was the Employment Promotion Act (AFG) of June 22, 1990, which promoted vocational training, further training and retraining as measures to stabilize the East German labor market and combat rising unemployment after reunification. The law came into force on 1 July 1990 together with the German economic, monetary, and social union. See SAPMO ZB 20049, "Arbeitsförderungsgesetz (AFG) vom 22. Juni 1990," Gesetzblatt der Deutschen Demokratischen Republik 1990, part 1, no. 36 (June 28, 1990): 403–45.
543 Advertisement "Lernen Sie, den PC zu beherrschen!," *Berliner Zeitung*, May 8, 1992, 22 (see Figure 6, left).
544 Advertisement "Sichern Sie sich Ihre Chance zum beruflichen Neueinstieg," *Berliner Zeitung*, June 12, 1993, 65 (see Figure 6, right).

der to success;" some of them aimed specifically at the unemployed in need of retraining (see Figure 6).

Figure 6: Advertisements in the Berliner Zeitung from the early 1990s for a "Basic course for computer users" (left) and a training course in "Clerical administration and EDP accounting" (right). Sources: Advertisement "Lernen Sie, den PC zu beherrschen!," Berliner Zeitung, May 8, 1992, 22; Advertisement "Sichern Sie sich Ihre Chance zum beruflichen Neueinstieg," Berliner Zeitung, June 12, 1993, 65.

In contrast to such imaginaries, in which computer training was deemed a valuable asset, there were also critical voices that publicly called into question the utility of computer and programming training for all. After the fall of the Berlin Wall, it suddenly became possible to openly and retrospectively challenge the SED's education policy with regard to the comprehensive introduction of computer instruction for trainees in all occupations, regardless of whether their work involved the use of a computer at all now or in the foreseeable future. In February 1990, a vocational teacher of informatics wrote in the newspaper *Neue Zeit*:

> We have introduced the subjects "Basics of Automation" and "Informatics" in the past years with a lot of money, time, effort, and propaganda. [. . .] Today we ask ourselves whether this training makes sense for professions such as hairdresser, specialized salesperson, etc. Those responsible remain silent or use the phrase of an all-round educated skilled worker as an excuse for their wrong decision. [. . .] As a teacher of informatics, I believe specialized training should be given to those apprentices who need it later in their profession. But here, too, there needs to be more of a practical orientation. Not everyone can and needs be able to create a program.[545]

---

545 "Muss jeder an Computer?" *Neue Zeit*, February 8, 1990, 5 (translated from German).

Clearly, the sociotechnical imaginary of the computer as a universal tool for enhancing human capabilities, and the accompanying belief that everyone, regardless of occupation, should receive a computer education and acquire basic programming skills, had lost its persuasive power. Or perhaps the SED's imaginary had never been able to convince the people who only now felt that it could be publicly challenged. The SED's once dominant sociotechnical imaginary of computer technology as the engine of *collective* social and economic progress for the socialist society of the GDR had been replaced by new imaginaries, some of them propagated by newly emerging providers of continuing and adult vocational education in the new information technologies, which instead foregrounded the promise of *individual* career success in a competitive labor market.

4.6 Chapter Conclusion

The first domestically produced computers in the GDR in the 1960s quickly led to a growing demand for personnel that was skilled in the operation, maintenance, and repair of these machines. Initially, large mainframe computers were mainly installed in research institutions and computing centers in higher education and industry. The development and use of computers had been the domain of few highly skilled experts in mathematics and engineering. However, with the rapid advances in computer technology, the computers became more powerful, and in turn, new applications for computers opened in science, the economy, and state administration. The mass production of the Robotron 300 mainframe created a need for a growing number of workers skilled in computer technology in state ministries, combines, and research and computing centers in higher education. In consequence, computer technology could no longer remain the exclusive domain of specialists with an academic degree. New job profiles for the operation and maintenance of mainframe computers in research, industry, and state administration were created. The GDR's system of vocational education and training was thus tasked with qualifying workers for the jobs of "technical computers" and "skilled workers for data processing." However, to be able to quickly respond to the new skill needs, companies and combines in the GDR initially resorted to ad-hoc training measures to meet their own immediate skill needs. This resulted in a disorganized landscape of a wide variety of short training programs, each of which only qualified the workers for the available technology and specific work tasks in the respective company where they were being trained.

The GDR's vocational training policymakers therefore saw the need for more centralized oversight and control of the education and training of skilled workers for computing, electronic data processing, and the maintenance and repair of the computer machines and systems. In 1967, the job profile of "skilled worker for electronic data processing" was created and included in the GDR's official system of skilled workers' occupations. At the same time, vocational training documents were drawn up for training in the new, officially recognized occupation, which were provided for a binding and homogenous basis for the vocational training and education of skilled workers in electronic data processing throughout the GDR. The aim was to provide these workers with a sound and broad technical qualification that would enable them to work with computer technology in a variety of sectors of the GDR economy. Equipped with a solid basic understanding of modern computer technology, which the new vocational curricula were designed to provide, it was expected that skilled workers in electronic data processing would be able to quickly familiarize themselves with different types of computers and more easily become acquainted with yet unknown new computer technology that would be developed in the future. In the 1970s, the GDR's computer industry adopted the technology of integrated circuits, which allowed for the development of production of even more powerful, faster, and smaller computers.

In 1977, the GDR's political leadership launched an extensive "microelectronics program" aimed at building up the domestic microelectronics industry and modernizing the country's industry. Dwindling export figures of the economically important machine tool industry of the GDR had alarmed economic policymakers: The economic competition on the world market was increasingly oriented towards machine tools and systems equipped with microelectronics to allow for computerized control, which the GDR was unable to produce. The goal of the microelectronics program was thus to build up substantial industrial capacities to produce computerized machines to export, as well as the use of computer technology in the GDR's own economy, to modernize and automate industrial production. This strategy was guided by the SED's sociotechnical imaginary which envisioned microelectronics and computer technology as a powerful means of increasing economic productivity, and the competitiveness of the GDR industry on the world market. Moreover, microelectronics had become a decisive arena in the struggle between the socialist and the capitalist systems, not just in terms of proving technological superiority and innovative prowess, but also as a matter of economic power and competitiveness. The SED's sociotechnical imaginary stressed this aspect by envisioning a future in which the socialist leadership of the GDR would succeed in exploiting the full potential of new computer technology, thereby proving the superiority of the socialist ideology and economic system over capitalism.

In line with Honecker's rhetoric of the "unity of social and economic policy," the authoritarian SED-leadership promised the people of the GDR, that the positive effects of the development and use of microelectronics and modern computer technology would not only be felt in terms of economic prosperity, but that it would also lead to social progress in the form of better working and living conditions for everyone. The propagation of this aspect of the GDR's sociotechnical imaginary served to legitimize the state's substantial investment in the emerging microelectronics industry at the expense of other sectors of the economy, and to mobilize the working people of the GDR to actively support the SED's economic strategy by acquiring the necessary skills and professional qualifications, accepting fundamental changes in their workplaces, and making sacrifices for the common good by accepting shift work, and by working even harder and more diligently.

Through the strategy of reinventing Western technology, the GDR succeeded in developing its own microprocessor, the U880 – modelled after Zilog's Z80 –, which formed the basis for a range of microcomputers that the GDR mass-produced from the mid-1980s onwards. The development of a domestically produced microprocessor enabled a multitude of new applications of computer technology in the GDR's economy. Computer technology was no longer confined to computer centers, but also found its way onto the desk of workers in offices in the form of microcomputers and enabled new applications in industry for the flexible automation of production. The newly developed microcomputers were much cheaper to produce, took up less space and offered significantly more computing power. The development of ready-made software for industrial and office applications, as well as operating systems with graphical user interfaces, made them easier to use and required far fewer skills from users than earlier generations of computers, which usually required extensive programming skills or even involved the use of punched cards.

The innovation of the microprocessor fuelled the dominant sociotechnical imaginary of the computer as a universally functional technology that could be used for a wide variety of purposes, and which, above all, would help to increase productivity in all branches of the socialist economy. The SED's vision of a computerized future for the socialist society of the GDR, in turn, inspired an educational policy aimed at providing the workforce with comprehensive training in the professional use of computer technology. The young generation of future workers was to be prepared ahead of time for the anticipated widespread use of computer technology in the GDR's economy.

The large-scale reform of vocational curricula in the 1980s account for the advances in new information technology thus included the introduction of the subject "Basics of Automation" which was meant to familiarize apprentices in all skilled occupations with the basics of modern computer and automation technology and informatics. Considering the anticipated spread of computers into all sectors of the economy, it was no longer only predominantly skilled workers in industry and computing centers that were expected to be working on computers and computerized machines, but also those working in clerical jobs and the service industry. Building on the foundation of a basic computer education in general vocational education, trainees would subsequently receive more in-depth training in the specific computer technology and applications used in their workplace. However, the design and implementation of revised vocational curricula posed a significant challenge for education policymakers, as in most jobs and workplaces, old and new technology were used concurrently. Workers thus needed to be qualified for the use of both. But regardless of whether an occupation was already computerized or planned to be computerized in the foreseeable future, all workers were to receive the mandatory basic computer instruction as part of their general vocational education. This was because the basic computer education not only served as preparation for the specific qualification requirements of their future job. Rather, the introduction for a mandatory basic vocational computer education *for all* was also based on the ideological stance that under socialism, all workers – even those who were not or only indirectly affected by the new computer technology – were meant to acquire a basic understanding of the new technology to support and participate in the realisation of the SED's sociotechnical imaginary of socialist computerization. The dominant imaginary envisioned computer technology as a tool in the hands of the working people. In this sense, it promised that all workers, and not just an exclusive economic or technical elite, would decide on the development of computer technology and control its use. Consequently, all workers in the GDR needed to be educated and trained in computer technology to become "masters" and "co-creators" of modern computerized production, and active contributors to the "scientific-technological revolution" that would bring about the desired computerized future as envisioned by the SED's sociotechnical imaginary.

In the second half of the 1980s, and in the context of the recent advances in microchip technology, CAD/CAM technology became a focal point of economic and education policy. The development and deployment of computerized machines, automated systems, and industrial robots in the GDR industry gave rise to visions of the fully automated factory floor, which would enable unprecedented productivity gains and significantly improve the quality of production. As a result, computerized industrial production was expected to also improve living standards in the GDR, by providing the people with consumer goods in greater quantities and of better quality. Moreover, the SED's sociotechnical imaginary promised an improvement in working conditions, as computerized machines would relieve workers of physically demanding, strenuous or mentally unstimulating and repetitive tasks. The microchip had paved the way for the development of modern computerized control systems and automation technology, thereby appearing to bring the realization of this aspect of the SED's sociotechnical imaginary into close reach.

However, the promise of social progress, which the authoritarian leadership promised would go hand in hand with economic progress, clashed with the reality of computerized workplaces in the GDR economy. While the use of computer-controlled machines and systems had led in part to a reduction in strenuous and monotonous work tasks, it had also led to the paradoxical effect of creating new tasks and jobs characterized by inferior or even dehumanising working conditions. In particular, jobs where workers had to take on the function of human "links" in the production process due to incomplete computerization often consisted in in the performance of repetitive and monotonous tasks and thus offered little mental stimulation. Moreover, frequent technical failures led to interruptions in production, which in turn caused periods of great stress for the staff to make up for lost time. To maximize economic productivity gains, the expensive computerized machine systems had to be in constant operation both during day and night. As a result of this policy, there was a significant increase in shift work, which led to physical, psychological, and social grievances among the workers concerned. From a Marxist-Leninist point of view, such adverse effects of the computerization of the workplace would have had to be deemed unacceptable in a socialist society. At least in part, the negative effects had resulted from a prioritization of the goal of economic productivity increased over the aim of social progress, such as in the case of increased shift work. The SED leadership, however, attributed the undesirable consequences by and large to the still incomplete computerization. It assured the people that the negative effects were merely a transitory phenomenon, which would be eradicated as these undesirable work tasks, too, would be taken over by computerized machines.

Through the structures of the FDGB, the SED's propaganda on technical and labor policy reached all the way down to the level of the individual companies. FDGB functionaries were responsible for managing paradoxical effects on the local level with ideological propaganda, by reinforcing the SED's sociotechnical imaginary and motivating workers to undergo further education and retraining in response to the new skill needs as a result of the development of new computer technology. Their task was to persuade workers to accept the profound changes in their work content and organization of their workplace resulting from the introduction of computer technology. In the pursuit of the anticipated economic and social progress for the GDR's society as a whole, the authoritarian socialist regime expected socialist workers to accept any burdens arising from these changes and to make individual sacrifices. Female workers became a focal point of the FDGB's campaigns in the 1980s in this regard, as they were disproportionately affected from negative effects of the computerization of workplaces in the GDR. Taking on shift work and having to undergo further education or retraining proved particularly problematic for women who were facing the double burden of household and care work on the one hand, and a salaried job on the other.

Moreover, women were at a particular risk of ending up in low-skilled jobs they were overqualified for, such as dull data entry jobs.

In some cases, workers who had been familiarized with the exciting possibilities and advantages of modern computer technology in during their vocational education and training experienced disappointment as the necessary technical equipment they had been trained on was simply not available in their workplace. The fall of the Berlin Wall revived the hope of these people that Western computer technology would now become available to them and that they would finally be able to apply their acquired knowledge and skills in their jobs. With the demise of the SED government, however, the sociotechnical imaginary that had dominated up until then also lost its clout. Nevertheless, a solid education and training in computer technology remained a sought-after asset, albeit no longer oriented towards realizing the SED's imaginary of computer technology as a guarantor of collective social and economic progress under socialism. Instead, computer education and training became a crucial part of the vision and promise of individual career success under capitalism and was advertised as a decisive advantage in the competition for jobs in reunified Germany.

5 Informatics For All? Computer Literacy in Schools

The introduction of computers into general schooling in many industrialized followed a twofold approach. On the one hand, computers were introduced as a *subject of learning*. As a result of rapid technological progress, computers had acquired such a significant status in the industrialized world by the 1980s, that educational policy makers felt that a basic computer education ought to be part of the general education of all citizens.[546] Accordingly, pupils had to acquire the necessary skills and knowledge to use a computer and learn about the role of new information technology in automating and modernizing economic production, as well as to reflect on the ways in which the new technology was changing society. On the other hand, in many industrialized countries, computers were also introduced as a *tool for teaching and learning*. In the 1980s, global players such as the OECD advocated for the use of computers in schools to promote quality and efficiency of education.[547] In this context, the computer was touted as a new kind of educational technology that would lead policymakers and educators around the world to rethink and transform the way children learn and teachers teach.[548]

In the GDR, the introduction of computer education into the general education system was initially met with resistance, from no less than the Minister of National Education herself, Margot Honecker. The first failed attempt to introduce computer education into the GDR's system of general education dates to 1970. At the time, the GDR computer scientist Gerhard Merkel had been appointed head of a newly established forecasting group on the "Basic Directions of the Automation of Material and Mental Processes" in the Council of Ministers of the GDR. The group's task was to prepare a forecast with a time horizon of 1980, which was subsequently discussed and adopted by the Politburo of the Central Committee of the SED in November 1970. However, Merkel later recalled that the proposed measures to introduce computer education into general education, which originally had been part of the proposal, had been strongly opposed by Ed-

---

[546] See, for example, the case studies on the introduction of computers into education in Europe in the volume edited by Flury and Geiss, *How Computers Entered the Classroom*.
[547] Barbara Hof and Regula Bürgi, "The OECD as an arena for debate on the future uses of computers in schools," *Globalisation, Societies and Education* 19, no. 2 (2021): 154–66.
[548] Carmen Flury and Michael Geiss, "Computers in Europe's Classrooms: An Introduction," in *How Computers Entered the Classroom, 1960–2000: Historical Perspectives*, ed. Carmen Flury and Michael Geiss (Berlin/Boston: De Gruyter Oldenbourg, 2023), 4.

ucation Minister Margot Honecker.[549] According to the proposed bill, computer knowledge was to become part of the general education of all citizens in the GDR. However, Merkel recounts that Margot Honecker raised an objection during the vote on the bill, claiming that the forecasting group's proposals were not compatible with the principles of communist education. Merkel protested, but the chairman of the Council of Ministers, Willi Stoph, argued that the bill could not be allowed to fail because of the part on education. Stoph advised the group to remove it from the proposal altogether. Merkel took this instruction quite literally and simply removed pages 115 to 132 from the bill, unnoticed by the decision-makers.[550] And indeed, the relevant pages are missing from the approved bill, which is kept in the Federal Archives in Berlin-Lichterfelde.[551]

Margot Honecker's intervention in 1970 may have prevented the early introduction of computers into general education in the GDR. It was not until the mid-1980s that her objection was overruled by a Politburo decision to introduce computers into general education. The leadership of the SED, at least according to what it conveyed to the wider population as party consensus in the 1980s, seemed convinced that it was indeed possible to introduce computers into the GDR's general education system without tampering with its basic socialist principles and philosophy. In fact, it was seen as an indispensable measure to prepare the young generation for the sociotechnical future of socialist computerization as envisioned by the SED.

The GDR's comprehensive education system, with the polytechnic school at its core, was established in 1965 after a period of reform and retained this structure without major changes until the end of the GDR in 1989 (see Figure 7).[552] Throughout the 1970s and 1980s, educational policy in the GDR focused on stabilizing and improving the established structure and organization of the school system without bringing in fundamentally new or disruptive impulses. Especially in general schooling, the structures adopted by the act on the integrated socialist education system in 1965 were largely adhered to, despite the challenges and new educational demands resulting from scientific progress and technological change.[553]

---

549 Merkel, "Computerentwicklungen in der DDR," 46.
550 Merkel, "Computerentwicklungen in der DDR," 46.
551 The following note was added to the table of contents of the forecast paper: "Die Seiten 115–132 sind in dieser Ausfertigung nicht enthalten" (eng.: The pages 115–132 are not included in this version). See BArch DF 4/23554, "Prognose: Grundrichtungen der Automatisierung materieller und geistiger Prozesse" (revised version), dated September 1970, 5.
552 cf. Act on the Integrated Socialist Education System (ger.: Gesetz über das einheitliche sozialistische Bildungssystem), passed on February 25, 1965.
553 Hans-Jürgen Fuchs and Eberhard Petermann, *Bildungspolitik in der DDR 1966–1990: Dokumente* (Berlin: Osteuropa-Institut der Freien Universität Berlin, 1991), 20.

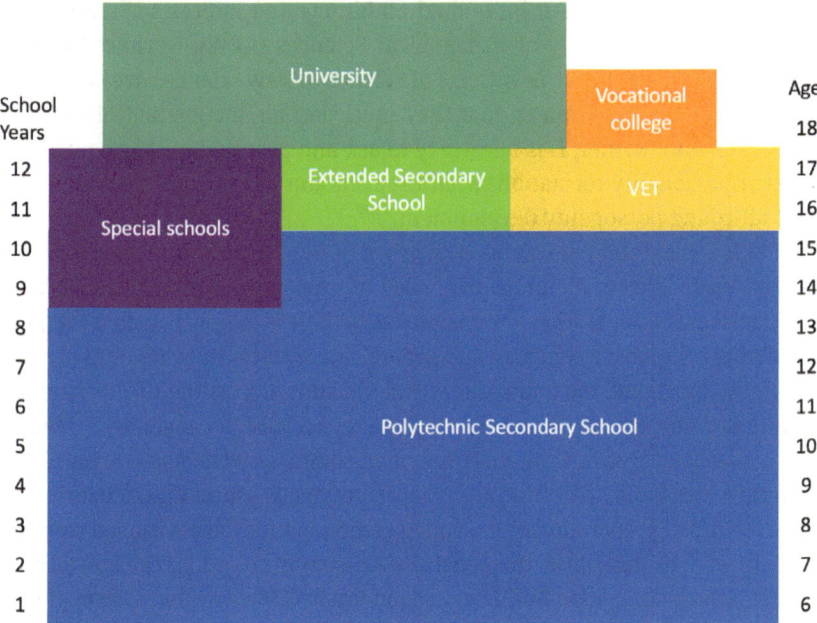

Figure 7: The system of education in the GDR since 1965, with the comprehensive ten-class general educational Polytechnic Secondary School (POS) at its core.

As a consequence, the introduction of computers was not seen as a measure to be taken with the aim of revolutionizing or fundamentally altering the educational system. Rather, computers were considered by education policymakers to support and optimize the educational structures and practices already in place. The curricular reforms of the 1980s followed the explicit premise of preserving what had been tried and tested, while at the same time being oriented towards future perspectives and requirements, in particular concerning scientific and technological advances.[554] Accordingly, computers had to be introduced in a way that did not profoundly challenge the status quo in general schooling and that was in line with the core values of socialist education. This meant that the educational use of computers had to first and foremost serve the purpose of forming the "socialist personality" – the "all-round education" of their mind, hearts, and hands, as well as the inculcation of values and virtues which were oriented towards socie-

---

554 Zentralkomitee der SED, *Bericht an den XI. Parteitag*, 106–107.

tal requirements rather than individual ambitions and interests.[555] The president of the GDR's Academy of Pedagogical Sciences (APW), Gerhart Neuner, stated in 1985: "The fact that new fields of science or new sciences are coming to the fore does not in itself mean that they must become the content of socialist general education. Rather, it is necessary to ask and prove what they contribute to socialist personality formation, what they amount to in terms of the foundations of all-round personality development."[556]

While the integration of computer instruction into higher and vocational education had resulted from the immediate need to train experts and skilled workers for the development, operation and maintenance of new computer technology, the integration of computer technology into general education seemed less straight forward. On the one hand, the main aim of general education in the GDR was to lay the foundations for the "all-round development of socialist personalities,"[557] hence Neuner's question regarding the contribution of computer technology to this objective. On the other hand, there was also a more pragmatic aspect involved, oriented towards the broad preparation of the young generation for the envisioned computerized future.[558] In this sense, the general education provided by the Polytechnic Secondary School also had to fulfill its function in the GDR education system of preparing the basis for the subsequent vocational and higher education, which in-

---

[555] Angela Brock, "The Making of the Socialist Personality: Education and Socialisation in the German Democratic Republic 1958–1978" (PhD diss., University College London, 2005), 2, https://discovery.ucl.ac.uk/id/eprint/1363641/ (accessed May 31, 2024); Brock, "Producing the 'Socialist Personality'?."

[556] Gerhart Neuner, "Zur weiteren Entwicklung der Allgemeinbildung in unserer allgemeinbildenden polytechnischen Oberschule," *Information*, no. 1 (1985): 20, as cited in Lutz Engelmann, *Zur Vermittlung und Aneignung informatischen Wissens und Könnens als Bestandteil sozialistischer Allgemeinbildung – dargestellt am Beispiel eines fakultativen Kurses "Informatik" für die Klassen 9 und 10 (Dissertation A)*, Band 1 (Berlin: Akademie der Pädagogischen Wissenschaften der DDR, 1988), 7 (translated from German) (translated from German).

[557] 'All-round personality development' (ger.: allseitige Persönlichkeitsentwicklung) essentially refers to the humanist educational ideal of a universal form of education aimed at the full development of all human faculties. In GDR pedagogy, the term also drew heavily on Karl Marx, who derived the call for all-round development from his analysis of capitalist production, which entailed an increasing one-sidedness of human activity. Borrowed from Soviet pedagogy, the all-round education of socialist personalities included five aspects: mental, moral, polytechnic, aesthetic, and physical development of the individual. Since the early 1970s, "all-round development" was increasingly defined not as an abstract, timeless category, but rather as a historical-concrete ideal in line with current social conditions and requirements, and in particular the state doctrine of the SED (Hans-Joachim Hausten, *Allgemeinbildung und Persönlichkeitsentwicklung: Ein Beitrag zur Aufarbeitung der DDR-Pädagogik* (Frankfurt am Main: Peter Lang, 2003), 34–36).

[558] Engelmann, *Vermittlung und Aneignung informatischen Wissens*, 8.

creasingly included knowledge and skills in informatics and new computer technology. It was therefore considered necessary for pupils to acquire a basic understanding of new computer technology and informatics, the specific perspectives and ways of thinking and working associated with computing, and basic skills in the use of computers that they would require in their later life and professional career. It is in this delicate balance between general personal development on the one hand, and vocational preparation for a computerized future on the other, that the educational policy efforts to integrate computer education into the general school system of the GDR took place.

The decisive impetus for the SED to consider introducing computers into education on a broad basis, was rooted in the party's economic policy strategy for the 1980s and the following decades, guided by the sociotechnical imaginary of a far-reaching computerization of the GDR's society and economy. Accordingly, to bring about the desired future, the people of the GDR needed to be skilled in the use of computers and persuaded by the assertion that there was no alternative to the potency of computer technology to achieve greater economic and social prosperity in the GDR. Education thus not only served to impart the skills, knowledge and competencies that were anticipated as necessary in the envisioned technological future, but also to stabilize the dominant sociotechnical imaginary of the political leadership. However, especially regarding general education, the more "instrumental" purposes of introducing computer instruction into schools as a pre-vocational preparation for an envisioned future of computerized workplaces, needed to be legitimized and carefully balanced with a rhetoric that foregrounded the value of computer technology in supporting and promoting the development of general mental abilities of pupils.[559]

In the GDR, the introduction of computers into general schooling focused primarily on instruction *about* the new technology, how it was to be used and the ways in which it would contribute to social and economic progress in the GDR, as envisioned by the SED's sociotechnical imaginary. By contrast, the use of computers as an educational technology, that is, as a tool of instruction with the aim of improving education, was of secondary importance to the political leadership. In this sense, the aspect of the dominant sociotechnical imaginary that envisioned the computer as a tool for increasing economic productivity and competitiveness by computerizing industry and training the workforce to use computers at work took precedence over the more pedagogically oriented vision of using computers to develop human intellectual capacity and creativity, and to stimulate learning. Consequently, in the 1980s, the available resources in educational policymaking for the

---

559 Engelmann, *Vermittlung und Aneignung informatischen Wissens*, 7–8.

introduction of computer technology into general education – both with regard to research and development capacities as well as technical equipment – were concentrated primarily on the task of introducing a computer education to prepare pupils for the computerized future rather than establishing the computer as an educational technology.[560] Teachers and educational scientists identified benefits of computer use in education – as a tool for knowledge retrieval, for instructional demonstrations, and experimentation, or as a teaching machine.[561] It was these pedagogical experts who faced the challenge of reconciling the predominantly economic and political motives behind computer integration into schools with the broader pedagogical goals and philosophy of general education. While pedagogues showed awareness of and interest in the pedagogical implications and possibilities of computer use,[562] these aspects seem to have played a subordinate role in the eyes of the political leadership. Ultimately, teachers and educational scientists had to conform to the SED's political agenda, which was underpinned by a sociotechnical imaginary that defined computers primarily as workplace tools for increasing economic productivity, and not as an educational technology.

---

[560] Against the backdrop of scarce human resources, the APW needed to focus on the development and implementation of new curricula for general education as mandated by the SED's political decision. Preliminary research on the use of microcomputer and software as an educational technology to support teaching and learning in a variety of subjects was largely concentrated in universities for teacher education, notably in Halle and Dresden (BBF APW 11296/4, "Bericht über Stand und Probleme bei der Realisierung des Informatikbeschlusses anlässlich der Kontrollberatung beim Präsidenten der APW am 4.5.1987," dated April 29, 1987, 10–13).

[561] Immo O. Kerner, "Internationaler Trend des Computereinsatzes in allgemeinbildenden Schulen und Bemerkungen zur Computerdidaktik," in *1. Seminar 'Computer als Mittel und Gegenstand der Ausbildung', Tharandt 1982. Seminarberichte Teil 1* (Dresden: Pädagogischen Hochschule Dresden, 1982), 8–9.

[562] Immo Kerner, Professor of Computer Science at the University of Teacher Education in Dresden, noted in 1982 that the "cybernetic and algorithmic aspect" of the pedagogical process of teaching and learning had long been known in the GDR, referring to efforts in the field of programmed instruction in the 1950s and 1960s. These attempts had been deemed as failed and ultimately abandoned in 1980. However, while they had been based on the medium of printed materials, new computer technology was expected to open up new ways of supporting the educational process, in particular through its "variability and flexibility" and because it allowed for a global view of the learner's previous learning trajectory, as well as the collection of data on pupils and the program for the purposes of evaluation and assessment (Kerner, "Internationaler Trend des Computereinsatzes," 4).

5.1 Electronic Pocket Calculators and Early Curricular Reforms

In the late 1970s and early 1980s, curricular reforms in general education were initiated in response to recent scientific and technological advances. Their purpose was to assess which school subjects and learning contents had gained in importance or needed to be adapted, and which curricular materials had to be revised in response to technological change and new scientific findings. At the same time, the notion of the "scientific-technological revolution" became increasingly influential in educational policymaking, in particular through the SED's demand to further intensify the polytechnic character of the secondary school. In 1976, Lothar Oppermann and Rudi Oelschlägel, responsible for the department for national education within the SED's Central Committee, stated: "In order to combine the scientific and technological revolution with the advantages of socialism, it is necessary to further strengthen the polytechnic character of secondary education. Without equipping young people with a high level of general education with a polytechnic character, without linking school with life, the scientific and technological revolution cannot be realized."[563] In other words, the SED wanted to orient general schooling more towards the world of work, by strengthening its pre-vocational function.[564] In this sense, polytechnic education served to prepare and familiarize pupils with the requirements of the sphere of economic production. Consequently, its contents had to be adapted in response to the fundamental changes that the economy was currently undergoing and was expected to undergo in the future, as envisioned in the SED's sociotechnical imaginary of socialist computerization.

At the occasion of the 8th Pedagogical Congress in 1978, educators and pedagogues in the GDR were informed of the practical consequences of recent scientific and technological developments for general education in mathematics and physics, as well as for polytechnic education and elective courses.[565] In her speech at the congress, Margot Honecker, Minister for National Education, also brought up the issue of computer technology in general education, specifically by raising the question of the role of electronic pocket calculators in schools. She called for

---

[563] Lothar Oppermann and Rudi Oelschlägel, "Ideologisch-theoretische Probleme der weiteren Ausprägung des polytechnischen Charakters unserer Oberschule," Einheit 31, no. 11 (1976): 1244 (translated from German).
[564] Fuchs and Petermann, *Bildungspolitik in der DDR*, 29.
[565] Ministerrat der DDR/Ministerium für Volksbildung, *VIII. Pädagogischer Kongress der Deutschen Demokratischen Republik vom 18. bis 20. Oktober 1978: Protokoll* (Berlin: Volk und Wissen, 1979).

an in-depth investigation into the matter to clarify from which grade level, for which subject areas, and with what aim the use of electronic pocket calculators could prove to be useful – and where and under which circumstances their use was inappropriate.[566] A student teacher, who attended the congress, later remembered a "distinct murmur" in the hall as the Minister of National Education talked about the possible introduction of electronic pocket calculators into general schooling,[567] thereby for the first time publicly reawakening the specter of automation and computer technology in the classroom since the failed attempts at developing and implementing teaching machines in the 1960s and 1970s.[568] None of the teachers and pedagogues present openly expressed their scepticism or opposition, but there certainly seems to have been some resistance among them to the very idea of introducing electronic calculators into schools. In contrast to the teaching machines, which had attempted to automate certain functions of the teacher, the electronic pocket calculator now promised to automate an important function performed by the *learner* in mathematics instruction. This raised concerns among mathematics teachers that children who used an electronic pocket calculator would become dependent on the device, and forget, or no longer learn how to calculate with no other help than their own minds.[569]

Following Margot Honecker's remarks at the 8th Pedagogical, a theoretical and practical investigation on the issue was launched. The Academy of Pedagogical Sciences (APW) of the GDR was instructed to conduct a research project to empirically test the use of the electronic calculator in school classes.[570] Between 1979 and 1983, 23 classes were equipped with three different models of electronic calculators for the use in the subjects of mathematics, physics, and chemistry. According to the researchers' findings, the use of the calculator automated some of the students' time-consuming written calculations, thus freeing up classroom time for the core content

---

566 Margot Honecker, *Der gesellschaftliche Auftrag unserer Schule. Referat von Margot Honecker, Minister für Volksbildung, auf dem VIII. Pädagogischen Kongress* (Berlin: Verlag Zeit im Bild, 1987), 44.
567 Organisationsbüro IX. Pädaogischer Kongress, *Bulletin no. 3 IX. Pädagogischer Kongress: Diskussionsbeiträge vom zweiten Konferenztag (14. Juni 1989)* (Berlin, 1989), 19.
568 For the history of teaching machines and programmed instruction in the GDR, see Zabel, "Lehrmaschinen und der Programmierte Unterricht."
569 BBF APW 11295/3, "Standpunkte zur Einbeziehung von Grundelementen der Informatik in den Mathematikunterricht der allgemeinbildenden Oberschulen," dated June 1984, 3; Organisationsbüro IX. Pädagogischer Kongress, *Bulletin no. 3 IX. Pädagogischer Kongress*, 19.
570 Tom Schnabel, "Der Einsatz der Taschenrechner," in *Kleincomputer in der DDR* (Thesis, Humboldt-Universität zu Berlin, 1999).Online: http://waste.informatik.hu-berlin.de/Diplom/robotron/diplom/texte/einsatz/taschenrechner.html (accessed May 31, 2024).

of the subjects in which the calculator was used.[571] Instruction in mathematics, natural sciences, and the polytechnical subjects would be rendered more effective and connected to real life. In addition, the introduction of the electronic calculator into general education was considered as a decisive step in the mission to introduce elements of informatics into schools.[572] In this sense, the use of electronic pocket calculators in mathematics and other subjects was seen as a means of preparing pupils for the general educational requirements placed on the working population by the development of science and technology,[573] and, more specifically, of familiarizing pupils with the "ways of thinking" associated with informatics.[574]

Because of the positive results of the school pilot project, electronic calculators were introduced in the 11th grade of the extended polytechnic secondary school in September 1984, and in the 7th grade of the polytechnic secondary school in September 1985. This made it necessary to take measures in the training of teacher students in the subject of mathematics, as well as in the further training of teachers already in service. Even though the SR1 was introduced as the official school calculator, other electronic pocket calculators were allowed. This meant that teachers needed to be prepared for the use of various types of calculators that pupils might bring to class. Simultaneously, new teaching and learning materials had to be developed to account for the profound implications which arose from the pocket calculator's use in the subject of mathematics. Pupils first needed to learn how to enter data into the calculator, perform basic arithmetic operations and to interpret the result. The revised mathematics textbooks, thus, included symbolic representations of the SR1's display and keyboard, and explained the calculator's various functionalities. A colorful image of the SR1 featured prominently on the cover of the new mathematics textbook that was published in 1985 for grade 7, in which pupils from now on would first be introduced to this new learning tool.

Fuelled by the concerns that the use of the electronic calculator would negatively affect pupils' numeracy skills, the materials to prepare mathematics teacher for the implementation of pocket calculators emphasized that particular attention should be paid to the practice of mental arithmetic.[575] The basic argument was that

---

571 "Pädagogen lassen sich täglich etwas einfallen," *Berliner Zeitung*, March 19, 1985, 3; "Taschenrechner für die Schule," Berliner Zeitung, March 23, 1985, 3.
572 BBF APW 11295/3, "Standpunkte zur Einbeziehung von Grundelementen der Informatik in den Mathematikunterricht der allgemeinbildenden Oberschulen," dated June 1984, 5.
573 "Pädagogen lassen sich täglich etwas einfallen," *Berliner Zeitung*, March 19, 1985, 3.
574 Immo O. Kerner and Ulrich Winkler, *Taschenrechner in der Schule. Lehrmaterial zur Aus- und Weiterbildung von Mathematiklehrern an der PH Dresden* (Dresden: Pädagogische Hochschule K. F. W. Wander Dresden, 1985), 54.
575 Kerner and Winkler, *Taschenrechner in der Schule*, 38.

pupils should never rely completely on the calculator, but rather to be able to estimate the expected result in advance and then check the calculator's output against it. Such a rationale was an expression of a variety of concerns on the part of pedagogues and educators about how pupils would interact with and be affected by this new technology. The concerns included a "blind trust in the device" and the output it produced, an unwillingness to check and verify the results it computed, and a loss of key math abilities, such as the understanding of quantities and numerical concepts.[576] The teachers were thus required not only to instruct the pupil on the technical functions and handling of the calculator, but also to educate them to use the new device with a critical attitude towards the results they obtained. While it was assumed that the calculator worked flawlessly – apart from the occasional technical glitch or battery failure – pupils had to be made aware that it was their input that mattered, and that it could lead to a false result if they were careless when typing.[577] The underlying notion was that these new information technology tools required their users to be conscientious, attentive, and self-disciplined to be able to put these devices to good and effective use.

In this sense, the introduction of electronic pocket calculators into schools were also intended to prepare children for the use of other computer technology later in their educational career and working life, such as microcomputers and computerized machines. By introducing pupils to the electronic pocket calculators, teachers were explicitly tasked to promote the development of certain working attitudes, routines, and mindsets in their pupils, that were deemed essential for the sensible use of information technology. As an example, teachers were encouraged to let their class draw a flowchart of the calculation process when solving more complex math problems.[578] Such a visualization of the individual steps leading to the result was meant to make the learners' understanding of the problem and their problem-solving process more transparent. But it also prepared pupils for the writing of programming flowcharts, which was seen as an important part of informatics and working with microcomputers, for example in after-school computer clubs and workshops.

In parallel, pedagogues and educational policymakers also addressed the question of the extent to which school curricula had to be adapted or supplemented to take account of the growing economic policy significance of microelectronics since the Politburo's decision on microelectronics from 1977. A first attempt was taken by the APW to introduce microelectronics as a subject of instruction into the curricula

---

[576] Kerner and Winkler, *Taschenrechner in der Schule*, 56.
[577] Kerner and Winkler, *Taschenrechner in der Schule*, 56.
[578] Kerner and Winkler, *Taschenrechner in der Schule*, 57.

of extended secondary schools (EOS) in the early 1980s.[579] A draft syllabus for an elective course called "microelectronics" was developed and tested in schools with two separate groups of pupils. The course comprised 50 hours, of which 20 hours were devoted to the basics of microelectronic technologies and digital integrated circuits. Another 18 hours were allocated for a practical work project, in which pupils should collaboratively complete one of two possible assignments. The first option consisted in the development and construction of a speedometer that could be used for industrial robots or CNC machines. The second option involved the development and construction of a digital multimeter, which could be used in a data acquisition system. The remaining hours of the course were dedicated to teaching the importance of microelectronics for the GDR's national economy, and to cover some supplementary hardware of microelectronic systems. Thus, the draft syllabus was strongly focused on the fabrication of microelectronic hardware components and its specific physical-technical aspects. Notably, the syllabus also explicitly stated the goal to develop the pupils' technical interests and inclinations in accordance with societal requirements. Moreover, the pupils were to be motivated to enroll in a microelectronics-related technical or economic science degree program or pursue a career in military.[580]

In October 1982, the curriculum advisory committee of the APW decided to stop the work on the draft syllabus.[581] The committee concluded that the draft syllabus could not be implemented, because physics teachers were not adequately qualified to teach it, and the EOS lacked the technical and material requirements. Moreover, the aims and focus of integrating the scientific and technological advancements in microelectronics into general schooling had shifted in the meantime. Instead of emphasizing physical and technical aspects, the curriculum advisory committee was now convinced that pupils rather needed to be familiarized with the application areas of microelectronics and user solutions. The committee consequently drew the conclusion, that the contents of the planned course could not be successfully taught under the given conditions and did not need to be taught within the scope of general education.[582] Instead of implementing further new programs, the committee argued that efforts should be concentrated on

---

[579] BBF APW 9901, "Entwurf des Rahmenprogramms 'Mikroelektronik' für den fakultativen Unterricht der EOS vom 20.8.1982."

[580] BBF APW 9901, "Entwurf des Rahmenprogramms 'Mikroelektronik' für den fakultativen Unterricht der EOS vom 20.8.1982," 2–3.

[581] BBF APW 9901, "Protokoll der Beratung zum Entwurf des Planes für den fakultativen Lehrgang 'Mikroelektronik' am 18.11.1982," 1.

[582] BBF APW 9901, "Protokoll der Beratung zum Entwurf des Planes für den fakultativen Lehrgang 'Mikroelektronik' am 18.11.1982," 2.

the implementation of the already approved syllabus for an optional class in "Electronics." which was to be introduced in the EOS in September 1983.[583] The syllabus encompassed the basics of electronic measurement technology, semiconductor diode circuits, amplifier circuits, the basics of digital technology, and the electrical measurement of non-electrical parameters. Together with existing after-school "Microelectronics" study groups for pupils, the "Electronics" course was expected to allow the APW to gain insights of what worked in practice and to determine more precisely, what should be taught about microelectronics in general schooling.[584]

Concurrently, the APW was also working on the revision of curricula in the subject "Introduction to Socialist Production" (ger.: Einführung in die Sozialistische Produktion, ESP) for polytechnic secondary schools (POS). This subject formed, together with the subject "Productive Work" (ger.: Produktive Arbeit, PA), the polytechnic core of the POS. It was aimed at raising the pupils' awareness of the use and applications of knowledge in the national economy, technology and production, while linking this knowledge to regional conditions, and the current political course of the SED as well as the achievements of the "scientific-technological revolution."[585] In PA lessons, pupils from grade 7 and above spent time in local businesses, combines and farms, where they were meant to apply what they had learned in the subject ESP, and gain hands-on work experience in industry, construction, or agriculture, under the guidance of pedagogically trained skilled workers.[586]

The APW revised the curricula for "Introduction to Socialist Production" for grades 9 and 10, by adding new content to cover the topics of "information electrics" and "microelectronics".[587] The content in "information electrics" focused on various electronic components and their use in the automation of production, in electronic data processing and measurement technology, as well as the transfer and exchange of information in technological processes. With regard to microelectronics, pupils were taught the key applications of microelectronics, and were familiarized with future trends and perspectives of computerized control of machines and microelectronic components such as the microprocessor for the auto-

---

583 BBF APW 9901, "Protokoll der Beratung zum Entwurf des Planes für den fakultativen Lehrgang 'Mikroelektronik' am 18.11.1982," 2.
584 BBF APW 9901, "Protokoll der Beratung zum Entwurf des Planes für den fakultativen Lehrgang 'Mikroelektronik' am 18.11.1982," 2.
585 Hausten, *Allgemeinbildung und Persönlichkeitsentwicklung*, 43.
586 Christoph Kodron, "Polytechnische Bildung und Erziehung in der Deutschen Demokratischen Republik," *International Review of Education* 24, no. 2 (1978): 213–14.
587 BBF APW 9901, "Auszug aus dem weiterentwickelten ESP-Lehrplan (Klassen 9/10)," undated.

mation of production in the GDR, as envisioned by the SED.[588] For the same grades, a new syllabus for PA in the metalworking and electronics industry was drafted to include tasks for pupils that involved automated machine tools and systems.[589] The syllabus stated that pupils were to execute simple tasks in preparing, operating, monitoring, and maintaining metal-cutting machines autonomously after prior instruction.[590] Where available, pupils were also meant to assist in operating, monitoring, maintaining automated machine tools, and learn about modern industrial robot technology. Working with automated machine tools and systems was meant to give pupils "realistic insights into fundamental trends in the development of scientific-technological progress and the working conditions in automated fields of production."[591]

The APW's curricular revisions regarding the introduction of microelectronics as a new field of teaching and learning in polytechnic instruction thus focused primarily on the electrical fundamentals of the new technology, and its applications in automated and computerized machines in industrial production. Computers as machines for information processing and mathematical computation, as they were used in research facilities and computing centers, did not feature in the revised curricula. The APW also followed an "incremental" approach in its curricular reforms, which did not change the subject structure but integrated the new content into existing subjects. However, this was to change fundamentally by the mid-1980s, as microcomputers began to take on an increasingly important role in education at an international level.

In July 1981, two pedagogues of the APW had been sent as delegates to the World Conference on Computers in Education (WCCE) in Lausanne, Switzerland. The conference was organized on behalf of the IFIP's Technical Committee on Education, which assured in its invitation to the conference that microprocessors were going to permeate classrooms just as they did factories and offices, and that the language and methods of informatics would fundamentally alter teaching of all subjects.[592] The two GDR delegates reported that advanced capitalist countries were developing microcomputers, which were expected to flood the market in

---

588 BBF APW 9901, "Auszug aus dem weiterentwickelten ESP-Lehrplan (Klassen 9/10)," undated, 8.
589 BBF APW 9901, "Plan für die produktive Arbeit der Schüler in Betrieben der metallverarbeitenden Industrie und der Elektrotechnik/Elektronik, Klassen 9 und 10 – Entwurf," undated.
590 BBF APW 9901, "Plan für die produktive Arbeit der Schüler in Betrieben der metallverarbeitenden Industrie und der Elektrotechnik/Elektronik, Klassen 9 und 10 – Entwurf," undated, 9.
591 BBF APW 9901, "Plan für die produktive Arbeit der Schüler in Betrieben der metallverarbeitenden Industrie und der Elektrotechnik/Elektronik, Klassen 9 und 10 – Entwurf," undated, 21.
592 "WCCE-81, Computers in Education, 3rd World Conference, Lausanne, Switzerland, 27–31 July, 1981," European Journal of Engineering Education 5, no. 1 (1981): 285.

large numbers in the years to come.⁵⁹³ At the conference, attendees from all over the globe had discussed the various ways in which microcomputers could be used for teaching and learning, in particular in higher education and secondary schools. The two GDR delegates reported that large research institutions had been set up in many economically advanced capitalist countries to develop programs for computer-assisted instruction.⁵⁹⁴ The GDR delegates felt that these were programs of an entirely new quality, with little resemblance in content or structure to those that had been developed under the rubric of programmed instruction in the 1960s, the so-called "cybernetic pedagogy."⁵⁹⁵ In response to the report, pedagogues at the APW concluded that the development of microcomputer technology would open up previously unimagined possibilities in education.⁵⁹⁶ In their view, the questions about programmed instruction from the 1960s had to be answered in the negative because of technological constraints and ideological concerns, were now being asked in a completely new way, under completely different circumstances. In line with the sociotechnical imaginaries of computer technology in education articulated at the WCCE, APW educators envisioned the microcomputer as a means of streamlining school administration and as a complex instructional tool that would allow for meaningful differentiation and individualization of learning and teaching with minimal effort. It was also noted that computers in schools would help pupils to develop user skills likely to be required in modern production soon, and therefore, they needed to be made part of general education.⁵⁹⁷

Hence, they urged that exploratory and basic research on the subject be initiated immediately. To add political weight to the demand, they pointed out that the Soviet Union was also commencing research into the role of microcomputers

---

593 BBF APW 11295/2, "Spezieller Bericht zur 3. WCCE unter der Sicht von Volksbildungsfragen," dated August 11, 1981, 1.
594 BBF APW 11295/2, "Spezieller Bericht zur 3. WCCE unter der Sicht von Volksbildungsfragen," dated August 11, 1981, 2.
595 By mentioning "cybernetic pedagogy," the GDR delegates referenced the work of Helmar Frank, a West German cybernetician and director of the "Research and Development Center for Objectified Teaching and Learning Methods" (ger.: Forschungs- und Entwicklungszentrum für objektivierte Lehr- und Lernverfahren, FEoLL) in Paderborn. For a more detailed account of cybernetic pedagogy and the contributions of Helmar Frank, see Barbara Hof, "From Harvard via Moscow to West Berlin: Educational Technology, Programmed Instruction and the Commercialisation of Learning after 1957." *History of Education* 47, no. 4 (2018): 445–65; Barbara Hof, "The Turtle and the Mouse: How Constructivist Learning Theory Shaped Artificial Intelligence and Educational Technology in the 1960s." *History of Education* 50, no. 1 (2021): 93–111.
596 BBF APW 11295/2, "Vorschlag zum weiteren Vorgehen auf dem Gebiet der Entwicklung neuer technischer Mittel für den Unterricht," dated September 1981.
597 BBF APW 11295/2, "Vorschlag zum weiteren Vorgehen auf dem Gebiet der Entwicklung neuer technischer Mittel für den Unterricht," dated September 1981, 2.

in education.[598] A working group was formed under Gerhart Neuner, the APW's president, to investigate the possible use of microcomputers in general education.[599] It was concluded that priority should be given to research on new computer technology, its theoretical foundations and practical use as a *subject of teaching* and part of socialist general education, rather than as a *tool of instruction*.[600] The reasoning behind this was as follows: The SED's economic strategy in the second half of the 1980s required that school graduates be prepared for subsequent vocational training and higher education, so the general education system had to act quickly to introduce the relevant knowledge and skills associated with the new information technology into school curricula. The use of the microcomputer as a tool of instruction, by contrast, would first require extensive empirical and experimental research to answer the question of how microcomputer hardware and software could be used effectively in teaching and learning, and to prove its superiority over traditional means and methods of instruction.[601]

Following the formulation of these initial positions, in March 1985 the APW produced a draft position paper entitled "Theses on School Computer Science."[602] In the paper, the APW pedagogues argued that the new discipline of informatics extended beyond the importance of "classic" disciplines such as mathematics, physics, chemistry, or biology, due to its broad impact: Virtually all professions would use information technology in the future as a tool to solve their problems. However, the APW also clearly stated that informatics should not be introduced as an additional subject. The additional subject of informatics would not only overburden both learners and the curricula, but also require the training of specialized teachers for informatics, which in turn, would lead to an "alibi function" for all other teachers, which would feel that they were not responsible for teaching informatics in their classes.[603] Instead, the pedagogues of the APW argued for an interdisciplinary approach to teaching informatics in general education, in

---

[598] BBF APW 11295/2, "Vorschlag zum weiteren Vorgehen auf dem Gebiet der Entwicklung neuer technischer Mittel für den Unterricht," dated September 1981, 2.
[599] BBF APW 11295/2, "Entwurf: Entscheidungsgrundlagen und Vorschläge zur Planung von Forschungen 1984–1990 über die Nutzung von Kleincomputern in der allgemeinbildenden Schule," dated April 1984, 1.
[600] BBF APW 11295/2, "Entwurf: Entscheidungsgrundlagen und Vorschläge zur Planung von Forschungen 1984–1990 über die Nutzung von Kleincomputern in der allgemeinbildenden Schule," dated April 1984, 4.
[601] BBF APW 11295/2, "Entwurf: Entscheidungsgrundlagen und Vorschläge zur Planung von Forschungen 1984–1990 über die Nutzung von Kleincomputern in der allgemeinbildenden Schule," dated April 1984, 5.
[602] BBF APW 16.315–16.316, "Thesen zur Schulinformatik," dated March 14, 1985.
[603] BBF APW 16.315–16.316, "Thesen zur Schulinformatik," dated March 14, 1985, 2.

particular by introducing the microcomputer as a tool of instruction in most of the subject – as an educational technology. The microcomputer would help to increase the vividness of instruction with graphical representation on the computer screen. Moreover, the use of microcomputers in school instruction was promised to help shift the emphasis from purely conceptual knowledge to process-oriented, problem-solving knowledge.[604] Importantly, it was stated in the paper that, in contrast to earlier efforts in programmed instruction, the teacher was not to be "eliminated" from the process of instruction with the introduction of computers as a tool of teaching and learning.[605] On the contrary, the role of teachers in communicating with students was expected to become more important.

However, computer-assisted teaching and learning that went beyond the mere visualization or simulation of certain principles and processes in mathematics and science on the computer screen by the teacher, or the individual work of pupils on the computer for problem-solving exercises, was controversial even among the educators of the APW. In particular, the possible future use of educational software, in which students would learn and practise independently on the computer, was a matter of disagreement. Such "programmed instruction" on the computer, the critics at the APW argued, carried the risk that "good software" would make teachers at least partially redundant and essentially deprive them of their direct educational influence on pupils.[606] The use of computers for "programmed learning," involving the use of sophisticated learning software for pupils, was therefore considered to be a distant possibility at the time, requiring clarification of numerous unresolved issues beforehand.

5.2 A Curriculum for Computer Instruction in General Education

In November 1985, the Politburo of the SED's Central Committee and the Council of Ministers of the GDR decided on the consequences to be drawn from the development of informatics and information processing technology in the field of education.[607] This decision marked the beginning of an extensive program to introduce computer technology into the educational system of the GDR, backed by an explicit

---

604 BBF APW 16.315–16.316, "Thesen zur Schulinformatik," dated March 14, 1985, 3.
605 BBF APW 16.315–16.316, "Thesen zur Schulinformatik," dated March 14, 1985, 4.
606 BBF APW 16.315–16.316, "Protokoll einer Konsultation bei Prof. Dr. Peschel, Leiter des Forschungsbereichs Mathematik und Informatik der APW, am 4.4.85," 5.
607 SAPMO DY 30/J IV 2/2/2138, "Standpunkte zu Konsequenzen aus der Entwicklung der Informatik und informationsverarbeitenden Technik für das Bildungswesen" (Attachment no. 7 to the

and detailed mandate from the political leadership. It outlined how new information technology was to be made part of curricula in general education, vocational education, and training and higher education. Before the decision, the first tentative steps towards an educational response to technological change in general schooling had already been taken as part of ongoing curricular reforms since 1978, and the APW's tentative steps in exploring the possibilities of introducing microcomputers into education. However, with a corresponding decision from the highest political level, a coherent strategy encompassing the entire education system was to be developed and implemented. Accordingly, the decision encompassed the planned measures to be taken by the Ministry of Higher Education, the State Secretariat for Vocational Education, and the Ministry of National Education regarding the introduction of information technology into education and training in the GDR.

The timing of this decision was no coincidence. Only a year earlier, the KC 85 microcomputer, originally designed as a home computer, had been presented at the spring trade fair in Leipzig.[608] The KC 85 was the first domestically produced computer in the GDR that was reasonably sized for classroom use, much cheaper to manufacture than earlier models, and could be produced in large numbers. The SED delegation visiting the fair, including Erich Honecker, decided that the KC 85 microcomputer should be used for educational purposes in schools and other institutions, rather than as a consumer product for people to play computer games in their homes.[609] With the technical innovation of the first domestically produced off-the-shelf microcomputer, the material basis was now potentially available to equip educational institutions on a broad scale. At the same time, the new microcomputers reinvigorated the SED's sociotechnical imaginary of a computerized society, the realization of which now seemed to have moved a significant step closer. The widespread introduction of computers into all areas of the economy and society now seemed within reach, which in the eyes of the political leadership made it necessary to train the general population to use this technology effectively and in accordance with socialist values and the party's political goals.

The impact of the efforts of educators and scientists at the beginning of the 1980s to persuade the political leadership of the SED to take a landmark decision for the rapid introduction of computers into the elementary school system should not be underestimated. In February 1984, the working group on information proc-

---

minutes no. 45 of November 12, 1985), identical in its wording with the corresponding decision of the Council of Ministers of November 14, 1985.
608 René Meyer, *Computer in der DDR* (Erfurt: Landeszentrale für politische Bildung Thüringen, 2019), 52–57.
609 Meyer, *Computer in der DDR*, 57

essing and cybernetics of the GDR's Academy of Sciences devised a position paper on the introduction of basic elements of informatics into the GDR's system of general education, signed by the working group's head, computer pioneer Nikolaus Joachim Lehmann.[610] The position paper outlined the "revolutionary" impact of microelectronics on economy, technology, science, and society based on increasingly cheap and mass-produced computer technology. The systematic exploitation and promotion of this new technology through state support and intervention in the leading industrialized countries, the authors alleged, only further intensified the sociopolitical poignancy of this development. Consequently, the authors urged that new information technology should be introduced and used on all levels of the GDR's system of education. The working group considered this to be essential, particularly with regard to general education, because informatics would require new forms of process-oriented thinking and working, which they argued were particularly easy to inculcate at a young age.[611] Instead of fundamentally altering curricular contents, general education would have to teach algorithmic and process-oriented thinking in all mathematical and scientific disciplines, and to impart knowledge on the computer and its importance for the development of the economy and society. The working group thus urged the GDR's political decision makers to include informatics as soon as possible in general education, and to immediately take all economically justifiable and possible steps to accelerate this process.[612]

While it is hardly possible to estimate the impact of such initiatives by pedagogues and scientists on the SED's policymaking in education, these stakeholder groups surely did their part in bringing the issue and their demands to the attention of the GDR's Council of Ministers. The SED's landmark policy decision on the consequences to be drawn from the development of new information technology in education was thus likely to have been precipitated both by the technological breakthrough of the first domestically produced microcomputers and by the repeated calls for action from scientists and educators committed to bringing computer technology into mainstream education.

---

610 BBF APW 11295/3, "Stellungnahme der Arbeitsgruppe Informationsverarbeitung/Kybernetik der AdW der DDR zur Einbeziehung der Grundlagen der Informatik in die Volksbildung," dated February 16, 1984.
611 BBF APW 11295/3, "Stellungnahme der Arbeitsgruppe Informationsverarbeitung/Kybernetik der AdW der DDR zur Einbeziehung der Grundlagen der Informatik in die Volksbildung," dated February 16, 1984, 2.
612 BBF APW 11295/3, "Stellungnahme der Arbeitsgruppe Informationsverarbeitung/Kybernetik der AdW der DDR zur Einbeziehung der Grundlagen der Informatik in die Volksbildung," dated February 16, 1984, 2.

In devising an educational strategy to prepare its workforce for the use of new information technology, the Politburo's decision taken in 1985 explicitly stated that the GDR should make use of the experiences and insights of the Soviet Union and other socialist countries in their responses to technological change. But it was also noted that the consequences of the introduction of new information technology were expected to play out differently in the specific structures of different educational systems,[613] which left a certain amount of leeway for the GDR to explore independent paths.

At the same time, the decision demanded for the GDR's educational strategy to be clearly distinguished from the technicist approach of capitalist countries. The SED criticized the "unprecedented marketization of schooling" and the "unscientific procedures" through which capitalist countries were "modernizing" their educational systems.[614] Drawing upon the criticism voiced by scientists and pedagogues in Western capitalist countries, it was pointed out how these countries were lacking a sound pedagogical concept for the introduction of information technologies, and that as a result, schools were turned into markets for hardware and software under the pressures of large corporations.[615] The approach of the GDR, in contrast, would have to be integrated into an overarching socialist concept of education and be "scientifically sound and differentiated."[616] Curricular revisions had to be justified and planned through methodological research, tested experimentally in schools and approved by a curriculum commission and pedagogical councils – a process that could easily take up several years. As a result, an immediate redesign of the compulsory curriculum within a short time frame was neither deemed feasible, nor was it intended in the GDR's system-theoretical approach to education policymaking. Non-compulsory and extracurricular education, by contrast, allowed for more short-term curricular innovation in the GDR's system of education. The setting up of optional courses and after-school computer clubs offered a way to respond more quickly to the changing educational demands in the context of technological change, without rushing to disrupt the carefully balanced curricular system of compulsory education.

---

613 SAPMO DY 30/J IV 2/2/2138, "Standpunkte zu Konsequenzen aus der Entwicklung der Informatik und informationsverarbeitenden Technik für das Bildungswesen," 4.
614 SAPMO DY 30/J IV 2/2/2138, "Standpunkte zu Konsequenzen aus der Entwicklung der Informatik und informationsverarbeitenden Technik für das Bildungswesen," 5.
615 SAPMO DY 30/J IV 2/2/2138, "Standpunkte zu Konsequenzen aus der Entwicklung der Informatik und informationsverarbeitenden Technik für das Bildungswesen," 5.
616 SAPMO DY 30/J IV 2/2/2138, "Standpunkte zu Konsequenzen aus der Entwicklung der Informatik und informationsverarbeitenden Technik für das Bildungswesen," 4.

In general schooling, the immediate consequences to be drawn from the development of information technology seemed less clear and straight-forward than in higher and vocational training and education, where the use of modern computers and computerized machines demanded for specific knowledge and skills. In its decision, the Politburo merely stated that it had yet to be determined to what extent general education needed to include a basic understanding of information and communication technology, and which character traits and personal qualities would gain in importance because of the increasingly widespread use of these technologies.[617] It was expected that the rapid technological development would generally place higher demands on the command of basic knowledge and skills, especially in mathematics, sciences, and technology, and required creative and scientific thinking and work practices. In addition, it was argued that certain "moral qualities" gained in importance, such as "mental agility, accuracy, diligence, discipline, sense of community, perseverance and organizational skills" as well as the "pleasure and interest in dealing with scientific and technical subjects."[618] The polytechnic secondary school in the GDR was thus tasked not only with providing students with relevant technical knowledge and skills, but also with instilling in them certain "soft skills" in the form of behaviors and attitudes that were found to be of value in computer users who would become part of the envisioned future workforce.

The decision of the Politburo and the Council of Ministers included a more detailed and concrete action plan that had been prepared beforehand by the Ministry of National Education.[619] The action plan elaborated on the implications of the development of new information technology on general schooling and the concrete measured to be taken in the following years. The main goal was that all pupils were taught the basics of informatics and information technology, explicitly to build a foundation for their later vocational education and training, and to prepare them for the requirements of modern production.[620] In contrast to earlier efforts to introduce content on new information technology into school curricula, which had focused on teaching the physical principles and microelectronic

---

[617] SAPMO DY 30/J IV 2/2/2138, "Standpunkte zu Konsequenzen aus der Entwicklung der Informatik und informationsverarbeitenden Technik für das Bildungswesen," 6.
[618] SAPMO DY 30/J IV 2/2/2138, "Standpunkte zu Konsequenzen aus der Entwicklung der Informatik und informationsverarbeitenden Technik für das Bildungswesen," 23.
[619] SAPMO DY 30/J IV 2/2/2138, "Maßnahmeplan des Ministeriums für Volksbildung zur Realisierung der Konzeption 'Standpunkte zu Konsequenzen aus der Entwicklung der Informatik und informationsverarbeitenden Technik für das Bildungswesen'."
[620] SAPMO DY 30/J IV 2/2/2138, "Maßnahmeplan des Ministeriums für Volksbildung zur Realisierung der Konzeption 'Standpunkte zu Konsequenzen aus der Entwicklung der Informatik und informationsverarbeitenden Technik für das Bildungswesen'," 1.

foundations of the new information technology, the new curriculum content was to focus on practical user skills and knowledge of computer technology. Pupils were now expected to learn the skills necessary to use a computer, develop a basic understanding of the principal structures and functions of information processing technology, and acquire knowledge and skills in problem-solving involving algorithms and in developing computer programmes.[621] This also required for pupils to practice "precise command of language." as well as "logical, algorithmic and technical-constructive thinking."[622] In addition, and very much in line with the SED's sociotechnical imaginary, teaching computer technology was supposed to instill in students the belief in the "controllability of modern technology," which included an understanding of the "fundamental social effects and the use of this technology for the benefit of mankind."[623]

While this conceptualization of a mandatory computer curriculum as part of general education was aimed at providing *all* pupils with basic skills and knowledge of computer technology, a more differentiated and specialized curriculum was envisioned for pupils in EOS and in the so-called "special schools" for mathematics, natural sciences, and technology.[624] These schools primarily catered to gifted and high-achieving pupils, who usually went on to study at a university or a technical college. In consequence, the role of these schools was not only to provide a sound general education, but also to prepare pupils for higher education, where applications of computer technology and informatics played an important role in an increasing number of disciplines. The Ministry of National Education's action plan prioritized curricular innovations targeted at this group of pupils and listed measures that should be promptly initiated and implemented. A possible reason for why such a prioritization in the introduction of computer curricula in general schooling was necessary, can be found in the limited resources that were made available to put the action plan into effect. The APW simply did not have

---

[621] SAPMO DY 30/J IV 2/2/2138, "Maßnahmeplan des Ministeriums für Volksbildung zur Realisierung der Konzeption 'Standpunkte zu Konsequenzen aus der Entwicklung der Informatik und informationsverarbeitenden Technik für das Bildungswesen'," 1.
[622] SAPMO DY 30/J IV 2/2/2138, "Maßnahmeplan des Ministeriums für Volksbildung zur Realisierung der Konzeption 'Standpunkte zu Konsequenzen aus der Entwicklung der Informatik und informationsverarbeitenden Technik für das Bildungswesen'," 2.
[623] SAPMO DY 30/J IV 2/2/2138, "Maßnahmeplan des Ministeriums für Volksbildung zur Realisierung der Konzeption 'Standpunkte zu Konsequenzen aus der Entwicklung der Informatik und informationsverarbeitenden Technik für das Bildungswesen'," 2.
[624] SAPMO DY 30/J IV 2/2/2138, "Maßnahmeplan des Ministeriums für Volksbildung zur Realisierung der Konzeption 'Standpunkte zu Konsequenzen aus der Entwicklung der Informatik und informationsverarbeitenden Technik für das Bildungswesen'," 3–6.

enough personnel resources to develop, test, and implement as many new curricula and curricular materials as required for alle the different school levels at the same time.[625] In addition, the implementation of computer education curricula required for an appropriate qualification of teachers, and schools needed to be equipped with or have local access to computers and peripheral equipment. The Action Plan estimated that by 1990 a total of 1,600 to 1,700 microcomputers deemed suitable for use in schools would be needed for general education alone, as well as the same number of television screens and 1,000 tape recorders.[626] Prioritizing "special schools" and EOS both in terms of the implementation of new curricula as well as the equipment with microcomputers was aimed at ensuring that as soon as possible, new students entering higher education would already be equipped with a solid foundation of basic computer knowledge and skills.

According to the action plan, informatics was not meant to be introduced as a new compulsory subject in general schooling – apart from the "special schools." where the introduction of a compulsory subject of informatics in its own right was planned to provide the high performing pupils attending these schools with a more in-depth and academically oriented education in informatics.[627] This decision followed the basic premise formulated by the APW that computer science and technology content should generally be integrated into existing subjects.[628] Consequently, information technology education was to be incorporated in the polytechnic subjects ESP and PA. In addition, the introduction of elective courses on information technology in both the POS and the EOS was planned to cater to pupils with a particular interest and talent in the field of information technology and informatics.

The APW was tasked with developing and implementing the curricular materials for new computer education courses in general education in line with the Ministry for National Education's action plan. To this end, researchers and pedagogues at the APW carefully developed, tested, evaluated, and revised new curricula and instructional materials which were then to be gradually implemented in schools over the course of the following years.

---

625 BBF APW 11296/4, "Lehrplanberatungen Informatik 1985–1987."
626 SAPMO DY 30/J IV 2/2/2138, "Maßnahmeplan des Ministeriums für Volksbildung zur Realisierung der Konzeption 'Standpunkte zu Konsequenzen aus der Entwicklung der Informatik und informationsverarbeitenden Technik für das Bildungswesen'," 10.
627 SAPMO DY 30/J IV 2/2/2138, "Maßnahmeplan des Ministeriums für Volksbildung zur Realisierung der Konzeption 'Standpunkte zu Konsequenzen aus der Entwicklung der Informatik und informationsverarbeitenden Technik für das Bildungswesen'," 3.
628 BBF APW 16.315–16.316, "Thesen zur Schulinformatik," dated March 14, 1985.

5.2 A Curriculum for Computer Instruction in General Education — 177

The starting point and core of computer education in the POS was a 30-hour block course that was to be introduced in the subject ESP in grade 9. For the development of the experimental syllabus for the basic computer course in the POS, a research group was formed under the leadership of the Institute for Polytechnic Education (IPB) of the APW.[629] In addition to developing the experimental syllabus, the research group was also tasked with preparing the necessary teaching software and the corresponding teacher and student materials.

The aim of the basic computer course was to provide all pupils in general education with a uniform, basic computer education and training that would serve as a solid foundation for more in-depth and advanced computer education in the EOS and in VET.[630] The content covered the diverse fields of application in the GDR's economy where computer technology was currently and would prospectively be used, as well as its anticipated socioeconomic effects. Furthermore, pupils were to be familiarized with text and data processing software, as well as computer graphics and computerized process control, and to receive practical training on a microcomputer to solve mathematical, scientific, and technical problems.[631] For this purpose, computer labs were to be set up in the so-called "Polytechnic Centers," which were educational facilities shared by multiple local schools for the purpose of polytechnic instruction.[632]

The basic computer course thus essentially consisted in a user education, foregrounding the hands-on use of ready-made software and the manifold applications of computer technology in industry and socialist businesses. However, the syllabus also included algorithmic problem solving, and the development of simple programs.[633] Pupils were meant to acquire some basic programming skills in the language BASIC, but only to enable them to use the computer for solving specific problems as suggested in the curricular materials. Accordingly, the learning of syntax and commands was limited to the very basics that were needed to solve

---

[629] BBF APW 11125/2, "Hinweise und Erläuterungen für den Experimentallehrplan zum Grundkurs 'Informatik' als Stoffgebiet im ESP-Unterricht," dated June 10, 1988, 4.

[630] Ministerrat der DDR/Ministerium für Volksbildung, *Lehrplan der zehnklassigen allgemeinbildenden polytechnischen Oberschule: Technik, Variante mit Stoffgebiet 'Grundkurs Informatik', Klassen 9 und 10* (Berlin: Volk und Wissen, 1990), 5.

[631] BBF APW 11125/2, "Hinweise und Erläuterungen für den Experimentallehrplan zum Grundkurs 'Informatik' als Stoffgebiet im ESP-Unterricht," dated June 10, 1988, 6.

[632] BBF APW 11125/2, "Hinweise und Erläuterungen für den Experimentallehrplan zum Grundkurs 'Informatik' als Stoffgebiet im ESP-Unterricht," dated June 10, 1988, 3.

[633] Ministerrat der DDR/Ministerium für Volksbildung, *Lehrplan Variante mit Stoffgebiet 'Grundkurs Informatik'*, 14–15.

the problems in class.[634] From a broader societal perspective, it was simply deemed unnecessary that everyone learned in more detail how to program a computer. According to Volker Kempe, director of the ZKI, an estimated 95% of workers would only ever use ready-made software, while merely 5% needed to be trained as programming experts capable of developing such software.[635] In-depth programming training was therefore considered to be outside the scope of general education in the POS and was instead envisaged as part of the computer courses to be introduced in the mentioned "special schools."

The researchers and pedagogues at the APW also compared the newly drafted syllabus for a basic computer education course in general education with conceptual developments of computer education in selected socialist and capitalist countries.[636] They concluded that their own efforts were in line with the international trend, which emphasized a more application-oriented rather than theoretical approach to computer education, and foregrounded the development of algorithmic thinking and problem-solving skills.[637] The experience of other countries also indicated that the content of computer education curricula was subject to accelerated change due to rapid technological development. The research group therefore noted that it may be necessary to adapt and develop the proposed syllabus soon after its introduction to keep up with the pace of technological change in science and the world of work.[638]

The first draft of the experimental syllabus was subsequently sent out for evaluation to selected external experts in informatics, computer technology, and pedagogy at universities and teacher training colleges. Their main criticism of the draft syllabus was that the specified objectives and contents could hardly be realized in the limited time available for teaching (30 hours), because of the proposed learning content which they considered to be demanding and rather

---

634 Ministerrat der DDR/Ministerium für Volksbildung, *Lehrplan Variante mit Stoffgebiet 'Grundkurs Informatik'*, 15.
635 BBF APW 16.318–16.319, Handwritten note regarding the expertise by Prof Bormann dated February 24, 1986, dated March 11, 1986, 2.
636 In the case of the newly developed syllabus for the basic computer course, comparisons were made to similar developments in the Soviet Union, Czechoslovakia, Bulgaria, Poland, and Hungary, as well as Austria, Norway, France, and the USA (BBF APW 11125/2, "Hinweise und Erläuterungen für den Experimentallehrplan zum Grundkurs 'Informatik' als Stoffgebiet im ESP-Unterricht," dated June 10, 1988, 9).
637 BBF APW 11125/2, "Hinweise und Erläuterungen für den Experimentallehrplan zum Grundkurs 'Informatik' als Stoffgebiet im ESP-Unterricht," dated June 10, 1988, 9.
638 BBF APW 11125/2, "Hinweise und Erläuterungen für den Experimentallehrplan zum Grundkurs 'Informatik' als Stoffgebiet im ESP-Unterricht," dated June 10, 1988, 9–10.

ambitious.[639] The experts also called for an increase in the proportion of practical work on the computer. Instead of learning a programming language or simply using ready-made software, they argued that more emphasis ought to be placed on "computational thinking and working." This was seen as important because the training of programming and software user skills was ascribed a largely instrumental and pre-vocational character in preparing for the later use of a microcomputer in the workplace, whereas the process of "computational thinking" and algorithmic problem solving could more readily be characterized as having general educational value, with applications beyond the purpose-bound work on a computer.

The intention was for pupils to broadly understand programming as a "description of the solution to a task," whereby human behavior was to be taken as a starting point to design and describe workflows.[640] In this sense, any given everyday task could be described in the form of an algorithmic program flowchart (see Figure 8). Thus, in order to develop "algorithmic" thinking and problem-solving skills, pupils were to describe any given technical or mathematical workflow to solve a problem and visualize it in the form of a programming flowchart (ger.: Programmablaufplan, PAP).[641] The next step consisted in translating the problem solution into a formal programming language, which was to be understood as a "universal mediator," or in other words, simply a pragmatically chosen set of defined terms and operations that would be "understood" by a computer.[642] Learning a programming language such as BASIC was therefore not to be understood as an end in itself, but rather a means to an end; namely to translate the algorithmic problem solution into a computer program that could be processed by a microcomputer. In the development of the syllabus for the basic computer course, researchers and pedagogues at the APW thus placed the educational focus firmly on the first step, the development of an "algorithmic structure," which was intended to teach students "algorithmic skills" independently of a problem-oriented programming language, while its translation into a BASIC program was seen as a necessary but subordinate step in the development of "algorithmic" ways of thinking.[643]

---

639 BBF APW 11125/2, "Hinweise und Erläuterungen für den Experimentallehrplan zum Grundkurs 'Informatik' als Stoffgebiet im ESP-Unterricht," dated June 10, 1988, 7.
640 BBF APW 11125/2, "Hinweise und Erläuterungen für den Experimentallehrplan zum Grundkurs 'Informatik' als Stoffgebiet im ESP-Unterricht," dated June 10, 1988, 7–8.
641 APW der DDR, *Lehrbuchergänzung ESP Klasse 9* (Berlin: Volk und Wissen), 15–17.
642 BBF APW 11125/2, "Hinweise und Erläuterungen für den Experimentallehrplan zum Grundkurs 'Informatik' als Stoffgebiet im ESP-Unterricht," dated June 10, 1988, 7–8.
643 BBF APW 16.318–16.319, Handwritten note regarding the expertise by Prof Bormann dated February 24, 1986, dated March 11, 1986, 1.

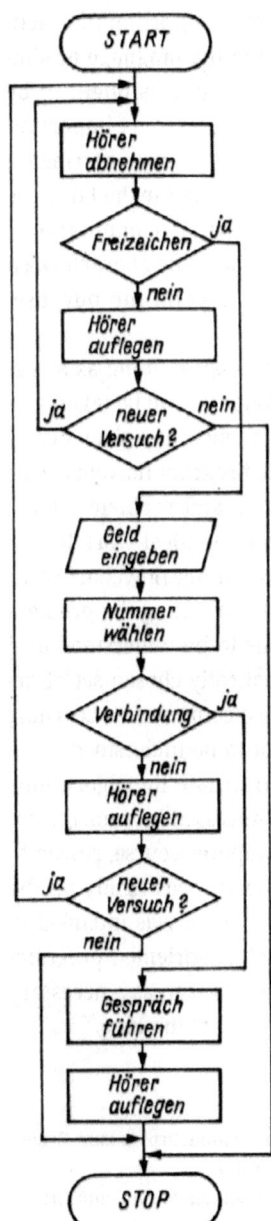

Figure 8: Example of a program flowchart for a phone call.
Source: APW der DDR, Lehrbuch-ergänzung ESP Klasse 9, 16.

In June 1987, the experimental syllabus was approved for pilot testing. In line with the premise that curricular innovations in the GDR school system had to be "scientifically sound,"[644] the pilot phase of the introduction of the new syllabus was accompanied by extensive educational research and evaluation. To assess the practical feasibility of the experimental syllabus, it was empirically tested in the 1987/88 school year in 55 computer laboratories in polytechnic centers in all districts of the GDR.[645] The school pilot program was accompanied by surveys of teacher and pupils by means of questionnaires, observations of teaching lessons, and discussions with the teachers involved. In addition, a computer program was specifically designed to test whether the students had acquired the knowledge and skills detailed in the syllabus.

The evaluation of the pilot test led to the conclusion, that the syllabus for the basic computer course was feasible in principle.[646] However, the lack of textbook supplements and methodological recommendations for classroom instruction, which were still in development, as well as the teachers' scant experience with the new teaching contents posed major obstacles in realizing the curricular goals. In particular, the work with the electronic coupling modules in the "Computerized Control" unit of the syllabus proved difficult for teachers to implement because of the technically highly demanding in-class activities and exercises, but also because of the technical difficulties that were often encountered. The teachers also complained in general about the high susceptibility of the technical equipment to malfunctions and the tedious work with software on cassettes. The microcomputers installed in the polytechnic centers were originally designed as home computers and were not intended for school use. The heavy use of the microcomputers, which were in operation for up to 14 hours a day, resulted in frequent technical breakdowns and significant downtime for repairs.[647]

The survey of students' prior knowledge revealed that 60.3% had never used a computer before. This result highlights the fact that for most children, the SED's sociotechnical imaginary of computerization had been a mere rhetoric and distant dream of the future, detached from their own reality of life. The research group interpreted this result as a validation of their efforts, as it underlined the

---

644 SAPMO DY 30/J IV 2/2/2138, "Standpunkte zu Konsequenzen aus der Entwicklung der Informatik und informationsverarbeitenden Technik für das Bildungswesen," 4.
645 BBF APW 11125/2, "Hinweise und Erläuterungen für den Experimentallehrplan zum Grundkurs 'Informatik' als Stoffgebiet im ESP-Unterricht," dated June 10, 1988, 10.
646 BBF APW 11125/2, "Hinweise und Erläuterungen für den Experimentallehrplan zum Grundkurs 'Informatik' als Stoffgebiet im ESP-Unterricht," dated June 10, 1988, 13.
647 BBF APW 11125/2, "Hinweise und Erläuterungen für den Experimentallehrplan zum Grundkurs 'Informatik' als Stoffgebiet im ESP-Unterricht," dated June 10, 1988, 15.

importance of a compulsory basic computer course so that all pupils could acquire at least some fundamental skills and experience in the use of computers.[648] The basic computer course meant for many children that they could see a computer up close for the first time and even be able to use it practically, which was expected to help them to develop a positive attitude towards computers.[649] Moreover, the researchers noted that the computer course fostered a number of desirable personality traits and behaviors in children, that were considered important for working with computer technology: Reliability, accuracy, perseverance, the ability to concentrate, as well as a disciplined and careful handling of hardware and software. However, only half of the pupils were able to use the computer and to solve simple tasks in a fully autonomous way after completing the course. They struggled with problem analysis and developing algorithms, whereas the implementation of their solutions in the programming language BASIC hardly caused any issues. Nevertheless, the pilot test was considered a success and it was decided to introduce the experimental syllabus for the basic computer course as part of the 9th grade ESP lessons starting in September1989.

The newly introduced computer course in the subject ESP provided pupils with some basic computer knowledge and skills and taught them about the various ways in which computer technology was used in the GDR economy. But educators in the GDR were convinced that, to fully grasp the extent to which new information technologies were changing economic production, pupils needed to experience it first-hand. Only if they could take part in practical work using new computer technology, solving real problems, and producing real goods, would they gain a deeper understanding of the importance of computers and microchip technology in modernizing economic production and increasing productivity. The real experience of computer technology in action was thus expected to convey the SED's sociotechnical imaginary of the computer as a powerful means of economic and social progress in a much more convincing and impressive way than simply hearing about the promises of a desired computerized future in class or writing a computer program for practice only. This was precisely the function of the school subject PA. As part of the polytechnic education of students in the GDR, the PA classes added a practical component to the subject of ESP, as students were assigned to a workplace in industry, agriculture, or construction to assist and work alongside workers under genuine production conditions. Already in 1983, new curricula for the subject PA had been introduced for grades 9 and 10,

---

648 BBF APW 11125/2, "Hinweise und Erläuterungen für den Experimentallehrplan zum Grundkurs 'Informatik' als Stoffgebiet im ESP-Unterricht," dated June 10, 1988, 11.
649 BBF APW 11125/2, "Hinweise und Erläuterungen für den Experimentallehrplan zum Grundkurs 'Informatik' als Stoffgebiet im ESP-Unterricht," dated June 10, 1988, 12–13.

specifically about the metalworking and electronics industry, which specified the deployment of pupils in workplaces involving automated machines and systems. The curricular reforms in the second half of the 1980s continued this line of development and were aimed at updating the PA syllabus to account for the ongoing and prospective digital transformation of workplaces in economic production. Thus, a syllabus for a new transversal field of work entitled "Information processing and computer technology" was designed and scheduled to be implemented in September 1989.[650]

The syllabus would apply to all pupils in grades 9 and 10, regardless of whether their work assignment was in industry, agriculture, or construction. The general goals and tasks of PA remained the same, but the work unit on "Assistance in the use of automated machine tools and systems" was expanded. Before, this work unit had comprised of tasks at automated workplaces in direct production, for example automatic process control and the work with CNC-machines. The addition of the new transversal work area to the syllabus was intended to also allow for the deployment of pupils in a wider variety of workplaces where information technology was used, including computerized workstations in all phases of the production process, as well as automated machines and industrial robots, to familiarize them with a wider range of its applications and functions throughout the entire operational process in a company.[651] However, while pupils were meant to gain some work experience and insights into the reality of work in socialist production as part of their education, the role of the POS was not to provide them with any sort of specialized vocational education and training.[652] Instead of providing pupils with an in-depth training to work on computerized machines, the main goal was a moral-ideological one: To familiarize pupils with a core process of the scientific-technological revolution – the transfer of informational processes to devices, machines and systems – and thereby to inculcate and reinforce the SED's sociotechnical imaginary of computer technology.[653] The aim was for the chil-

---

[650] BBF APW 11125/2, "Hinweise und Erläuterungen für das übergreifende Arbeitsgebiet 'Informationsverarbeitung und Rechentechnik' in der produktiven Arbeit der Schüler der Klassen 9 und 10 in Kombinaten und Betrieben der Industrie, des Bauwesens und der Landwirtschaft," dated June 10, 1988, 1.
[651] Gottfried Loos, "Produktive Arbeit an informationsverarbeitender Technik," *Polytechnische Bildung und Erziehung* 29, no. 4 (1987): 124; Günter Rummler, "Heranführen der Schüler an die automatisierte Produktion," *Polytechnische Bildung und Erziehung* 28, no. 11 (1986): 403.
[652] Rummler, "Heranführen der Schüler," 403.
[653] BBF APW 11125/2, "Hinweise und Erläuterungen für das übergreifende Arbeitsgebiet 'Informationsverarbeitung und Rechentechnik' in der produktiven Arbeit der Schüler der Klassen 9 und 10 in Kombinaten und Betrieben der Industrie, des Bauwesens und der Landwirtschaft," dated June 10, 1988, 3.

dren to experience vividly and at first hand the use of microelectronics and modern information processing technology in real-life economic production, while at the same time being made to believe that these new technologies were accelerating economic growth, and would continue to do so over the next decades, thereby strengthening the performance of the national economy.[654]

The development of the new syllabus in PA for grades 9 and 10 followed a similar pattern as the basic computer course. However, as the subject of PA also involved pre-vocational aspects, representatives from combines and businesses in industry, construction, and agriculture were included in the efforts to design the new syllabus. A draft was sent out for review to the ministries of the various economic sectors, the State Secretariat and the Central Institute for Vocational Education, the Central Council of the FDJ,[655] as well as the Central Executive Committee of the Education Trade Union.[656] In their response, the consulted experts criticized the overemphasis on training for work skills. Both the representatives of the State Secretariat and the Central Institute for Vocational Education argued that the objective of lessons in PA was not to practice vocational skills or to qualify pupils to carry out specific work tasks and jobs independently, but rather to contribute to their general education and socialist upbringing.[657]

Industry representatives demanded that pupil assignments at workplaces involving information technology would be required to last for up to one year, as these work tasks were more demanding and required a longer initial training period during which pupils had to be closely supervised, thus resulting in a higher workload for the staff.[658] However, this request was denied by the pedagogues at

---

[654] BBF APW 11125/2, "Hinweise und Erläuterungen für das übergreifende Arbeitsgebiet 'Informationsverarbeitung und Rechentechnik' in der produktiven Arbeit der Schüler der Klassen 9 und 10 in Kombinaten und Betrieben der Industrie, des Bauwesens und der Landwirtschaft," dated June 10, 1988, 3.

[655] Free German Youth (ger.: Freie Deutsche Jugend, FDJ). The FDJ was the Socialist Unity Party's official youth movement in the GDR for young people between the ages of 14 and 25. Its main goal was for the GDR's political leadership to exert political and ideological influence on the youth.

[656] BBF APW 11125/2, "Hinweise und Erläuterungen für das übergreifende Arbeitsgebiet 'Informationsverarbeitung und Rechentechnik' in der produktiven Arbeit der Schüler der Klassen 9 und 10 in Kombinaten und Betrieben der Industrie, des Bauwesens und der Landwirtschaft," dated June 10, 1988, 4.

[657] BBF APW 11125/2, "Hinweise und Erläuterungen für das übergreifende Arbeitsgebiet 'Informationsverarbeitung und Rechentechnik' in der produktiven Arbeit der Schüler der Klassen 9 und 10 in Kombinaten und Betrieben der Industrie, des Bauwesens und der Landwirtschaft," dated June 10, 1988, 9.

[658] BBF APW 11125/2, "Hinweise und Erläuterungen für das übergreifende Arbeitsgebiet 'Informationsverarbeitung und Rechentechnik' in der produktiven Arbeit der Schüler der Klassen 9

the APW, who emphasized the need for pupils to switch work positions regularly so they would experience the contrast of working both in computerized workplaces, as well as more traditional workplaces that did not involve the use of information technology. Gaining an understanding of how traditional technology functioned was seen as foundational in order for pupils to understand how these modern information processing systems and computerized machines worked, which had now taken over some of the tasks in the production process.[659] In addition, gaining an understanding of traditional, non-automated production was meant to help students appreciate how "revolutionary" the use of microelectronics and information technology in production was, and how conventional measurement and control technology was gradually being replaced by computer-based automation systems, and thereby making the production process less immediately tangible to the human senses.[660] For this reason, it was seen as important that pupils developed a "mental model" of the invisible physical and informational processes that took place inside the automated systems as part of their ESP instruction, before they worked on such machines during their PA lessons.

Opinions also differed as to which tasks involving computer technology could be carried out by pupils, and which should be left to skilled workers and specialists. The reviewers of the draft syllabus were especially concerned by the fact that the syllabus mentioned that pupils would be working autonomously on computerized systems and machines. Instead, they recommended to change the wording of the recommended activities of pupils to "helping" or "assisting" if the work involved computer technology.[661] Such concerns were addressed by a paragraph that was added to the syllabus which stated that the respective tasks were recommended for PA work assignments of pupils in the case that the specific conditions in the company permitted it.[662] In this way, the companies were given a certain amount of leeway to decide for themselves which workplaces the pupils were as-

---

und 10 in Kombinaten und Betrieben der Industrie, des Bauwesens und der Landwirtschaft," dated June 10, 1988, 16.
659 Klaus-Dieter Brömmel, "Schüler an die moderne Technik – Aspekt meiner Leitungstätigkeit," *Polytechnische Bildung und Erziehung* 29, no. 1 (1987): 11.
660 Rummler, "Heranführen der Schüler," 405.
661 BBF APW 11125/2, "Hinweise und Erläuterungen für das übergreifende Arbeitsgebiet 'Informationsverarbeitung und Rechentechnik' in der produktiven Arbeit der Schüler der Klassen 9 und 10 in Kombinaten und Betrieben der Industrie, des Bauwesens und der Landwirtschaft," dated June 10, 1988, 9–10.
662 BBF APW 11125/2, "Hinweise und Erläuterungen für das übergreifende Arbeitsgebiet 'Informationsverarbeitung und Rechentechnik' in der produktiven Arbeit der Schüler der Klassen 9 und 10 in Kombinaten und Betrieben der Industrie, des Bauwesens und der Landwirtschaft," dated June 10, 1988, 15.

signed to, and which machines or systems were off-limits to them. When it came to the decision, which pupils were assigned to work with the most modern and sophisticated kinds of technology, prior skills and knowledge mattered greatly. It was usually the more talented and tech-savvy pupils that were selected to work with the modern, intellectually demanding technical systems and robots. This differentiation in familiarizing pupils with new technology in the workplace according to their prior knowledge, skills and talents was not considered as inherently problematic within the GDR's unified school system, but rather as an important part of promoting technically gifted pupils, to "achieve excellence in the field of scientific and technological progress."[663]

As more and more production systems in industry were automated, the portion of manual activities in the production process that could be carried out by pupils declined. During their PA lessons, pupils therefore would also have to learn that productive work increasingly not only encompassed manual labor, but also intellectual work in the form of monitoring and controlling automated production lines. It was noted, however, that not all work assignments and tasks offered the same quality of learning conditions for pupils. Computer workplaces upstream, downstream, or adjacent to the direct production process, such as in research and development, project planning and design, quality assurance and control, logistics, maintenance, as well as in operational management, administration and accounting were considered by the APW pedagogues to offer more favorable conditions for the deployment of pupils than workplaces on automated machines in the immediate production process itself.[664] The latter often involved monotonous and intellectually less demanding activities that offered little opportunity for the pupils to develop the desired algorithmic thinking skills and a deeper understanding of information processing technology. Also, such tasks did not sit well with the sociotechnical imaginary that the political leadership and the pedagogues in the GDR sought to convey to the pupils. The often-repeated vision of a society in which machines took over monotonous and physically demanding work to set workers free to engage in creative, more intellectually demanding tasks conflicted with the reality of industrial production that still involved a significant share of uninspiring, repetitive, and tedious work. While it would have been convenient for workers to offload some of these tasks onto the pupils, pedagogues demanded that pupils were also engaged in intellectually and technically more challenging activities.[665] Varied and exciting tasks on modern computerized

---

663 Brömmel, "Schüler an die moderne Technik," 12.
664 Loos, "Produktive Arbeit," 125.
665 This demand conflicted with the standpoint taken by representatives of certain companies that some of the activities listed in the draft syllabus, such as setting up and maintaining computerized

machines were, of course, far better suited to demonstrate and reinforce the political narrative and sociotechnical imaginary of an enthralling and inspiring "scientific-technological revolution" driven by computer and automation technology. After all, the students were to be convinced of the SED's promise that the new technology would bring sweeping social and economic progress, and therefore, confronting them with the full reality of fragmented modernization with ambivalent effects on working conditions would only tarnish the bright sociotechnical imaginary of socialist computerization.

Pilot testing of the new syllabus took place in 62 companies in industry, construction, and agriculture. The main issue was that the availability of workplaces that were equipped with new information technology varied greatly between different companies. In some industries, such as chemicals, textiles and clothing, the degree of automation was much higher than in the metalworking industry, electrical engineering, and electronics.[666] Across all industries and economic sectors, computer-assisted workplaces were particularly widespread in operational management, planning and accounting.[667] These workplaces generally involved the work on microcomputers in offices and computer labs. In these workplaces, as laid out in the newly introduced computer education curricula, pupils were not meant to develop their own programs, but rather work with existing user software for text and data processing, graphics programs and solving calculation tasks. This required for pupils to have at least some basic computer and software user skills.[668] It was thus recommended for companies to mainly assign pupils in grade 10 to these computer workplaces, as they had already completed the 9th grade basic computer course and were not expected to require any further preparatory courses.[669] However, the computers used

---

machines and systems, and entering data and programs, should be reserved for specialists (BBF APW 11125/2, "Hinweise und Erläuterungen für das übergreifende Arbeitsgebiet 'Informationsverarbeitung und Rechentechnik' in der produktiven Arbeit der Schüler der Klassen 9 und 10 in Kombinaten und Betrieben der Industrie, des Bauwesens und der Landwirtschaft," dated June 10, 1988, 15).

666 BBF APW 11125/2, "Hinweise und Erläuterungen für das übergreifende Arbeitsgebiet 'Informationsverarbeitung und Rechentechnik' in der produktiven Arbeit der Schüler der Klassen 9 und 10 in Kombinaten und Betrieben der Industrie, des Bauwesens und der Landwirtschaft," dated June 10, 1988, 6.

667 BBF APW 11125/2, "Hinweise und Erläuterungen für das übergreifende Arbeitsgebiet 'Informationsverarbeitung und Rechentechnik' in der produktiven Arbeit der Schüler der Klassen 9 und 10 in Kombinaten und Betrieben der Industrie, des Bauwesens und der Landwirtschaft," dated June 10, 1988, 14.

668 Ministerrat der DDR/Ministerium für Volksbildung, *Arbeitsbereich "Informationsverarbeitung und Rechentechnik" für die produktive Arbeit der Schüler der Klassen 9 und 10* (Berlin: Volk und Wissen, 1989).

669 BBF APW 11125/2, "Hinweise und Erläuterungen für das übergreifende Arbeitsgebiet 'Informationsverarbeitung und Rechentechnik' in der produktiven Arbeit der Schüler der Klassen 9

in industry and offices were very different from the KC 85 microcomputers used in schools: They had a different keyboard, used different commands, were more powerful and could therefore perform a wider range of tasks, and used floppy disks to store and retrieve files from external storage, as opposed to the magnetic tape cassettes that pupils used in school.[670]

In September 1989, the curricula for PA in grades 9 and 10 were supplemented by the newly developed syllabus for work deployments in the field of "Information processing and computer technology." While the syllabus suggested a wide variety of activities for pupils involving the use of new information technology in workplaces, it was left to the responsible authorities of the individual companies to decide which work tasks could be at least partially delegated to pupils.[671] Particular importance was attached to the inculcation of a sense of high appreciation for the "high material and idealistic value" of workplaces involving modern computer technology. Pupils should be taught to handle hardware and software with the utmost care, to work conscientiously, with concentration and discipline, and to meticulously follow the rules and instructions given.[672]

The mandatory basic computer education course in the subject ESP, as well as the new syllabus in PA which allowed for the assignment of pupils to workplaces involving modern computer technology, were meant to equip all pupils with a uniform level of basic computer knowledge and skills as part of their general education in the POS. In 1988, the APW reported that 81.7% of pupils stated that they had used computer technology before: 41.6% in computer clubs and extracurricular study groups, 33.6% in the subject ESP, and 24.8% at home. Additionally, 44% had mainly worked with ready-made user software, another 18.9% had at least modified ready-made software in some way, and the 34.9% had even developed their own small programs.[673] The familiarization with computer technology among pupils in grade 9 and 10 was thus by no means "uniform," both in quantitative and

---

und 10 in Kombinaten und Betrieben der Industrie, des Bauwesens und der Landwirtschaft," dated June 10, 1988, 10.
670 BBF PL 89–15–51, "Erfahrungen und Ergebnisse des Einsatzes der Schüler der 9. und 10. Klassen im Arbeitsgebiet 'Informationsverarbeitung und Rechentechnik' in der Differenzierungsrichtung metallverarbeitende Industrie," dated 1989, 7.
671 Ministerrat der DDR/Ministerium für Volksbildung, *Arbeitsbereich "Informationsverarbeitung und Rechentechnik,"* 2.
672 Ministerrat der DDR/Ministerium für Volksbildung, *Arbeitsbereich "Informationsverarbeitung und Rechentechnik,"* 2.
673 BBF APW 11125/2, "Hinweise und Erläuterungen für das übergreifende Arbeitsgebiet 'Informationsverarbeitung und Rechentechnik' in der produktiven Arbeit der Schüler der Klassen 9 und 10 in Kombinaten und Betrieben der Industrie, des Bauwesens und der Landwirtschaft," dated June 10, 1988, 14.

qualitative terms. As polytechnic centers were only gradually equipped with microcomputers, not all schools were able to implement the new basic computer course just yet, which prospectively would provide all pupils with at least a basic level of hands-on computer experience. At the time, however, most pupils still acquired computer knowledge and skills outside of mandatory school instruction, in particular with regard to more advanced computing that involved working on the development of their own computer programs.

The GDR's Minister of National Education, Margot Honecker, stated in an interview with Radio DDR in 1987 that the intention behind the curricular reforms was not to turn *all* children into computer experts.[674] While she acknowledged the great importance of computer technology for the further development of GDR society and economy, she adamantly insisted that "[. . .] we are not living in the computer age [. . .]. We are living in an age when we are continuing to build and shape socialist society, a society in which the decisive role is played by the human being."[675] In her opinion, a basic level of computer education had a legitimate place in general education, but its importance was not to be overemphasized compared to other subjects. It was considered necessary for only a fraction of all pupils to acquire an advanced level of computer knowledge and skills and be put on the educational path to become computer experts. Crucially, then, the considerable differences in young people's computer experience and knowledge were not an accidental phenomenon but were to a certain extent deliberate and even encouraged through the GDR's education policy.

One fundamental factor in differentiating general education was anchored in the structure of the general GDR education system, namely the continuation of the POS through the two grades of the EOS for selected pupils. While the majority of pupils entered vocational training after the POS, a proportion of about 10% spent two more years at the extended polytechnic secondary school (EOS) to graduate with the Abitur, which allowed them to enter higher education.[676] Consequently, pupils in EOS (as well as in "special schools") were to receive a more extensive informatics and programming instruction, which would prepare them for their subsequent studies in higher education, regardless of the subject they would go on to study.[677] Thus,

---

674 "Was in der DDR getan wird, um eine kluge, vorwärtsdrängende Jugend heranzubilden," *Neues Deutschland*, December 21, 1987, 3.
675 "Was in der DDR getan wird, um eine kluge, vorwärtsdrängende Jugend heranzubilden," *Neues Deutschland*, December 21, 1987, 3 (translated from German).
676 Helmut Köhler and Manfred Stock, *Bildung nach Plan? Bildungs- und Beschäftigungssystem in der DDR 1949 bis 1989* (Wiesbaden: Springer Fachmedien, 2004), 39.
677 BBF APW 11828/1, "Hinweise und Erläuterungen zum Lehrplan für den obligatorischen Unterricht in den Klassen 11 der Erweiterten Oberschulen im Fach Informatik," dated September 1988, 2.

while the basic computer course in the POS aimed to provide a general understanding of the new computer technology for the vast majority of pupils who were imagined as computer *users* in the future computerized GDR, informatics instruction in the EOS was designed to set a smaller proportion of pupils on the path to *mastering* or even *further advancing* computer technology as part of their future careers. It put more emphasis on fundamental concepts of algorithm theory as well as to solve problems independently using self-developed programmes and ready-made software.[678]

To this end, it was decided in March 1987, to also introduce an additional, mandatory *subject* of "Informatics" into grade 11 of the EOS, with two hours of instruction per week.[679] This conception clearly deviated from the APW's principle of not introducing informatics as an independent subject in general education schools. The new subject was meant to be gradually implemented between 1988 and 1990 in every EOS as soon as it had access to the required technical equipment and qualified teaching staff.[680] The course was intended to build upon the mandatory basic informatics course in grade 9 of the POS, allowing for every pupils that graduated and would enter higher education to have received a minimum of 90 hours of informatics instruction in total.[681] However, the planned new subject was not part of the prospective curriculum as it had been envisioned in the Politburo's decision of 1985, and therefore did not feature in the APW's predefined research program for the second half of the 1980s. As a result, the APW struggled with the tight schedule and insufficient human resources to test the experimental syllabus and develop the necessary curriculum materials until the planned implementation in 1988.[682] A new syllabus for the subject "Informatics" was eventually introduced in

---

678 Ministerrat der DDR/Ministerium für Volksbildung, *Experimentallehrplan Informatik Klasse 11* (Berlin: Volk und Wissen, 1989), 5–6.
679 BBF APW 11828/2, "Auszüge aus den Vorschlägen zur Weiterentwicklung des Inhaltskonzepts und der Bedingungen für die Abiturstufe vom 3.3.1987," undated, 1–2; BBF APW 11296/4, "Bericht über Stand und Probleme bei der Realisierung des Informatikbeschlusses anlässlich der Kontrollberatung beim Präsidenten der APW am 4.5.1987," dated April 29, 1987, 5.
680 BBF APW 11296/4, "Bericht über Stand und Probleme bei der Realisierung des Informatikbeschlusses anlässlich der Kontrollberatung beim Präsidenten der APW am 4.5.1987," dated April 29, 1987.
681 If, as planned by the APW, all the new curricula were implemented and all schools equipped with computers by the mid-1990s, all pupils were to receive 30 hours of basic computer instruction in grade 9 of the POS, and 60 hours of computer instruction for those attending the EOS. In addition to this compulsory minimum of 90 hours, pupils were free to take up to 160 hours of elective courses in new information technology in grades 9 to 12.
682 BBF APW 11296/4, "Bericht über Stand und Probleme bei der Realisierung des Informatikbeschlusses anlässlich der Kontrollberatung beim Präsidenten der APW am 4.5.1987," dated April 29, 1987, 5.

1989, albeit on an experimental basis.[683] The largest share of lessons in the subject covered the basics of using a computer, and specifically algorithms, for problem solving (28 hours). The remainder of the course involved practical problem solving on the computer, with students developing their own programs (18 hours) and using ready-made problem-oriented software for word processing, data processing, computer graphics, and process automation (14 hours). The syllabus for the subject "Informatics" thus reflected a conceptualization of the computer as the central working tool of informatics, and programming as its central operating principle and as the basis for the "universality" of new computer technology.[684] The latter was to be demonstrated to the pupils in the context of extensive practical training on the computer.

In addition to the EOS, elective courses were another key factor in providing differentiated computer education for students with higher levels of academic achievement and a specific interest in new information technologies. The 1980s saw a significant expansion of the provision of elective courses in grades 9 and 10 of the POS.[685] This concerned mainly subjects related to "key technologies" such as computing, which were considered highly relevant to the SED's political course of mastering the scientific-technological revolution. Elective courses offered a way to include new contents into general schooling much faster than through curricular revisions in mandatory subjects.[686] Moreover, they allowed for a certain degree of flexibility in the GDR's system of general schooling, by introducing a higher degree of differentiation into the POS, which was otherwise geared to uniformity. The fundamental educational objective behind elective courses was to identify and foster individual talents and inclinations among the pupils and promote their aspirations to pursue a vocational career or a higher education in the specific field.[687] In this sense, the non-mandatory computer courses in schools were designed to encourage high achievers to follow the career path of becoming the computer experts urgently needed to implement the SED's ambitious computerization plans.[688]

---

683 Ministerrat der DDR/Ministerium für Volksbildung, *Experimentallehrplan Informatik Klasse 11*.
684 BBF APW 11828/2, "Stellungnahme zum Entwurf des Lehrplanes Informatik für die Klasse 11 der POS," dated May 21, 1987, 2.
685 Oskar Anweiler, *Schulpolitik und Schulsystem in der DDR* (Opladen: Leske+Budrich, 1988), 121.
686 The first elective courses in microelectronics and computing were introduced in selected POS as early as 1984 by committed teachers who were enthusiastic about computer technology (see, for example, BBF PL 87-2-10, "Der Einsatz von Computern bei der selbständigen Erarbeitung von Programmen in BASIC und Maschinensprache im fakultativen Unterricht," dated 1987).
687 Anweiler, *Schulpolitik und Schulsystem*, 121.
688 BBF PL 87-06-12, "Ergebnisse und Probleme in der Arbeit mit Schülern der Klassen 9 und 10 im Rahmen des fakultativen Unterrichts (FK/R) in Informatik," dated 1987, 6.

In the POS, two elective courses, "Informatics"[689] and "Information Processing – Process Automation,"[690] were developed and included in the National Ministry of Education's framework program for non-mandatory courses in informatics for grades 9 and 10 in 1989. With regard to the EOS, the APW also prepared curriculum materials for two different elective courses: A course entitled "Development of Programs,"[691] and the course "Microcomputing Technology in Socialist Production."[692] This way, both in the POS and the EOS, elective courses on new information technology were offered to pupils with, on the one hand, a more mathematical and algorithmic orientation (courses "Informatics" and "Development of Programs"), and on the other, a more polytechnical orientation (courses "Information Processing – Process Automation" and "Microcomputing Technology in Socialist Production"). As the interest in these courses often exceeded their capacity, the teachers had to select who would be admitted to the course. Typically, they chose students who had good academic records and wanted to further their education in new computer and automation technology as part of their future vocational training or higher education.[693] The elective courses were designed more openly than mandatory school instruction and required less strict orientation to a fixed syllabus. They provided ample space for the students' project work based on their individual interests and talents. However, this also placed high demands on the course instructors, who had to have a very broad knowledge and in-depth experience in computing to guide and advise the pupils in their various projects.

The division between mathematically oriented and algorithm-centered courses on computer technology reflected the dual nature of the nascent discipline of computing in the GDR during the 1980s, rooted in both mathematics and engineering (see Table 5). It was also reflected in the distribution of roles and responsibilities within the APW in the development of the new curricula for computer education. The main institutes involved were the Institute for Polytechnic Education (ger.: Institut für Polytechnische Bildung, IPB) and the Institute for Mathematics and Science Education

---

[689] Ministerrat der DDR/Ministerium für Volksbildung, *Rahmenprogramm für den fakultativen Kurs Informatik in den Klassen 9 und 10* (Berlin: Volk und Wissen, 1989).
[690] Ministerrat der DDR/Ministerium für Volksbildung, *Rahmenprogramm für den fakultativen Kurs Informationsverarbeitung – Prozessautomatisierung in den Klassen 9 und 10* (Berlin: Volk und Wissen, 1989).
[691] APW der DDR, *Experimentallehrplan für den fakultativen Unterricht in Informatik in der erweiterten Oberschule: Lehrgang Entwickeln von Programmen* (Berlin: Volk und Wissen, 1988).
[692] BBF APW 11828/1, "Protokoll über die Beratung der Leitgruppe Informatik am 27.1.1989," dated February 13, 1989, 4.
[693] The selection of participants was at the discretion of the teachers. Some teachers gave preference to pupils who wanted to study informatics, mathematics, and physics, or who wanted to become teachers of mathematics, physics, and polytechnics (see, for example BBF PL 88–10–19, "Unsere Erfahrungen beim Aufbau und der Nutzung eines Computerkabinetts zur Erprobung des Lehrgangs 'Informatik' im Rahmen des fakultativen Unterrichts der Klassen 11 und 12," dated 1988, 4).

(ger.: Institut für Mathematischen und Naturwissenschaftlichen Unterricht, IMN). The IPB was responsible for the development and implementation of the basic computer course in grade 9, the curricular revisions in the subjects ESP, PA and WPA, as well as the elective courses "Information Processing – Process Automation" in grades 9 and 10, "Microcomputer Technology in Socialist Production" in grade 12. The IMN, by contrast, was only responsible for the elective courses "Informatics" in grades 9 and 10, and "Development of Programs" in grade 12, which put more focus on algorithmic thinking and problem solving. The later decision to also introduce informatics as a mandatory subject in grade 11, thus, posed a welcome opportunity for the IMN to strengthen its role in shaping the computer education in general schooling.

Table 5: Overview of computer education in general schooling in the GDR by the end of the 1980s (POS: grades 9 and 10; EOS: grades 11 and 12).

Grade	Mandatory education	Elective courses[694]
12	Work projects at computerized workplaces in the subject "Scientific-Practical Work"[695]	*Development of programs* (mathematically/algorithm-oriented, 50 hours)
		Microcomputer Technology in Socialist Production (technically oriented, 50 hours)
11	Mandatory subject *Informatics* (60 hours)	
10	Cross-thematic teaching about new information technology in ESP and PA	*Informatics* (mathematically oriented)
9	Mandatory basic computer education block course *Informatics* in ESP (30 hours)	*Information processing – Process automation* (technically oriented)
	Transversal field of work *Information processing and computer technology* in PA	Each course consisted of 100–110 hours, spread over two school years (grades 9 and 10).)

Sources: BBF APW 11828/2, "Auszüge aus den Vorschlägen zur Weiterentwicklung des Inhaltskonzepts und der Bedingungen für die Abiturstufe vom 3.3.1987," undated; BBF APW 11828/3, "Erste Erfahrungen aus der Erprobung des obligatorischen Informatikunterrichts in Klasse 11," undated, 1

---

[694] Before the introduction of the mandatory subject of 'Informatics' in 1988, two elective courses were offered in selected EOS on an experimental basis, spanning over two years (grades 11 and 12): *Informatics* (110h) and *Microcomputer Technology in Socialist Production* (110h) (BBF APW 11828/2, "Auszüge aus den Vorschlägen zur Weiterentwicklung des Inhaltskonzepts und der Bedingungen für die Abiturstufe vom 3.3.1987," undated, 3).

[695] Scientific-Practical Work (ger.: Wissenschaftlich-Praktische Arbeit, WPA) was a practical polytechnic subject in EOS designed to familiarize pupils with the working reality of socialist production, corresponding to PA in the POS.

The two compulsory informatics courses in grades 9 and 11 were meant to provide all pupils with a common minimum level of computing knowledge and algorithmic problem-solving skills, but both also included a section on computer-aided process control and automation. It seemed important to the pedagogues at the APW to familiarize pupils with a wide range of possible applications of new computer technology as it was envisioned in the GDR, both in terms of a computing machine using algorithms to solve numerical problems, and as a programmable machine to automate and control industrial production processes.

The binding curricula for compulsory computer science teaching were scheduled to come into force in September 1989 and replace the previous experimental curricula.[696] However, at the beginning of 1989, 60% of the EOS and only about 15% of the polytechnic centers were equipped with microcomputers.[697] Thus, due to the incomplete equipment of schools with computers, in early 1989 70% of all pupils in class 11 of the EOS were taught in compulsory subject of informatics, but only 30% of all pupils in class 9 of the POS were able to take the basic computer science course as part of their ESP instruction.[698] It was planned to complete the equipment of all EOS with 7 microcomputers each by 1990, and of all polytechnic centers with 9 computers each by 1993.[699]

5.3 Teacher Training for Computer Education

The introduction of new curricula to introduce computer technology into general schooling not only required a huge effort to equip all schools with computers, but also to train a large number of teachers for the new subject content and the use of the new technology. In the GDR, secondary school teachers (grades 5–12) were trained both at universities of teacher education (ger.: Pädagogische Hochschulen) and other universities. Each institution only offered a limited number of predefined subject combinations. At the Dresden University of Teacher Education, for example, mathematics teachers were only trained in the subject combinations of mathematics and physics or mathematics and geography. At other insti-

---

696 BBF APW 11828/3, "Erste Erfahrungen aus der Erprobung des obligatorischen Informatikunterrichts in Klasse 11," undated, 1.
697 BBF APW 11828/1, "Protokoll über die Beratung der Leitgruppe Informatik am 27.1.1989," dated February 13, 1989, 7.
698 BBF APW 11828/1, "Protokoll über die Beratung der Leitgruppe Informatik am 27.1.1989," dated February 13, 1989, 2–4.
699 BBF APW 11828/1, "Protokoll über die Beratung der Leitgruppe Informatik am 27.1.1989," dated February 13, 1989, 3–4.

tutions, mathematics could also be combined with other subjects.[700] In 1982, a fifth year was added to the previous four-year diploma teacher study program course, mainly to accommodate the extended school placement for student teachers, which provided for a longer period of practical training in the classroom.[701] After five years of study, students graduated with the academic degree of diploma teacher for their specific subjects, which allowed them to teach both at POS or EOS.

In the GDR, the newly introduced courses in informatics and computer education in general schooling were mainly taught by teachers of mathematics, polytechnic instruction and, to a lesser extent, physics.[702] These different disciplinary backgrounds of the educators who taught informatics and computer education in schools also played a role in shaping the subject profile of the new learning content. The more mathematically oriented courses were predominantly taught by mathematics teachers, the technically oriented elective courses by teachers for polytechnic instruction.

Already since 1971, diploma teachers for mathematics were taught the basics of informatics and computer technology as part of their studies, more than a decade before informatics was even introduced into school curricula.[703] The initial training of mathematics teachers had helped to acquaint them with new computer technology and the basic concepts of informatics, but had not prepared them for the specific requirements of teaching computer education in schools according to the new curricula, which were introduced in the second half of the 1980s. Both, in terms of content and in terms of the didactic design of computer science lessons, the previous training was no longer aligned with the new requirements and for many teachers it was several years in the past without having been applied in their teaching.

In 1986, preparation began for the first teachers to participate in testing the new curricula in selected schools. Because of the ambitious schedule, this had to be done at the same time as the new curriculum materials were being prepared. As a result, there was little time for thorough training and in-depth familiariza-

---

700 Steffen Friedrich and Bettina Timmermann, "Lehrerbildung Informatik – Basis für die Informatik als Allgemeinbildung," in *Informatik in der DDR – Grundlagen und Anwendungen*, ed. Birgit Demuth (Bonn: Gesellschaft für Informatik, 2008), 201.
701 Thomas Arnold, "Lehrkräfte(aus)bildung im Deutschland des 20. Jahrhunderts – Kontinuitäten und Brüche," *Seminar*, no. 3 (2021): 12.
702 BArch DR 201/572, "Einschätzung zur Realisierung des Programms der Qualifizierung von Polytechniklehrern auf dem Gebiet Informatik und Informationstechnik (vom Januar 1987)," dated November 1987, 2.
703 Immo O. Kerner, *Studienmaterial zur Informatikausbildung für Lehrerstudenten* (Dresden: Pädagogische Hochschule K. F. W. Wander Dresden, 1988), preface.

tion of teachers with the new curriculum requirements and materials. For this purpose, short training courses in informatics and computer courses were organized, drawing upon locally available training resources at universities, technical schools, in combines and enterprises, as well as courses of the GDR's chamber of technology (ger.: Kammer der Technik, KdT) and vocational training courses.[704] However, most of the computer courses on offer focused on imparting computing and programming skills, but were not tailored to the specific needs of teachers, or the contents of the new school curricula in informatics. Thus, the universities for teacher education in Erfurt and Güstrow set up one-week training courses for ESP teachers, those in Potsdam and Dresden offered courses in informatics for mathematics teachers. A one-week crash course, however, soon proved inadequate. For this reason, the universities offered an additional second one-week course starting in 1987 for a more advanced preparation of teachers.[705]

Responsible for setting up and coordinating local further training measures were the "District Cabinets for Education and Training of Teachers and Educators" (ger.: Bezirkskabinette für Unterricht und Weiterbildung der Lehrer und Erzieher, BUW).[706] The central body to support the BUW's efforts was the Central Institute for Further Training of Teachers and Educators (ger.: Zentralinstitut für Weiterbildung der Lehrer und Erzieher, ZIW). The ZIW functioned as a link between the APW's work to revise and design new curricula, and the BUWs, who were tasked with preparing teachers for the implementation of new curricular materials.[707] The ZIW delegated staff to participate in the APW's workgroups on informatics to stay up to date with new curricular developments, and in response devised central framework programs for the further education of teachers. It was also entrusted with the operational control over the realization of further education programs and concepts in the districts of the GDR, thus overseeing the work of the BUWs.[708]

To support the BUWs efforts in setting up local further training courses, the ZIW devised two framework programs for informatics in 1987, consisting of a total of 60 hours of instruction and training – one for polytechnic teachers, one for mathematics teachers.[709] The program could be divided into a 30-hour basic course, which was to

---

[704] BBF APW 11296/4, "Information über Stand und Probleme bei der Realisierung des Informatikbeschlusses," undated, 8–9.
[705] BBF APW 11296/4, "Information über Stand und Probleme bei der Realisierung des Informatikbeschlusses," undated, 9–10.
[706] BArch DR 201/572, "Konzeption für die weitere Qualifizierung von Lehrern und Betreuern auf dem Gebiet Informatik und Informationstechnik (1. Entwurf)," dated May 3, 1988, 4.
[707] Koch and Linström, "Pädagogischen Lesungen," 40–41.
[708] BArch DR 201/572, "Konzeption für die weitere Qualifizierung von Lehrern und Betreuern auf dem Gebiet Informatik und Informationstechnik (1. Entwurf)," dated May 3, 1988, 5.
[709] BArch DR 201/572, "Informatik in der Weiterbildung," dated May 21, 1987, 3.

introduce teachers to the use of new computer technology and programming in BASIC, and a 30-hour advanced course, aimed at teachers who were about to be deployed for the teaching of an informatics course, to familiarize them with the content and aims of the newly developed informatics curricula.[710] The advanced course The universities of teacher education were advised to adhere to the ZIW's program in the provision of their further teacher training courses in the field of informatics.

The compulsory informatics courses in grades 9 and 11 posed a challenge for mathematics teachers, as they also included a rather technical section on process automation. This involved the practical use of electronic components and coupling modules, such as the "Pupils' Experiment Device" (ger.: Schülerexperimentiergerät, SEG) for teaching electronics and automation technology. These modules could be used, for example, for computerized temperature measurement and control, or for experiments involving the numerical control of a machine model.[711] The elective courses with a technical focus were even more challenging to teach, requiring in-depth knowledge of microelectronics and computer hardware to conduct advanced experiments and supervise the pupils' projects that went beyond what was covered in the ZIW's 60-hour program. To meet this need, so-called "special courses" on technological process control and automation were offered by the universities of teacher education in Güstrow, Erfurt, and Potsdam, as well as at the Humboldt University of Berlin and the Martin Luther University of Halle-Wittenberg.[712] However, the ZIW criticized that the high academic standard of the special courses overwhelmed many of the participating teachers, who ended up having to repeat the course.[713] The same was true of the basic informatics courses for in-service teacher education offered at universities, which presupposed a high level mathematical knowledge and skills.[714]

Overall, the new qualification requirements imposed on teachers by the introduction of the new informatics curricula in general education demanded a considerable effort and commitment from them. The training courses were usually held during the spring or summer holidays so that teachers did not miss a day of

---

710 BArch DR 201/572, "Konzeption für die weitere Qualifizierung von Lehrern und Betreuern auf dem Gebiet Informatik und Informationstechnik," dated June 29, 1988, 2.
711 BArch 201/570, "Rahmenprogramm für die Qualifizierung auf dem Gebiet Informatik und Informationsverarbeitende Technik – Polytechnik," dated 1989, 9–10.
712 BArch DR 201/662, "Konzeption (Entwurf) für die weitere Qualifizierung von Lehrern und Betreuern auf dem Gebiet Informatik und Informationstechnik," dated May 18, 1986, 3; BArch DR 201/572, "Einschätzung zur Qualifizierung von ESP-Lehrern und Betreuern auf dem Gebiet der Informatik und Informationstechnik," dated December 12, 1989, 2.
713 BArch DR 201/572, "Einschätzung zur Qualifizierung von ESP-Lehrern und Betreuern auf dem Gebiet der Informatik und Informationstechnik," dated December 12, 1989, 2.
714 BArch DR 201/572, "Einschätzung zur Qualifizierung von ESP-Lehrern und Betreuern auf dem Gebiet der Informatik und Informationstechnik," dated December 12, 1989, 4.

school. In addition, the ZIW urged teachers to devote more of their free time to training, as the 60-hour informatics program was far from sufficient to provide a sound level of qualification. It was emphasized that the training courses were only a starting point, after which teachers were expected to deepen their education on their own account, for example, by studying the new curriculum materials, textbooks, and journal articles on informatics and computing. To this end, the two teacher magazines *Polytechnische Bildung und Erziehung* (eng.: Polytechnic Education) and *Mathematik in der Schule* (eng.: Mathematics in School) published articles on the new curricular developments in order to inform teachers of mathematics and polytechnic education about the contents and aims of the new informatics curricula, to familiarize them with microcomputers and programming, or to share reports by informatics teachers on their own teaching experiences and suggestions on how to teach certain sections of the new curricula effectively.[715]

It was planned to qualify all ESP teachers through further education courses in informatics until 1992, to support the implementation of the compulsory basic informatics course.[716] This extension of in-service training efforts from the "pioneers" among the teachers who had volunteered to participate in the testing of the new curricula and who were personally interested in new computer technology to all ESP teachers brought with it the problem that a proportion of the teachers were not intrinsically motivated to grapple with the challenge of new information technology.[717] The ZIW reported that especially older ESP teachers were reluctant to undergo further training in information technology because they felt overwhelmed

---

715 See, for example, Siegfried Bohnsack, "Aufbau, Arbeitsweise, Programmierung des Kleincomputers KC 85/1," *Polytechnische Bildung und Erziehung* 29, no. 2/3 (1987): 57–60; Harald Diesel, "Integration der Informatik und der informationsverarbeitenden Technik in den polytechnischen Unterricht," *Polytechnische Bildung und Erziehung* 30, no. 5 (1988): 154–57; Jürgen Migdalek, "Zur Einführung von Elementen der Informatik in die Allgemeinbildung," *Polytechnische Bildung und Erziehung* 28, no. 4 (1986): 105–10; Jürgen Migdalek, "Grundlagen der Informatik im polytechnischen Unterricht," *Polytechnische Bildung und Erziehung* 28, no. 12 (1986): 429–31; Jürgen Migdalek, "Herausbildung eines Grundverständnisses der Informatik als fester Bestandteil des polytechnischen Unterrichts in den Klassen 9 und 10," *Polytechnische Bildung und Erziehung* 30, no. 5 (1988): 157–59; Hans-Jürgen Sprengel, "Kleincomputer und Programmierung (Teil 1)," *Mathematik in der Schule* 24, no. 12 (1986): 835–39; Hans-Jürgen Sprengel, "Kleincomputer und Programmierung (Teil 2)," *Mathematik in der Schule* 25, no. 1 (1987): 8–13; Hans-Jürgen Sprengel, "Kleincomputer und Programmierung (Teil 3)," *Mathematik in der Schule* 25, no. 2/3 (1987): 91–97; Dieter Müller, "Zur Einführung des obligatorischen Unterrichts im Fach Informatik an den erweiterten Oberschulen," *Mathematik in der Schule* 26, no. 9 (1988): 577–84; Lothar Flade, "Zum fakultativen Kurs 'Informatik' in den Klassen 9 und 10," *Mathematik in der Schule* 27, no. 7/8 (1989): 541–43.
716 Barch 201/570, "Weiterbildung der ESP-Lehrer," dated April 11, 1989.
717 BArch DR 201/572, "Einschätzung zur Qualifizierung von ESP-Lehrern und Betreuern auf dem Gebiet der Informatik und Informationstechnik," dated December 12, 1989, 3.

by the new qualification requirements. Having only recently undergone further education in the new and challenging subject of information electronics and automation technology, they now had to complete yet another demanding 60-hour course to acquire new knowledge and skills in informatics and computing.[718]

By contrast, in-service training for PA teaching staff[719] in the use of new information technologies remained rather limited. It was estimated that by the early 1990s, less than 5% of workplaces would be equipped with new information technology.[720] Moreover, pupils were only to be deployed at workplaces involving new computer technology if it would not disrupt operational productivity and if the pupils could actually meaningfully contribute to production with their work.[721] The further training of staff supervising pupils during their PA lessons was thus only initiated on a small scale at first, focusing on individual supervisors in workplaces where the deployment of pupils on computerized workplaces was deemed feasible and appropriate. The technical training of the main supervisors in the new computer technology was to take place through their participation in measures for the vocational qualification of adults in combines and companies, for example on CNC machines and automation technology, or in the qualification courses set up for ESP teachers to prepare them for teaching the basic course in computer education.[722] The pedagogical preparation for their new task was organized through the individual instruction of designated subject advisors for polytechnic education.[723] In addition, the PA supervisors, too, were instructed to study the curriculum materials in their free time in order to familiarize themselves with the objectives, content, and

---

718 BArch DR 201/572, "Einschätzung zur Realisierung des Programms der Qualifizierung von Polytechniklehrern auf dem Gebiet Informatik und Informationstechnik (vom Januar 1987)," dated November 1987, 1.
719 The 'productive work' of pupils in workplaces of the GDR economy was pedagogically guided and supervised by volunteers among the workers at the respective company who, in addition to their regular job, took on the task of supervising and instruction of pupils during their PA assignments.
720 BArch DR 201/662, "Konzeption (Entwurf) für die weitere Qualifizierung von Lehrern und Betreuer auf dem Gebiet Informatik und Informationstechnik," dated May 18, 1986, 4.
721 BArch DR 201/572, "Einschätzung zur Qualifizierung von ESP-Lehrern und Betreuern auf dem Gebiet der Informatik und Informationstechnik," dated December 12, 1989, 2.
722 BArch DR 201/572, "Einschätzung zur Qualifizierung von ESP-Lehrern und Betreuern auf dem Gebiet der Informatik und Informationstechnik," dated December 12, 1989, 4.
723 Subject advisors were teachers who had distinguished themselves through outstanding pedagogical and professional competence and political loyalty to the SED, who taught in schools as full-time teachers and additionally advised and trained teachers on a part-time basis. They initiated the exchange of experiences between teachers of a specific subject, provided teachers with study material and thus acted as multipliers of further training measures and good teaching practice (Koch and Linström, "Pädagogischen Lesungen," 40).

methodological approach of the new syllabus for PA lessons in workplaces where new computer technology was being used.

Since it was expected that computer and automation technology would become increasingly widespread in the GDR economy in the near future, it was expected that PA supervisors would need to be trained in the new information technology on a wider basis from the late 1980s onwards.[724] To this end, the ZIW devised a new supplement to the framework program for the further training of full-time PA supervisors in 1989, on the educationally effective planning and design of pupils' productive work in grades 9 and 10, in particular on workplaces involving information processing and computer technology.[725] As part of these courses, PA supervisors were instructed on how to select appropriate tasks for pupils using modern computer technology, and were invited to share their experiences with their colleagues of how they were entrusting pupils with tasks in computerized workplaces.[726]

In 1989, the framework of short-term in-service teacher education in informatics was complemented by more comprehensive qualification programs, by establishing a postgraduate program in informatics for the further training of experienced teachers, as well as teaching diploma programs in informatics for initial teacher training. For the SED leadership, it was of paramount importance that the education and training of teachers in informatics not only consisted in a technical familiarization with computers and theoretical instruction in informatics, but that it would link new computer technology to the SED's economic plan and vision of a desirable computerized future. In this sense, teacher education in informatics would have to include elements of the SED's sociotechnical imaginary of computer technology, especially those regarding the anticipated social and economic progress the imaginary promised for the future. In a position paper on the introduction of informatics and computer technology into general education, the party leadership stated that while teachers were neither expected to become economists nor computer specialists, they were required to know about key technologies and understand their link to socialist in-

---

724 BArch DR 201/572, "Konzeption zur Qualifizierung von hauptamtlichen Betreuern für den Grundkurs: Informatik," dated July 1986.
725 BArch DR 201/572, "Ergänzung zum Rahmenprogramm für die Weiterbildung der hauptamtlichen Betreuer im polytechnischen Unterricht. Thema: Die bildungs- und erziehungswirksame Planung und Gestaltung der Produktiven Arbeit der Schüler in den Klassen 9 und 10 (Entwurf – Diskussionsmaterial)," dated July 1988.
726 BArch DR 201/572, "Ergänzung zum Rahmenprogramm für die Weiterbildung der hauptamtlichen Betreuer im polytechnischen Unterricht. Thema: Die bildungs- und erziehungswirksame Planung und Gestaltung der Produktiven Arbeit der Schüler in den Klassen 9 und 10 (Entwurf – Diskussionsmaterial)," dated July 1988, 6.

tensification.[727] Teachers needed to be aware of the desired economic effects, know about changes in the nature of work, and understand the link between scientific-technological and social progress. For this reason, all education and training programs for teachers in informatics included a section on the economic and social importance of new computer technology in the GDR as set laid out in the SED's policies and vision of a computerized future of the country.

The postgraduate study program was introduced in February 1989, which allowed for teachers to obtain an additional teaching qualification in informatics.[728] The half-year program was aimed at subject teachers of mathematics, physics, or polytechnic instruction, who already had some teaching experience. Selected teachers were delegated by their local school or district authorities to study the course on a full-time basis and were released from their teaching duties for one school term.[729] The postgraduate study program was offered in two specialization tracks, corresponding to the dual approach of mathematically oriented informatics courses on the one hand, and more technically oriented informatics courses in schools on the other: "Informatics and Mathematics" and "Informatics and Process Automation."[730] Reflecting the goals of the school curricula for informatics, the postgraduate study program aimed at preparing teachers not only to teach informatics in general education based on the newly introduced curricula, but also to foster the development of personality traits deemed indispensable for working with new information technology.[731] In addition, teachers were also meant to raise pupils' enthusiasm for informatics as a science and for the use of new information technology. Finally, the teachers were expected to convey to the students the importance of informatics for the development of society under socialism and its key role in mastering the "scientific-technological revolution."[732] The common core of both tracks consisted in the three courses on the "Ideological-philosophical, historical, economic, and social aspects of informatics," "Algorithms, data, programs," and "Hardware and operating systems." While the latter

---

727 SAPMO DY 30/5831, "Probleme und Diskussionen zum Informatikbeschluss," undated, 2.
728 Ministerrat der DDR/Ministerium für Volksbildung/Ministerium für Hoch- und Fachschulwesen, *Studienplan für das postgraduale Studium zur Qualifizierung von Diplomlehrern auf dem Gebiet der Informatik an Universitäten und Hochschulen der DDR*. (Berlin: Volk und Wissen, 1987).
729 Friedrich and Timmermann, "Lehrerbildung Informatik," 203.
730 Ministerrat der DDR/Ministerium für Volksbildung/Ministerium für Hoch- und Fachschulwesen, *Studienplan für das postgraduale Studium*, 5.
731 Ministerrat der DDR/Ministerium für Volksbildung/Ministerium für Hoch- und Fachschulwesen, *Studienplan für das postgraduale Studium*, 4.
732 Ministerrat der DDR/Ministerium für Volksbildung/Ministerium für Hoch- und Fachschulwesen, *Studienplan für das postgraduale Studium*, 4.

two covered basic technical knowledge and skills in the use of computer technology and programming, the former was of a more ideologically charged nature. It covered the anticipated impact of new computer technology and informatics on science, economic production, and society. In particular, it related the scientific-technological development to the "global class struggle against imperialism," emphasizing the inseparable unity of Marxist-Leninist philosophy, socialist ideology and policy with the development of the new scientific discipline of informatics and the new computer technology in the GDR.[733] In line with the SED's imaginary of new computer technology, the syllabus reiterated the promise that the development and use of information technology in the GDR was the key to securing growing labor productivity on a lasting basis and the further development of the socialist society.[734] According to the study plan, this positive vision was to be contrasted with a discussion of the alleged societal implications of informatics and new information technology in capitalist countries, namely the "deepening of the fundamental contradiction of capitalism,"[735] the anti-democratic political objectives underpinning the development of working conditions, and the new forms of political control and manipulation arising from the use of modern computer technology.[736]

In addition to the common core modules, the mathematical track of the program included the courses "Mathematical Foundations of Informatics" and "Mathematical Modelling of Selected Processes." The technically oriented track included a course on "Automation of Production Processes." In both tracks, 45 hours out of a total of 330 hours were devoted to the "Methodology of informatics instruction."[737] In this regard, the program stipulated that the pedagogical requirements for standard software used in schools were to be addressed in the postgraduate teacher training course, as well as how pupils could be taught to select and confidently use

---

[733] Ministerrat der DDR/Ministerium für Volksbildung/Ministerium für Hoch- und Fachschulwesen, *Studienplan für das postgraduale Studium*, 12.
[734] Ministerrat der DDR/Ministerium für Volksbildung/Ministerium für Hoch- und Fachschulwesen, *Studienplan für das postgraduale Studium*, 13.
[735] The 'fundamental contradiction of capitalism' in Marxist theory refers to the transhistorical contradiction between the forces of production (labor power and means of production) and relations of production (ownership and control of the means of production) (Matt Vidal, "Work and Exploitation in Capitalism," in *The Oxford Handbook of Karl Marx*, ed. Matt Vidal et al. (New York: Oxford University Press, 2019), 243).
[736] Ministerrat der DDR/Ministerium für Volksbildung/Ministerium für Hoch- und Fachschulwesen, *Studienplan für das postgraduale Studium*, 13.
[737] Ministerrat der DDR/Ministerium für Volksbildung/Ministerium für Hoch- und Fachschulwesen, *Studienplan für das postgraduale Studium*, 39–40.

ready-made standard software.[738] In order to help pupils to develop the desired skills in "algorithmic" thinking and working as laid out in the informatics school curricula, teachers had to learn various ways of representing algorithms, such as verbal descriptions, program flow charts, structure diagrams, and computer programs.[739] The section on the methodology of computer instruction also listed the discussion of the advantages and disadvantages of different programming strategies. It emphasized the need for teachers to take care that pupils would develop a "good programming style," and the ability to critically evaluate and interpret computer output.[740]

From a didactic point of view, two fundamental principles were foregrounded regarding the teachers' planning of their computer education lessons: The guidance of pupils towards increasing autonomy and the extensive use of practical project-based activities. Teachers were provided with suggestions for the design of learning activities that were considered to promote the development of pupils' autonomy in working with new computer technology, such as solving complex tasks, presentations prepared and given by the pupils, or cooperative forms of problem solving and computer work.[741] In the didactic design of project assignments, teachers were instructed to focus on practice-oriented projects and to use both individual and collective forms of work. Teachers were also encouraged to consider pupils' interests and to involve them in choosing suitable projects. Furthermore, the course plan for the instruction of teachers accommodated the different phases of computer projects, from the initial problem definition to the final program documentation.[742]

So far, all efforts to prepare teachers for their new role as computer science teachers had focused exclusively on in-service training, as this was the fastest way to get qualified teachers ready for the introduction of the new informatics courses into schools. This changed in September 1989, when the informatics course was finally integrated into the system of initial teacher training in the GDR, thus further consolidating a permanent place for informatics in general education. Two new teaching diploma study programs were introduced, one for the

---

[738] Ministerrat der DDR/Ministerium für Volksbildung/Ministerium für Hoch- und Fachschulwesen, *Studienplan für das postgraduale Studium*, 34–35.
[739] Ministerrat der DDR/Ministerium für Volksbildung/Ministerium für Hoch- und Fachschulwesen, *Studienplan für das postgraduale Studium*, 34.
[740] Ministerrat der DDR/Ministerium für Volksbildung/Ministerium für Hoch- und Fachschulwesen *Studienplan für das postgraduale Studium*, 35.
[741] Ministerrat der DDR/Ministerium für Volksbildung/Ministerium für Hoch- und Fachschulwesen, *Studienplan für das postgraduale Studium*, 34.
[742] Ministerrat der DDR/Ministerium für Volksbildung/Ministerium für Hoch- und Fachschulwesen, *Studienplan für das postgraduale Studium*, 35.

subject combination mathematics and informatics, and one for the polytechnic instruction and informatics.

The teaching diploma program for the subject combination mathematics and informatics encompassed a comprehensive course in informatics of 480 hours, preparing teachers for the instruction of both compulsory and elective school courses in informatics.[743] Initially, the study program was only offered at the Technical University in Chemnitz, and the University of Teacher Education in Güstrow.[744] It introduced teachers to programming and software technology, computer and operating systems, the use of standard software used in school instruction, theoretical and applied informatics, and computer geometry. In addition, and in contrast to the introductory short courses, it also covered in depth the "Methodology of informatics instruction," thus for the first time addressing in detail didactic approaches to informatics instruction as part of general education, in particular regarding the practical use of microcomputers by the pupils. Included in the 480 hours of informatics education of the study program were 150 hours of practical training on a microcomputer for the practical training of teachers in programming and in the use of standard software.

The teaching diploma study program for polytechnic instruction and informatics prepared teachers in particular for teaching the compulsory computer course in the subject ESP, and the more technically oriented elective courses on computer and automation technology.[745] The program included 300 hours of instruction in informatics, as well as 150 hours on the topic of automation technology.

The informatics section of the program included an extensive education and training on assembler and machine code, programming in BASIC and PASCAL, as well as the methods of program development.[746] In contrast to the program for the subject combination of mathematics and informatics, it put more emphasis on the use of computer systems and software used in the GDR's industry and service sector, the modelling and simulation of technical systems and processes, and CAD/CAM. 120 hours were dedicated to practical computer use to familiarize

---

[743] Ministerrat der DDR/Ministerium für Volksbildung/Ministerium für Hoch- und Fachschulwesen, *Studienplan für die Ausbildung von Diplomlehrern der allgemeinbildenden polytechnischen Oberschulen in der Fachkombination Mathematik/Informatik an Universitäten und Hochschulen der DDR* (Berlin: Volk und Wissen, 1989).

[744] Friedrich and Timmermann, "Lehrerbildung Informatik," 201.

[745] Ministerrat der DDR/Ministerium für Volksbildung/Ministerium für Hoch- und Fachschulwesen, *Studienplan für die Ausbildung von Diplomlehrern der allgemeinbildenden polytechnischen Oberschulen in Polytechnik – Informatik an Universitäten und Hochschulen der DDR* (Berlin: Volk und Wissen, 1989).

[746] Ministerrat der DDR/Ministerium für Volksbildung/Ministerium für Hoch- und Fachschulwesen, *Studienplan Polytechnik*, 10.

teachers with the various types of microcomputers and peripherals, and for extensive training in the development, programming, and testing of structured programs, algorithmic problem-solving, the use of software tools, and working with data bases.[747]

The section on automation technology focused on the structure and functioning of automated systems and information processing for the control of technical processes in industrial production. It introduced teachers to the use of computer technology for process control and flexible automation, which, as the syllabus emphasized, was primarily aimed at increasing labor productivity and improving the working and living conditions of people in the GDR.[748] Practical training in the field of automation technology comprised 60 hours and focused on automated information processing in technical systems and computerized measurement and process control.[749] In order to prepare the teachers for teaching the module on automated process control, which formed part of the ESP basic computer course, the practical training was also meant to equip the teachers with knowledge of the use of the "SEG" models for technical experiments in the classroom with computerized control and measurement, and the coupling of various peripheral devices to the school computers.[750]

The introduction of the postgraduate course and the two diploma courses for initial teacher training in informatics paved the way for the replacement of the system of short-term informatics courses for teacher training by longer-term and more comprehensive courses for preparing teachers to teach both compulsory and optional courses in informatics and computer technology in schools. But perhaps more importantly, the new teacher training programs also reflected an important step towards the incorporation of didactic considerations for the teaching of informatics and computing in general education into teacher training in the GDR.

---

747 Ministerrat der DDR/Ministerium für Volksbildung/Ministerium für Hoch- und Fachschulwesen, *Lehrprogramme für die Ausbildung von Diplomlehrern der allgemeinbildenden polytechnischen Oberschulen im Fach Polytechnik – Informatik an Universitäten und Hochschulen der DDR* (Berlin: Volk und Wissen, 1989), 37.
748 Ministerrat der DDR/Ministerium für Volksbildung/Ministerium für Hoch- und Fachschulwesen, *Lehrprogramme*, 66.
749 Ministerrat der DDR/Ministerium für Volksbildung/Ministerium für Hoch- und Fachschulwesen, *Lehrprogramme*, 69.
750 Ministerrat der DDR/Ministerium für Volksbildung/Ministerium für Hoch- und Fachschulwesen, *Lehrprogramme*, 69–70.

5.4 A Dedicated Computer for Schools: The "Bildungscomputer" (BIC)

With the introduction of informatics and computer technology as a subject of teaching into general education, microcomputers had entered schools not just a subject, but also as a means of instruction. However, this was not the result of a conscious, deliberate effort to establish the computer as an educational technology, but rather a welcome side-effect of the introduction of informatics and computer education at schools. The principle of a practically oriented computer education required the pupils to use and write programs on the computer in the process of learning, as GDR pedagogues were convinced that only through practical use pupils would develop the desired skills of "algorithmic thinking" and problem-solving with the help of computer programs. Thus, at least in the context of informatics education, the microcomputer became an educational technology, a tool of instruction, that was considered important, if not indispensable, for achieving the educational and pedagogical goals set out in the curricula. One expression of this was the fact that some first steps towards a didactics of computer education and informatics were increasingly integrated into the training of schoolteachers, which dealt with effective ways of teaching and learning with the computer.

A second expression of the gradual establishment of the computer as an educational technology was the stance taken by educational policymakers that the "universal" information processing machine that the computer was, whatever its form and shape, was not necessarily suitable for use in schools, but that there was a need to develop a microcomputer tailored to the specific needs of school education. In fact, the Politburo decision on the "Consequences of the Development of Informatics and Information Processing Technology for the Education System" of November 1985 had also envisaged the development and domestic production of a special national school computer for use in secondary schools and vocational training. Accordingly, educational policymakers were tasked with working out the pedagogical and technical requirements for a dedicated school computer that would meet the specific conditions of use in general education, vocational education, and training.[751]

The idea of developing a purpose-built school computer was not unique to the GDR. Similar state-led initiatives to develop computer devices specifically for use in education were also launched in other socialist countries, such as the "Elwro Junior" in Poland, the "Didaktik Gama" in Czechoslovakia, or the "Pravetz 82" in Bul-

---

[751] SAPMO DY 30/J IV 2/2/2138, "Standpunkte zu Konsequenzen aus der Entwicklung der Informatik und informationsverarbeitenden Technik für das Bildungswesen," 31–32.

garia.[752] But also in capitalist countries, the state commissioned or incentivized the development of computers intended for use in schools during the 1980s, such as the "BBC Micro" in Britain, the "Compis" in Sweden, the "Smaky" in French-speaking Switzerland, or the "ICON" in Canada, and an unfinished school computer in Australia.[753] In the cases of Canada and French-speaking Switzerland, it becomes apparent that the state-led initiative to develop a computer specifically designed for educational use was fueled by the hope of not only equipping schools with a new tool of instruction, but also of supporting its local or national computer manufacturing industry.[754] However, most of these projects to establish a dedicated school computer in capitalist countries ultimately failed in the long run due to both the rapid technological advances of the 1980s and the economies of scale that favored general-purpose computers. New computers quickly surpassed the technical standards of school computers, offered better compatibility with popular software, and were sold at a lower price, thus making them more attractive to schools than the niche product of purpose-built educational computers.[755]

In the GDR, as the curricula for computer education were gradually implemented in the mid-1980s, classrooms were at first equipped with various types of KC 85 microcomputers, the first generation of domestically produced microcomputers in the GDR.[756] However, the KC 85 microcomputers had originally been developed for

---

[752] Patryk Wasiak and Jaroslav Švelch, "Designing Educational and Home Computers in State Socialism: The Polish and Czechoslovak Experience," *Journal of Design History* 36, no. 4 (December 30, 2023): 377–93; Petrov, *Balkan Cyberia*, 232–34.
[753] Alison Gazzard, *Now the Chips Are Down: The BBC Micro* (Cambridge, Massachusetts: The MIT Press, 2016); Rosalía Guerrero Cantarell and Carmen Flury, "Making the Computer Fit for School: Efforts to Develop a State-Mandated Educational Computer in Sweden and East Germany (1980s-1990s)," *Historical Studies in Education / Revue d'histoire De l'éducation* 35, no. 2 (2023): 5–29; Fabian Grütter, "The Smaky School Computer. Technology and Education in the Ruins of Switzerland's Watch Industry, 1973–1997," *Learning, Media and Technology* 49, no. 1 (May 24, 2023): 49–62; Robert J.D. Jones, "Shaping Educational Technology: Ontario's Educational Computing Initiative," *Educational & Training Technology International* 28, no. 2 (1991): 129–34; Ivor F. Goodson and J. Marshall Mangan, "Computers in Schools as Symbolic and Ideological Action: The Genealogy of the ICON," *The Curriculum Journal* 3, no. 3 (1992): 261–76; Arthur Tatnall, "The Australian educational computer that never was," *IEEE Annals of the History of Computing* 35, no. 1 (2013): 35–47.
[754] Jones, "Shaping Educational Technology," 131; Grütter, "The Smaky School Computer."
[755] Jones, "Shaping Educational Technology," 133; Guerrero Cantarell and Flury, "Making the Computer Fit for School," 21–22; Grütter, "The Smaky School Computer," 57.
[756] Two compatible types of computers with different technical parameters (KC 85/1 and KC 85/2) were developed simultaneously at the Robotron combine in Dresden and at VEB Mikroelektronik "Wilhelm Pieck" in Mühlhausen. However, when the development was finalized in 1984, the computers ended up being too expensive to be sold as consumer goods and were used mainly in educational institutions (Meyer, *Computer in der DDR*, 55–57).

other purposes, namely leisure and home computing, and were not deemed suitable by pedagogues and education policymakers for a "high quality and modern" computer instruction in schools.[757] These microcomputers simply did not provide enough computing power to meet the ambitious goals of the new informatics curricula in the medium term.[758] Moreover, the intensive use of computers in educational institutions led to frequent breakdowns of computer equipment.[759] With the GDR computer industry already stretched to its limits, schools were faced with very long repair times of up to a year without temporary replacements. In addition, the lack of homogeneous computer equipment posed a problem of efficiency in several respects: On the one hand, in terms of teaching, as everything had to be explained several times for the different systems used. On the other hand, it proved problematic in terms of teacher training and software development, as there were several computer systems that teachers had to learn about and for which different school software had to be developed. It was hoped that a new, purpose-built school computer would solve these problems, as teacher training and teaching materials could be aligned to a single, specified set of hardware that would be used for computing as part of both general and vocational education.

Based on the curriculum contents and goals for computer instruction in the GDR, the Academy of Pedagogical Sciences, together with the Ministries of Vocational and Higher Education, compiled a catalogue of pedagogical-technical requirements for the envisioned school computer, the so-called "Bildungscomputer" (BIC). However, the final technical specifications needed to be negotiated with the representatives of the Robotron combine, which was tasked with the development and production of the educational computer.[760] Barriers to fulfilling the demands of the pedagogues were either of a technical or financial nature: Technical parts were either not available in sufficient quantities to equip all schools in the foreseeable future, or were too expensive to arrive at a final

---

[757] Weise, *Erzeugnislinie Heimcomputer*, 33–34; Gert Keller and Gunter Kleinmichel, "Bildungscomputer robotron A 5105," *Mikroprozessortechnik* 2, no. 10 (1988): 292.

[758] Gert Keller and Gunter Kleinmichel, [Bildungscomputer (BIC) A 5105], *Neue Technik im Büro* 33, no. 2 (1989): 62–64.

[759] BBF APW 11125/2, "Hinweise und Erläuterungen für den Experimentallehrplan zum Grundkurs 'Informatik' als Stoffgebiet im ESP-Unterricht der Klasse 9," dated June 10, 1988, 15.

[760] BArch DR 3/26973, "Information und weitere Maßnahmen zur Entwicklung eines Ausbildungscomputers für die Volksbildung, die Berufsbildung und die Ausbildung an den Hoch- und Fachschulen," dated November 17, 1986.

product affordable to schools, i.e. within the 11,000 Marks price limit set by the Ministries of Education and Vocational Training.[761]

The development of the school computer was primarily focused on the realization of the computer education curricula of general and vocational schools. Its design and functionalities were thus geared to the curriculum goals, namely, to impart a basic understanding of how a computer works, to teach the fundamentals of programming, and for pupils to learn how to use standard applications confidently in solving mathematical and technical problems.[762] By contrast, the use of the educational computer as an educational technology in other subjects than mathematics and ESP played only a marginal role in determining the required technical parameters of the device.

The BIC was designed as a robust microcomputer, consisting of three components that were mechanically firmly attached to each other: The basic computer unit, a floppy disc storage unit and a monitor.[763] The new school computer was envisioned neither as a further development of the KC 85 microcomputers, nor as an upgraded variant of the PC 1715 widely used in workplaces throughout the GDR, but as an independent, new concept of a computer for teaching purposes, which was largely based on domestically produced electronics.[764] Educators demanded for a "robust and ergonomic design" and an adequate size to fit on a standard classroom table. In contrast to the KC 85 computers, which had monochrome screens and used tape cassettes to load and save programs, they asked for a color screen and a floppy disk drive. To fulfill the curriculum goals, the BIC also had to offer the possibility to connect various devices and peripherals such as a printer and a plotter, as well as pupils' experiment devices for the more engineering-oriented curriculum sections on process automation, and computer-aided measurement, control, and regulation. For programming instruction, the BIC was equipped with RBASIC (Robotron-BASIC), an interpreter for the programming language BASIC specifically developed by Robotron for the BIC. To cater for the needs of vocational education, the BIC also included the operating system SCP[765] which was commonly used on workplace computers in the GDR. This allowed for

---

761 BArch DR 2/11787, "Information über Ergebnisse der Verteidigung der Entwicklungsstufe K5 des Bildungscomputers A 5105 am 1.7.1988 im VEB Messelektronik 'Otto Schön' Dresden," dated July 7, 1988.
762 Keller and Kleinmichel, "Bildungscomputer robotron A 5105."
763 Keller and Kleinmichel, "Bildungscomputer robotron A 5105."
764 BArch DR 3/26973, "Information und weitere Maßnahmen zur Entwicklung eines Ausbildungscomputers für die Volksbildung, die Berufsbildung und die Ausbildung an den Hoch- und Fachschulen," dated November 17, 1986, 1.
765 SCP (Single User Control Program) was a widely used operating system in the GDR which ran on the most common workplace computers. It was developed by Robotron and closely mod-

the running of already available educational and office software for computer instruction in both vocational and general education, such as text processing and spreadsheet programs.[766]

It is notable that the catalogue of requirements that educators put forward for the new educational computer contained qualifiers such as "modern," or that it needed to be in line with "advanced international standards" and "trends."[767] Evidently, the BIC was seen as an opportunity to provide educational institutions with a state-of-the-art computer that would allow for the future-oriented technology education that had been envisioned, preparing the young generation for an imagined future in which the GDR would be at the international forefront of computer technology development and use. Contrasting the BIC with the previously available domestically produced microcomputers, there was a certain dissatisfaction with the apparent lag of GDR computers behind international technological standards, especially in relation to the Western capitalist countries.

Development of the BIC was completed in 1988, and the first 30 units were made available to the education sector for testing in October of that year.[768] Mass production was to begin in 1989. The Ministry of Education seemed pleased with the new machine, claiming that the BIC was vastly superior to the microcomputers previously used in education and was technologically up to date with "advanced international standards."[769]

The BIC met almost all the demands and requirements set out by the educators involved in its development. However, there were two points of contention: The monitor, and the fact that the BIC was running on an 8-bit microprocessor. A private, state independent computer magazine, newly founded by a computer enthusiast in the GDR, reviewed the BIC in the early 1990s and concluded: "Its design is in no way up to international standards. A particularly inglorious point is the 'monitor'."[770] Robotron had been unable to deliver on the demand of a color mon-

---

elled after (and compatible with) CP/M (Control Program for Microcomputers) developed in the 1970s by the US American software developer Digital Research.

766 Keller and Kleinmichel, "Bildungscomputer robotron A 5105."
767 BArch DR 3/26973, "Information und weitere Maßnahmen zur Entwicklung eines Ausbildungscomputers für die Volksbildung, die Berufsbildung und die Ausbildung an den Hoch- und Fachschulen," dated November 17, 1986, 2; Keller and Kleinmichel, "Bildungscomputer robotron A 5105," 292.
768 "Erste Bildungscomputer von Robotron übergeben," *Neues Deutschland*, October 4, 1988, 4.
769 BArch DR 3/26973, "Information und weitere Maßnahmen zur Entwicklung eines Ausbildungscomputers für die Volksbildung, die Berufsbildung und die Ausbildung an den Hoch- und Fachschulen," dated November 17, 1986, 1–2.
770 "BIC: hält er, was seine Entwickler versprachen?," *Bit POWER*, no. 1 (1990): 5–6 (translated from German).

itor in the appropriate size, due to the current "material-technological basis and technological possibilities" of the GDR. A development or even import of such a monitor was also not envisaged in the future either. Even a monochrome monitor posed serious challenges, as their demand for the production of personal computers could not be met at present.[771] In the end, the BIC was fitted with an inferior monochrome monitor. Unlike the rest of the machine, it did not receive the GDR seal of approval.[772]

The second point of critique was the 8-bit architecture of the BIC, which relied on the U 880 microprocessor in order to comply with the agreed price ceiling.[773] However, the Academy of Pedagogical Sciences had instead pleaded for the introduction of the new 16-bit ESER PC 1834, modified for the conditions and user needs of schools – a GDR microcomputer that was "at the global forefront in the field."[774] Based on knowledge gained from international conferences, study visits and literature, they had concluded that the international trend was clearly towards 16-bit microcomputers in both capitalist and socialist countries.[775] Bulgaria had introduced a 16-bit "Pravetz" computer for the use in schools already in 1987, and the Soviet Union, too, was currently developing its own 16-bit educational computer.[776] In the report on his visit to the European Conference on Computers in Education (ECCE) 1988 in Lausanne, Immo Kerner, Professor of Computer Science at the University of Rostock, wrote in blunt words: "The computer must be a 16-bit device. Other (8-bit) computers have no chance whatsoever internationally, even in the market sector of education. It is to be estimated that the GDR development of the school computer BIC will have a locally and also temporally limited impact. It can already be recommended to start preliminary investigations for the

---

771 BArch DR 2/11787, "Verständigung für die Beratung am 30.6.1987," undated, p 1.
772 BArch DR 2/11787, "Information über Ergebnisse der Verteidigung der Entwicklungsstufe K5 des Bildungscomputers A 5105 am 1.7.1988 im VEB Messelektronik 'Otto Schön' Dresden," dated July 7, 1988, 2.
773 Keller and Kleinmichel, "Bildungscomputer robotron A 5105."
774 BArch DQ 4/3400, Message from Gottfried Schneider, vice president of the APW, to deputy minister for education Harry Drechsler regarding "Standpunkte der ZFKP zur Frage des Einsatzes von 8- oder 16-bit-Prozessoren in der Lehrerbildung sowie im polytechnischen Unterricht vom 3.12.1987," dated January 14, 1988.
775 BArch DQ 4/3400, Message from Gottfried Schneider, vice president of the APW, to deputy minister for education Harry Drechsler regarding "Standpunkte der ZFKP zur Frage des Einsatzes von 8- oder 16-bit-Prozessoren in der Lehrerbildung sowie im polytechnischen Unterricht vom 3.12. 1987," dated January 14, 1988, 4.
776 BArch DQ 4/3400, Message from Gottfried Schneider, vice president of the APW, to deputy minister for education Harry Drechsler regarding "Standpunkte der ZFKP zur Frage des Einsatzes von 8- oder 16-bit-Prozessoren in der Lehrerbildung sowie im polytechnischen Unterricht vom 3.12. 1987," dated January 14, 1988, 2.

use of the IBM-compatible 16-bit computer EC 1834 in schools and other educational institutions."[777]

In contrast to the BIC, the newly developed 16-bit ESER PC 1834 would also ensure compatibility with IBM systems, which were increasingly setting the new global standard for personal computers and software development, as well as future software developed and used in the GDR industry. While acknowledging that the current computer education curriculum could be taught on an 8-bit computer, the education scientists argued that it was uncertain how educational needs would change in the 1990s and warned that the BIC was likely to be obsolete by then because of already foreseeable technological developments.[778]

The ministry of education agreed with the pedagogues' assessment of the rapid technological development, but vehemently rejected the notion of the BIC being already outdated from the start. The education policymakers reminded that while implementing the SED's ambitious economic strategy within a short time frame, the GDR industry's enormous demand for computers had been predominantly met by equipping enterprises and combines with 8-bit computers. Thus, they argued, even with the availability of a 16-bit computer in the form of the ESER PC 1834, which went into serial production in 1988, the current 8-bit standard would not change much for the time being: "It is not to be expected that this technology will be scrapped in the following years because of the transition to ESER PC 1834, especially since the numbers involved are considerable. These computers will continue to have their raison d'être in the near future."[779] Since computer education instruction at school primarily had a preparatory function for later entry into the computerized world of work, it seemed only logical that the lessons should be based on the computer standard that was prevalent in the workplaces of the GDR economy at the time and in the near future. An 8-bit computer such as the BIC was thus considered sufficient to introduce pupils to the use of computer technology, which they could later build on and eventually switch to higher order computers as part of a subject-specific or further vocational education.[780] The Ministry of Education also did not accept the pedagogues' argument

---

777 BBF APW 13.846, "Reisebericht Immo Kerner zur ECCE '88 in Lausanne," dated 1988, 2 (translated from German).
778 BArch DQ 4/3400, "Message from Gottfried Schneider, vice president of the APW, to deputy minister for education Harry Drechsler regarding "Standpunkte der ZFKP zur Frage des Einsatzes von 8- oder 16-bit-Prozessoren in der Lehrerbildung sowie im polytechnischen Unterricht vom 3.12. 1987," dated January 14, 1988, 4.
779 BArch DR 2/13566, "Zu Fragen des Einsatzes von 8- oder 16-bit-Computern im Bildungswesen," dated February 24, 1988, 1 (translated from German).
780 BArch DR 2/13566, "Zu Fragen des Einsatzes von 8- oder 16-bit-Computern im Bildungswesen," dated February 24, 1988, 3.

that an IBM-compatible computer would allow for the use of IBM software. They warned against adopting software from other countries, even from other states in the Socialist Bloc. Instead, because of the differences in educational systems and content, as well as the nature of the content to be taught, software was to be developed specifically for each application. The Ministry also pointed out that the GDR industry could not be expected to support the education sector in software development, as it was itself suffering from a lack of software development capacity.[781]

However, the decisive reason against a 16-bit architecture for the GDR's school computer was probably more of a financial nature: The authorities at the ministry of education outright admitted that the educational system of the GDR could simply not afford to equip schools with 16-bit computers, which would come at least at double the price of a BIC school computer: "From an education economics point of view, the principle that we can only afford what is available to us in terms of financial resources also applies to us. [. . .] The current introduction of the ESER-PC into the education system is not economically justifiable."[782]

Even without considering a more expensive 16-bit computer, the financial viability of the BIC was a major challenge. The computer configuration was specifically designed for use in the education sector, which severely limited the market for the new BIC, and the planned production volume was small compared to other computers. This situation was exacerbated by the decision of the Ministry of Electronics and Electrotechnology, not to approve the device as a home computer and thus to basically prevent its sale to private individuals.[783] In addition, it was still unclear whether the school computer could be a financially viable export product that could be sold to other socialist countries. For these reasons, the Ministry of Electronics and Electrotechnology and the Robotron representatives demanded a purchase guarantee from the central state education authorities in line with the production volume agreed between the manufacturer and the buyers.[784]

It soon became clear that this step had been taken with a wise sense of foresight. After a few initial tests in schools, the development of the BIC was finalized at the end of June 1989, and it entered serial production. The planned equipment of

---

[781] BArch DR 2/13566, "Zu Fragen des Einsatzes von 8- oder 16-bit-Computern im Bildungswesen," dated February 24, 1988, 2.
[782] BArch DR 2/13566, "Zu Fragen des Einsatzes von 8- oder 16-bit-Computern im Bildungswesen," dated February 24, 1988, 2–3 (translated from German).
[783] BArch DQ 4/3401, "Berichterstattung in der Dienstberatung des Staatssekretärs zum Stand der Vorbereitung des Einsatzes des Bildungscomputers in den Einrichtungen der Berufsbildung," dated May 23, 1989, 2.
[784] BArch DR 3/26973, "Information und weitere Maßnahmen zur Entwicklung eines Ausbildungscomputers für die Volksbildung, die Berufsbildung und die Ausbildung an den Hoch- und Fachschulen," dated November 17, 1986, 3.

the first schools was scheduled to start in the second half of the same year.[785] However, the fall of the Berlin Wall on 9 November 1989 was a major blow to the planned equipping of schools with the BIC. From one day to the next, the political and economic landscape of the GDR had changed, with yet unclear consequences. But for schools, it meant the sudden availability of cheaper, more powerful computers made in the West. In the eyes of educators, the BIC looked rather unattractive in comparison. The ministry for vocational education suspended the further supply of vocational training institutions with the BIC in May 1990. Going forward, the financial resources from the Ministry of Education's central fund for equipping municipal vocational schools with computer technology were to be used to purchase high-quality 16-bit computers instead. Meanwhile, the municipal vocational schools demanded the right to decide independently on the purchase of computer technology – but this request was denied by the central state authorities.[786]

The Ministry of National Education also discussed with Robotron representatives the conditions under which the supply of BICs to schools could be stopped. The Robotron representatives expressed their concern about the reluctance of many schools to accept further deliveries and offered to reduce the price of the BIC from 11,000 to 3,000 Marks, but to no avail. Without a guarantee of acceptance, they warned, production would cease in mid-April 1990. And production was indeed stopped in April 1990, after a total of merely 5,000 educational computers had been produced. The last 2,000 units that had never been delivered to schools were sold from remaining stocks under the name of "Alba PC" after the reunification of the two Germanies.[787]

5.5 Chapter Conclusion

Over the course of the 1980s, microcomputers were gradually established not only as a subject in general education, but also a means of instruction in the school systems of many industrialized countries. In the GDR, the former clearly outweighed the latter, as the SED's educational policy focused primarily on intro-

---

785 BArch DQ 4/3399, "Information in der Dienstberatung beim Staatssekretär am 23. Mai 1989 über die Ergebnisse und Probleme der objektkonkreten Abstimmungen mit den Ministerien u. a. zentralen Staatsorganen und mit den Räten der Bezirke zum Aufbau von BIC-Kabinetten 1989/90 sowie zur schrittweisen Ausstattung der Einrichtungen der Berufsbildung mit Bildungscomputern im Zeitraum 1991 bis 1995," dated May 19, 1989.
786 BArch DQ 4/3400, "Begründung zur Entscheidungsvorlage Ausstattung der KBS im Jahre 1990 mit Computertechnik," dated May 23, 1990, 3–4.
787 Meyer, *Computer in der DDR*, 62; Weise, *Erzeugnislinie Heimcomputer*, 69.

ducing computer technology as a subject of teaching and learning in general education to familiarize the broader population with the new technology and the hopes, promises, and expectations of its economic and social effects that were attached to it in line with the party leadership's sociotechnical imaginary. In the GDR, general education had a double function. On the one hand, it was to serve the personality development of everyone in the socialist society. On the other, it was to prepare pupils for their further education in the system of VET or higher education, equipping them with a solid foundation of knowledge and skills in key areas to build upon. In the introduction of computers into general education, these two main functions were linked to several aspects of the SED's sociotechnical imaginary of computers.

For one, the dominant imaginary posited that under socialism, all people were to actively participate in and contribute to the "scientific-technological revolution," in which computing was to play a key role. As part of the socialist educational philosophy, it was therefore required that *all* children would have the opportunity to receive a basic education and training in computing and informatics. This education was intended to help them not only to understand, but also to actively participate in the envisioned "revolutionary" process of modernizing the socialist economy of the GDR and in bringing about social progress through the use of new computer technology. For computer technology to be truly under the control of working people, each individual of the future workforce needed to have a working understanding of this new technology was and the societal goals it was intended to serve.

Secondly, the imaginary envisioned computer technology as an engine of economic growth, that would fundamentally change the world of work and thus require a new set of knowledge, skills, and attitudes of the future workforce. Accordingly, every child was to be introduced to computers, not only to be able to make effective use of this new work tool, but also to develop a positive attitude towards this technology, and a willingness to learn about it and engage with it. This rationale for the integration of computing and informatics into general education was of a more instrumental nature and called for an early preparation of pupils for the challenge of mastering computer technology from an education point of view. This could be done by equipping pupils with basic skills in computer use and programming, thus providing a foundation for computer education in vocational and higher education. However, it also meant that not all pupils needed the same depth of computer education. High performing pupils, who attended special schools, and pupils in EOS preparing for entering higher education, as well as talented pupils selected to participate in elective courses received a more comprehensive training in informatics, computer use, and programming. From an economic point of view, it was simply not deemed necessary to educate all pupils to a high level of competence in computing and programming. For the majority, basic user training and an introduction to the economic significance of computer technology had to suffice.

Third, computers were also envisioned as a powerful tool to expand human mental capacity. In this sense, computer knowledge and skills were perceived as a new aspect of personality development that needed to be considered in the pursuit of the "all-round" development of socialist personalities. Specifically, through the development of new skills such as "algorithmic" thinking and problem-solving, computers in schools were promised to contribute to the cognitive development of children. Moreover, the teaching of computer user skills was intended to open pupils to the myriad of new possibilities the computer, envisioned as a universal tool, was promised to offer.

However, the curricular reforms in general education that introduced computer technology as a subject of teaching were not only a function of the SED's sociotechnical imaginary, but in turn also served to reinforce and stabilize the SED's imaginary of computer technology. A general computer education "for all"[788] also served to establish a common understanding of computer technology, its purpose, and desired social effects. It was meant to mobilize the young people, and in particular, the future workforce, to embrace and actively support the SED's technology policy regarding the development and use of computer technology in the GDR, including its anticipated economic and social implications.

Finally, the introduction of computers into general education in the GDR followed an international trend. Most industrialized countries, both in the East and the West, were introducing computers into schools, and given the ideological struggle between the systems, socialist countries were unwilling to risk falling behind the capitalist states in the race for technological and educational advancement. Moreover, the people in the GDR were well aware of the availability of home computers in West Germany.[789] In the GDR, where home computers were a marginal phenomenon due to the scarce production capacity of the domestic microelectronics industry and the high costs of microcomputers, the introduction of computers in schools was a way of providing access to the new technology for a

---

[788] The POS as the core of the general education system in the GDR did not in fact reach all children, as it excluded certain groups of children, such as those with disabilities, who were relegated to special schools, which were generally not equipped with computers.
[789] See, for example, Schröder, *Auferstanden aus Platinen*, 75; Erdogan, *Avantgarde der Computernutzung*, 74–75.

major proportion of young people in the GDR.[790] The SED was keen to prove to its people that it could keep up with international trends and to demonstrate the innovative technical capabilities of the GDR under SED leadership, by providing access to computers in schools, and a computing and informatics instruction for all pupils within the realm of general education. In addition, since the curriculum's intention was to convey the SED's dominant sociotechnical imaginary of computer technology in the GDR, and to educate pupils about the desired social and economic effects of its use, computer education also served to win the support of young people for the SED's economic and technological policies.

In line with the economic importance attributed to computer technology by the SED's imaginary, envisioning the computer as a power tool to rationalize and automate the economy, computer education was, on the one hand, integrated into polytechnic instruction. This subject was aimed at familiarizing pupils with the socialist world of work and industrial production. On the other hand, the curricular reforms also encompassed the introduction of mathematically oriented elective courses and subject content into the new computer education curriculum. Thus, the dual orientation of informatics towards both the disciplines of engineering and mathematics that had been established in higher education more than a decade earlier,[791] was in some sense replicated in general education. This was also reflected in teacher training, as training to prepare teachers for their new role of teaching informatics in schools focused on mathematics and polytechnics teachers and, to a lesser extent, physics teachers. Initially, teacher preparation consisted of short training courses in computing and informatics, which were supplemented in 1989 by more comprehensive study programs for postgraduate teacher education and initial teacher training in informatics.

Computer technology had entered general education in the GDR already a few years before the introduction of the new informatics curriculum, in the form of electronic pocket calculators. The electronic pocket calculator had been introduced in the first half of the 1980s into mathematics curriculum with the explicit and deliberate role and function of an educational technology, aimed at making instruction in mathematics and natural sciences more efficient. This, however, led educators to become concerned that the calculator would completely take over arithmetic for students and that, consequently, they would unlearn it.

The microcomputer, by contrast, had initially not been envisioned by the SED leadership as an educational technology, but rather as a subject of instruction in

---

[790] Jürgen Danyel and Annette Schuhmann, "Wege in die digitale Moderne: Computerisierung als gesellschaftlicher Wandel," in *Geteilte Geschichte: Ost- und Westdeutschland 1970–2000*, ed. Frank Bösch (Göttingen: Vandenhoeck & Ruprecht, 2015), 301.
[791] See chapter 2 on 'Computer Science and Informatics Instruction in Higher Education'.

schools. The political rhetoric about ensuring that the workforce of the future socialist computerized society was computer literate overrode debates about the introduction of microcomputers into schools for purely educational benefits. The vision of the computer as an educational technology was marginalized against the strong economic rationale for the introduction of computers into education, which foregrounded the imaginary of the computer as a workplace tool to raise economic productivity. This resulted in curricular innovations that focused on an education *about*, rather than *with* computers, limiting the status and use of microcomputers as an educational technology in schools largely to informatics instruction during the 1980s.

However, amongst the pedagogues and educators that were associated with the APW, microcomputers were indeed envisioned as a pluripotent educational technology that could serve as a means of instruction in all subjects. With its possibilities of simulating and visualizing complex processes and functions and imagined, and the imagination of yet to be developed educational software that promised more personalized learning, the computer was assigned a potentially significant role in modernizing school-based instruction. In contrast to the SED's decision on educational policy regarding the introduction of computers into schools, the curricula and teaching aids developed by the APW included the idea that learning with computers allowed for the development of specific skills and ways of thinking, most notably the notion of algorithmic thinking and skills in computational problem-solving. Moreover, computer education was perceived by the educators and pedagogues of the APW as having positive effects on the learning processes and motivation of pupils in other school subjects, as well as on their personality development. However, these potential benefits of computer education were not the decisive factor for the introduction of computers in schools for the SED. Rather, they were seen as a secondary effect of the introduction of a practice-oriented computer curriculum and were used as an argument to justify the introduction of a teaching and learning subject with a clearly pre-vocational purpose into general education. In this sense, the microcomputer was envisioned as a crucial tool for effectively teaching the desired skills and knowledge in using the new technology and developing the "algorithmic" thinking and problem-solving skills, that would be required of the future workforce.

In parallel with the development of a new computer curriculum, educational policymakers in the GDR initiated the development of a dedicated school computer whose material and technical specifications were to be closely aligned with the requirements of classroom use. In this sense, the computer was seen not so much as a piece of hardware that could be used universally in any context, but rather as something that had to be at least partially reimagined and redesigned as an educational technology. For the pedagogues at the APW, the development of the BIC was initially tied to the hope of making the most modern computer technology available to schools that was in line with international standards and trends. However, this hope was dashed, and the GDR's school computer ended up being a short-lived phenomenon as schools turned to the Western computers that had become available after the fall of the Berlin Wall. This development illustrates the fragile nature of the educators' belief in the SED's sociotechnical imaginary, which promised that the GDR would be able to keep pace with technological developments in the West and prove the country's innovative prowess under the party's authoritarian leadership.

6 Extracurricular Computer Education

Leisure activities involving computers posed new opportunities for the young to learn about computer technology and what it had to offer, outside of the curriculum-regulated paths of school-based computer education. Computer classes in schools were dedicated to a dominant sociotechnical imaginary propagated by the political leadership that envisioned computer technology primarily to boost economic productivity the GDR's industry and as a pathway to renewed prosperity of the socialist society and its system. Learning with and about computers in the state-controlled educational system was thus inscribed with a mainly instrumental purpose and served to prepare pupils for their future life as productive members and workers in the socialist society. But outside of the formal educational system, computer enthusiasts found ways to engage with this new technology in their own ways and to appropriate it according to their individual needs and ideas.

However, the political leadership of the GDR also promoted the use of computers as a leisure activity with specific goals in mind. Youth policy was an essential part of the political-ideological work of the SED to consolidate class-oriented thinking and behavior among young people and to achieve their agreement with the political course of the SED.[792] The political-ideological influence and regimentation of the lives of children and young people began at an early age. The Pioneer Organization "Ernst Thälmann" was responsible for promoting the ideology and youth policy of the SED among younger schoolchildren.[793] From the age of 14, they were admitted to the Free German Youth (ger.: Freie Deutsche Jugend, FDJ).[794] The FDJ was the only state-approved and supported youth organization in the GDR. It dominated the lives of all young people, being the most important organizer of cultural, leisure, and holiday activities for young people in the GDR. The task of the FDJ was to introduce the youth to the Marxist-Leninist world view, to educate young people to become "class-conscious socialists," and to identify with the authoritarian leadership's policies and visions for the future of the GDR.

Supported by the activities and structures of the Pioneer Organization "Ernst Thälmann" and the FDJ, the SED's youth policy thus aimed, among other goals, to

---

792 Christa Mahrad, "Jugendpolitik in der DDR," in *Jugend im doppelten Deutschland*, ed. Walter Jaide and Barbara Hille (Opladen: Westdeutscher Verlag, 1977), 211.
793 Beate Kaiser, *Die Pionierorganisation Ernst Thälmann. Pädagogik, Ideologie und Politik: eine Regionalstudie zu Dresden 1945–1957 und 1980–1990* (Frankfurt am Main: Peter Lang, 2013).
794 Henry Krisch, *The German Democratic Republic: The Search for Identity* (Boulder, CO: Westview Press, 1985), 154–58.

mobilize the younger generation to identify with the SED's social and economic strategy with regard to new computer technology and to become personally invested in the realization of the sociotechnical imaginary promoted by the party. A major focus of the SED's youth policy was extracurricular education in the form of organized leisure-time activities, which was designed to channel the activities of children, adolescents, and young adults in a direction that was deemed desirable by the authoritarian government.

The system of extracurricular education allowed for a much broader impact of computers among the people by providing the public with access to this new technology, and in particular, by promoting and developing special talents among the young in the field of computing and informatics. As the socialist school system was unable to respond quickly to new developments in science and technology and was restricted by the obligation to provide all pupils with an equal minimal level of a basic understanding of computer technology, extracurricular education offered more flexibility to develop and nurture scientific and technological interests and talents.

On the one hand, extracurricular education was put in charge of responding to the desire of the youth to be involved in the societal process of computerization and address the fascination of young learners for computers.[795] On the other hand, it was also aimed at helping to satisfy the expected growing demand for workers who were skilled in the use of computer technology in order to realize the SED's economic strategy.[796] It complemented the formal school system in providing a larger part of the youth with a computer education, instilling certain ideological and moral convictions and attitudes, as well as sparking interest in science and technology.[797] In consequence, the politically motivated organization of extracurricular activities pursued the purpose of matching the needs of learners with societal requirements. In this sense, extracurricular activities were considered to contribute to the development of interests that were deemed relevant to society, to the promotion of creative thinking and practice, to the formation of certain politically desired personality traits and attitudes, and to the stimulation of learning.[798]

---

795 Hutterer, "Orientierungen," 7.
796 Hutterer, "Orientierungen," 5.
797 See § 17 of the GDR's Law on the Unified Socialist Education System (SAPMO ZB 20049, "Gesetz über das Einheitliche Sozialistische Bildungssystem vom 25. Februar 1965," Gesetzblatt der Deutschen Demokratischen Republik 1965, part 1, no. 6 (February 25, 1965): 83).
798 Hutterer, "Orientierungen," 9.

6.1 Popular Education and Extracurricular Learning about Computer Technology

The field of extracurricular computer education in the GDR encompassed a wide range of different activities. Public events served as a platform to display of computer technology and its uses to a broader public. Computers were featured prominently at the "Central Festival of Learning" (ger.: Zentrales Fest des Lernens) in 1986 at the Palace of the Republic in East Berlin, as well as at the "MMM" – the "Messe der Meister von Morgen" (eng.: Fair of the Masters of Tomorrow), a fair for young innovators to showcase their achievements and compete against other young researchers and developers for medals.[799] Various computer clubs participated in the central MMM in Leipzig with lectures on computers and programming, displays of their computer technology projects, and self-developed software to provide the broader public with insights into their activities.[800] The main goal of the MMM was to showcase the GDR's technological innovation capabilities and enthusiasm of the young generation to drive scientific and technological progress, thus increasing the credibility of the sociotechnical imaginary of a thriving computerized society and economy as a desirable and attainable future.

Computer technology also featured prominently in educational television and radio shows, aimed at teaching the broader population about this new technology, its use and purposes in the socialist economy and society of the GDR. In October 1986, the radio show "Effektives Programmieren in BASIC" (eng.: Effective Programming in BASIC) was aired on Radio DDR II, which was met with a highly positive response from listeners.[801] In January 1987, a new show under the title of "BASIC – 1x1 des Programmierens" (eng.: BASIC – Programming 101) was launched with 20 episodes, accompanied by free written materials for learners, which were later also published as a special issue by the popular science magazine *URANIA*.[802]

---

[799] "Lernfest – Freundenfest," *technikus*, no. 7 (1986): 17–19; "Computer und Software prägen MMM in Dresden," *Neues Deutschland*, July 10, 1987, 2.
[800] Norbert Spitzky, *Die 30. Zentrale Messe der Meister von morgen (ZMMM) im Urteil der Besucher* (Leipzig: Zentralinstitut für Jugendforschung, 1988), 35–36, https://nbn-resolving.org/urn:nbn:de:0168-ssoar-401358 (accessed May 31, 2024); Norbert Spitzky, Leonhard Kasek, and Michael Chrapa, *Information zur 31. Zentralen Messe der Meister von Morgen* (Leipzig: Zentralinstitut für Jugendforschung, 1989), 17, https://nbn-resolving.org/urn:nbn:de:0168-ssoar-402897 (accessed May 31, 2024).
[801] "REM und der DT 64-Computerklub – die Computermagazine des DDR-Rundfunks," *Funkamateur*, no. 5 (1989), http://www.kc85emu.de/scans/fa0589/REM.htm (accessed May 31, 2024).
[802] Ursula Grote and Horst Völz, *BASIC: Einmaleins des Programmierens. Eine Einführung in die Programmiersprache BASIC* (Leipzig/Jena: Urania-Verlag, 1987).

The programming course was authored by Horst Völz, a professor at the ZKI, and broadcast under the radio station's program "REM – das Computer Magazin."[803] The show was hosted by Joachim Baumann, a science editor of the school radio at Radio DDR II. In each of the weekly episodes, a short computer program was broadcasted over radio. The transmission used audible tones, which the radio listener could record on a cassette tape at home. The computer program could then be loaded to a computer with the help of a cassette player. The transmission was accompanied by an explanation of the program and instruction in some of the BASIC commands used.[804] The radio show was an astounding success, and the radio station was flooded with countless phone calls and reportedly over 20,000 letters by listeners – a response that clearly indicated "an explosion of interest in computers," as the show editors remarked.[805] In consequence, the BASIC programming course by Völz was also adopted by the GDR youth radio station DT 64 and broadcast as part of its "Computerclub" radio program.[806] Following on from the great success of the show and given the great interest of the audience in computing, an advanced BASIC programming course was broadcasted on both channels starting in January 1988. In 1989, Völz also created two additional computing courses which were broadcasted over Radio DDR and DT 64, one on machine code, and one on BASICODE, a cross-platform BASIC standard designed which allowed for the transmission of software by radio.[807]

With a similar aim of popularizing new computer technology and educating the people about how and what it could be used for, GDR television created a show dedicated to computer technology. "Computerstunde" aired for the first time on 26 May 1987.[808] The show was geared towards viewers with little to no prior knowledge of computer technology and aimed to help them overcome their fears and reservations about the new technology. It was not intended to replace a computer user course, but to lower the barrier to using computers. The show was hosted by Gabriele and Reinhard Lehmann from the Martin Luther University of Halle.[809] As

---

803 The name refers to the keyword REM, which is used to introduce comments in the BASIC programming language.
804 Lars Leppin and Tom Schnabel, "Informatik und Rechentechnik in der DDR [Studienarbeit]" (Humboldt-Universität zu Berlin, April 28, 1999), 45–47, https://archive.org/details/Leppin-Schnabel_1999_Informatik-in-der-DDR (accessed May 31, 2024).
805 "Den Computer kennenlernen," *Berliner Zeitung*, March 28, 1987, 3.
806 "REM und der DT 64-Computerklub – die Computermagazine des DDR-Rundfunks," *Funkamateur*, no. 5 (1989), http://www.kc85emu.de/scans/fa0589/REM.htm (accessed May 31, 2024).
807 Horst Völz, *Persönliches zur Rechentechnik in der DDR* (2015), 32, https://docplayer.org/68357290-Persoenliches-zur-rechentechnik-in-der-ddr.html (accessed May 31, 2024).
808 "'Computerstunde' im DDR-Fernsehen," *Neue Zeit*, May 22, 1987, 2.
809 Meyer, *Computer in der DDR*, 115.

part of the show, the two scientists demonstrated the practical use of a microcomputer and various software applications on a PC 1715 that had been set up in the studio. While the first episode focused on an introduction to computers and presented various domestically produced computers, the following episodes focused on ready-made software and what could be done with it. The second season of the show was concerned with teaching the foundations of programming in BASIC, an introduction to CAD/CAM workplaces and the preparation of GDR companies for the introduction of computers into production and office workplaces.[810] As was the case with the radio programs, the television program also attracted a large number of viewers. The television show also featured a quiz for viewers, as part of which the correct answers that had been sent in were entered into a draw for prizes, such as microcomputer kits, pocket calculators and quartz watches.[811] A staggering number of around 300,000 viewers took part in the quiz over the first six episodes.[812]

In addition to the mass media coverage, and the publication of popular science books and magazines on computing and programming, the GDR's institutions of popular education also stepped up their offers in courses and talks in order for the broader population to learn about new computer technology and informatics.[813] The GDR's society for the popularization of scientific and technological knowledge "Urania" hosted events and public lectures, often in cooperation with partners from science and industry.[814] Similarly, adult education centers offered courses in computer use and programming in BASIC. In contrast to the systematic computer education conveyed in vocational education and training, these courses were meant for the initial familiarization of the general public with the new technology, setting a particularly low barrier of entry in order to relieve people's fears and reservations of the new technology. The public lectures organized by Urania reflected current social and technological developments and created an opportunity for dialogue between the general population and the computer experts.[815] As such, the Urania events for popular education posed an opportunity for people to openly express their opinions and reservation against new technology. However, they often

---

810 "'Computerstunde' im DDR-Fernsehen," *Neue Zeit*, May 22, 1987, 2.
811 "Einblicke für jedermann: Neue Sendereihe 'Computerstunde' im Fernsehen," *Berliner Zeitung*, May 23, 1987, 7
812 "Grosses Zuschauerecho auf Fernseh-Computerstunde," *Neues Deutschland*, July 10, 1987, 4.
813 Meyer, *Computer in der DDR*, 110–12.
814 "Anhaltendes Interesse für URANIA-Computerlehrgänge," *Neues Deutschland*, October 10, 1987, 10; "Vorgeführt, was Computer und Roboter leisten," *Neues Deutschland*, 21.6.1986, 7.
815 Manfred Gartz, "Die Urania im Bildungssystem der DDR 1980/90," *REPORT Literatur- und Forschungsreport Weiterbildung*, no. 30 (1992): 101.

also served the propaganda, legitimization, and implementation of the SED's sociotechnical imaginary and technology policy.

Despite the many opportunities for the people of the GDR to learn about the new computer technology and to become acquainted with how and for what purposes it could be used, one essential element was lacking: Access to computer technology outside of school and the workplace to put the knowledge acquired into practice and to explore the possibilities of the computer on one's own terms.

The domestic production of microcomputers in the GDR was very limited due to scarce resources and an insufficient number of skilled workers who could build computers. Western computer technology was hard to come by because the COCOM embargo did not allow for exports of microchip-based technologies to socialist countries.[816] The few computers that were produced in the GDR were designated for the use in companies, educational institutions, the military, and research facilities.[817] Thus, both as a result of limited resources and the fact that home computing was not part of the dominant sociotechnical imaginary that guided the SED's decisions regarding the production and distribution of computers as a consumer good, microcomputers were not generally accessible to the broad public outside of the workplace and educational institutions for private use. The restricted access posed a significant barrier to certain practices and purposes of using computers which fell outside the political leadership's imaginary of what uses computer were intended for. The easiest option for teenagers and young adults to gain access to a microcomputer in their free time was to join one of the GDR's numerous extracurricular computer clubs and study groups. Officially acknowledged computer clubs were granted access to the computer labs in educational institutions and youth centers as part of the SED's efforts to make computers accessible to the broader population.

In the following, three forms of extracurricular activities are discussed, through which school-aged children, teenagers, and young adults gained access to and learned about computers. Each of these forms was situated in a different social setting with various degrees of autonomy regarding the purposes of computer use: 1. Individual engagement with computers at home; 2. membership in school-associated computer classes and societies for talented pupils (often taught by mathematics or physics teachers after school hours); and 3. the participation in computer clubs in youth centers, which were open to all.

---

816 Klenke, *Kampfauftrag Mikrochip*, 61; Donig "DDR-Computertechnik und COCOM-Embargo;" Klenke, "Globalisierung, Mikroelektronik und Scheitern," 424.
817 Danyel and Schuhmann, "Wege in die digitale Moderne," 301.

Individual tinkerers and autodidacts

Computer-curious individuals who did not have the opportunity or preferred not to join a state sponsored computer club experimented with hard- and software on their own – largely unsupervised, following their own visions and ideas of how this new technology could be put to good use. These hobbyists thus transgressed the SED's attitudes and approaches to computing, which were oriented towards economic modernization rather than visionary ideas of creative exploration. The socialist regime of the GDR saw computer technology primarily as a means to optimize and improve efficiency in industry and administration, thus raising the economy's competitiveness on the global market. Such an imaginary of the appropriate use of information technology did not entail the use of computers by the people in their private homes for their own purposes.

Computer enthusiasts in the GDR found creative ways to acquire or even build their own hardware for private use, backed by the prolific do-it-yourself culture of the GDR. The people of the GDR engaged in do-it-yourself practices not only as a hobby and pastime, but it also served as a strategy for coping with and overcoming the very limited choice and notorious shortage of consumer goods in the socialist society.[818] At first, building simple computers for learning purposes was mainly practiced in higher and vocational education institutions. But with the distribution of assembly kits – made of parts that did not meet the strict quality standards or from production above plan – computer enthusiasts with a basic understanding of electronics were enabled to build their own simple but functional home computer with the help of construction manuals, step-by-step guides in magazines, and books or the instruction of a more experienced mentor. This kind of "build-it-yourself" computer was not a phenomenon specific to the GDR. Kits for building simple personal computers but had already been produced and sold in the USA since the early 1970s – ten years before this kind of consumer product would become available in the GDR.[819] A popular example of a GDR single-board computer sold as a kit was the Z 1013, produced by the VEB Robotron-Elektronik Riesa between 1985 and 1990.[820] The Z 1013 kit had to be pre-ordered and was usually only delivered after a

---

818 Reinhild Kreis, "A 'Call to Tools': DIY between State Building and Consumption Practices in the GDR," *International Journal for History, Culture and Modernity* 6, no. 1 (2018): 49–75.
819 Haigh and Ceruzzi, *New History of Modern Computing*, 172–73.
820 Klaus-Dieter Weise, *Anlage 1 – Erläuterungen zu Begriffen im Dokument 'Erzeugnislinie Heimcomputer, Kleincomputer und Bildungscomputer des VEB Kombinat Robotron' der UAG Historie Robotron der Arbeitsgruppe Rechentechnik in den Technischen Sammlungen Dresden* (Dresden: Förderverein für die Technischen Sammlungen Dresden, 2005), 46–55, http://robotron.foerderverein-tsd.de/322/robotron322b.pdf (accessed May 31, 2024).

waiting time of one year. It included the hardware components, as well as operating instructions and a detailed manual. The Z 1013 kit itself was educational in nature, as the assembly of the Z 1013 involved the acquisition and development of skills and knowledge in applied microcomputing through experimentation and practice.[821] The manuals and various articles on the Z 1013 published in magazines for DIY enthusiasts and electronic hobbyists offered guidance for self-learners.[822] In a similar vein, the electronics hobbyist's magazine *Funkamateur* developed and published the construction manuals for the so-called "amateur computer" AC1.[823] The popular science and technology magazine *Jugend + Technik* developed its own self-build "JuTe-Computer," aimed especially at young readers and computer-curious teenagers.[824] In contrast to the Z 1013, however, both for the AC1 and the JuTe-Computer, the magazines merely printed and distributed construction manuals and did not sell any kits.[825] All electronic components, necessary materials and peripherals needed to be acquired independently. In addition, building an AC1 or JuTe-Computer involved the complicated and time-consuming assembly of the printed circuit boards.

The fact that home computers were largely only available in the form of self-assembly kits rather than prebuilt devices, initially resulted in a limited impact of computer technology on the broad public. Having to assemble your own hardware posed a high entry barrier to home computing, as it required access to the technical components and assembly equipment, as well as the necessary knowledge, skill, and time. But even after the first domestically produced microcomputers in the KC-85-series became available in the GDR, these pre-assembled computer systems were first and foremost distributed to public institutions and only sold in small quantities as consumer goods to individuals.[826] Moreover, they

---

821 "Mikrorechner zum Einsteigen," *technikus*, no. 7 (1986): 41.
822 "Mikrorechnerbausatz Z1013. Technische Daten – Bezug – Erweiterungen," *practic*, no. 2 (1987): 54; "Vom Bausatz zum PC," *practic*, no. 3 (1989): 135–138; "Mikrorechner-Bausatz aus dem VEB Robotron-Elekronik Riesa," *Funkamateur*, no. 12 (1984): 612–13.
823 Julia G. Erdogan, "Computerkids, Freaks, Hacker: Deutsche Hackerkulturen in internationaler Perspektive," in *Let's Historize it! Jugendmedien im 20. Jahrhundert*, ed. Aline Maldener and Clemens Zimmermann (Köln: Böhlau Verlag, 2018), 64.
824 "JU+TE Computer selbst gebaut," published in nine parts in the magazine Jugend + Technik, issues no. 7 (1987) to no. 2 (1988). The collected instructions were made available online by Volker Pohlers, *Dr. Helmut Hoyer: JU+TE Computer selbst gebaut* (2004), https://hc-ddr.hucki.net/wiki/lib/exe/fetch.php/tiny/jutecomp1.pdf (accessed May 31, 2024). In 1989, the instructions to build the 'TINY' computer were also published as a booklet: Helmut Hoyer and Norbert Klotz, *TINY, der kleine Selbstbau-Computer* (Berlin: Verlag Junge Welt, 1989).
825 Meyer, *Computer in der DDR*, 48–49.
826 Danyel and Schuhmann, "Wege in die digitale Moderne," 301.

were unaffordable for most, so that an extensive home computing culture comparable to the one in the USA or West Germany could not develop in the GDR.[827] Similar difficulties were encountered in attempts to acquire a Western computer because of the embargo that was established by the Western Bloc, which heavily restricted the export of computer technology to socialist countries.[828]

Nevertheless, there were ways and means for a citizen of the GDR to acquire a Western computer. One option was to receive it as a gift from relatives that lived in West Germany, or GDR citizens who were allowed to travel to the West. The GDR also maintained several hundred "Intershops" where Western products such as a Commodore 64 or an Atari 800 could be purchased but had to be paid for in West German Mark.[829] Another option was to buy a Western computer second-hand. Over the course of the 1980s, journals and special interest magazines featured private classified ads for computers and peripherals. Second-hand and home electronics shops in larger cities occasionally sold second-hand technology from the FRG.[830]

School computer clubs and study groups to promote talented pupils

An alternative to individual do-it-yourself projects and attempts to acquire a KC-85 or even a Western computer, was to gain access to computers in educational facilities and youth centers by joining a computer club or after-school study group. For the majority of young people, who could not afford or build their own home computer, extracurricular clubs were the only way they could use a microcomputer in their leisure time. The SED leadership supported the establishment and promotion of computer clubs and courses. The party's political and material support of computer clubs was aimed at increasing the impact of computer technology on the broader public, and in particular the young generation, by offering them hands-on learning experiences. However, the SED's support of extracurricular computer clubs can also be understood as an effort to channel young people's leisure activities with computer technology into formal settings, in order to gain oversight and control over their activities.

The first computer study groups and clubs in the GDR were initiated by teachers or engineers and computer experts in local combines in the early 1980s and established at schools, "stations for young scientists and technicians" or "pioneer

---

[827] Erdogan, *Avantgarde der Computernutzung*, 74–75.
[828] Donig, "DDR-Computertechnik und COCOM-Embargo."
[829] Meyer, *Computer in der DDR*, 71.
[830] Meyer, *Computer in der DDR*, 71.

houses."[831] The "stations for young scientists and technicians" were centers for after-school education in mathematics, natural sciences, and technology. They offered a variety of study groups for pupils with a particular talent or interest in a specific field or discipline. The "pioneer houses" were leisure centers for young people under the auspices of the FDJ, which provided a wide range of extracurricular courses and clubs for creative, scientific, technical, and social leisure activities.[832] Facilities and technical equipment for the clubs were often provided by regional and local education authorities and local companies.

The establishment of computer clubs and study groups directly associated with schools served to provide an environment for children and teenagers to learn about and use a computer in their leisure time, but in a more formal setting under the qualified pedagogic guidance of teachers and instructors. Some teachers and pedagogues in the GDR argued, that such a guidance of students' activities involving computers was indispensable, in particular during the phase when they first encountered this new technology and learned how to make use of it.[833] Without supervision and regulation by a teacher, they reasoned, the possibilities of the computer were rarely explored systematically.[834] This random approach to gaining an understanding of this new technology would then potentially lead to rather lopsided ideas about computers and informatics. Following this line of reasoning, extracurricular computer classes were set up in association with local schools and companies, which were in principle modelled after traditional school lessons with interchanging phases of theoretical instruction and practical work on the computer but did not have to adhere to an official syllabus or specific curricular goals.

The GDR's numerous school-associated extracurricular courses and study groups for physics and microelectronics targeted in particular students that performed well in mathematics or physics.[835] The extracurricular computer classes were often taught by teachers of these subjects or skilled workers from local companies with professional experience in microelectronics and computing.[836] The dif-

---

831 One of the earliest computer clubs in the GDR was the study group "Junge Konstrukteure" at the Pioneer Palace "Ernst Thälmann" in Berlin, founded in 1980 ("Das ist ihre Welt: Knobeln, Konstruieren und Bauen," *technicus* no. 4 (1985): 43).
832 Leonore Ansorg, *Kinder im Klassenkampf: Die Geschichte der Pionierorganisation von 1948 bis Ende der fünfziger Jahre* (Berlin: Akademie Verlag, 1997), 102–4.
833 See for example BBF PL 87–12–04, "Computerprogrammierung mit Schülern in der außerunterrichtlichen Tätigkeit," dated 1987, 3.
834 BBF PL 87–12–04, "Computerprogrammierung mit Schülern in der außerunterrichtlichen Tätigkeit," dated 1987, 3.
835 "Station im Kommen," *technikus*, no. 5 (1986): 2.
836 Extracurricular study clubs often formed partnerships with local combines, see for example "Paten, Pioniere – Partner," *technikus*, 6 (1986): 27.

ferent qualifications and inclinations of instructors resulted in different content orientation and objectives of the courses. Mathematics teachers often focused predominantly on numerical informatics and programming, while skilled workers with experience in electronics and automation concentrated more on hardware tinkering and the solving of problems inspired by practical applications in industry. Although the school-based computer clubs were highly practice-oriented, the learning setting was similar to that of classroom instruction. The activities were pre-structured and closely supervised by the teacher or study group leader. Thus, it depended heavily on the instructor to what extent the students were able to contribute and implement their own ideas, needs, and imaginations regarding the use of a computer. Some of these clubs and study groups focused primarily on computer hardware: Tinkering and experimenting with circuits, microelectronic components, and building their own devices as well as peripherals and additional equipment for computers. These technically oriented groups appealed mostly to older male teenagers, some of which had already experimented with electronics at home or seized the opportunity to build their own home computer with the help of the study group's instructor and the tools that were provided.[837] In contrast, a much larger and growing part of learners were exploring the computer as a ready-made technical device, its application and programming possibilities, the development of their own small programs as well as the use of a computer for learning, playing, data management, and problem solving in various fields of interest.[838] These study groups were primarily concerned with software development, often inspired by challenges to automate various tasks in economic production. Especially in scientifically and technologically oriented clubs and study groups, the fulfillment of specific work assignments in response to the needs of local companies was an important component of extracurricular computer activities.[839] For example, students would write programs to control robots and machine tool models used in industrial production, or optimize experiments on feeding, breeding, and animal behavior in agriculture.[840]

The exclusivity of access to a computer club or study group depended on the respective group leader or instructor. Since demand for extracurricular computer education by students in many cases exceeded the capacity being offered in many places, a selection had to be made as to which students were accepted. In other cases, mathematics or physics teachers had been setting up and were conducting

---

[837] Hutterer, "Orientierungen," 12.
[838] Hutterer, "Orientierungen," 12.
[839] "Betriebe geben den Schülern Aufträge," *Berliner Zeitung*, February 28, 1987, 11.
[840] Hutterer. "Orientierungen," 14.

the computer clubs and study groups in their free time to provide their highest achieving students with more extensive opportunities to develop their talents and capabilities – which the comprehensive school of the GDR without any internal or external differentiation did not permit.

One example of this were the computer lessons of the Potsdam Club of Young Mathematicians (ger.: Potsdamer Klub der Jungen Mathematiker, PKJM). The PKJM served as an extension of school mathematics for students with a special talent and interest in this area.[841] It focused its activities on the preparation of students for the Olympiad of Young Mathematicians and their later educational career.[842] The club comprised of mathematics classes taught by volunteering mathematics teachers at the pioneer house "Erich Weinert" in Potsdam and were attended by students during their free time. In September 1986, the PKJM acquired two microcomputers and another one in the following month, which allowed for the inclusion of informatics into the weekly lessons for classes 9 to 12.[843] The most talented students of grades 6 to 8 were taught informatics and were allowed to use the computer lab at the Potsdam University of Teacher Education "Karl Liebknecht" once a month for an additional two to three hours together with the club members of grades 11 and 12.[844] These students learned how to solve numerical problems on the computer by writing their own programs. Structured programming in BASIC was at the core of these computer classes, with the intent that the students would develop algorithmic thinking skills.[845]

The computer lessons of the PKJM are an example of how the historical link between the academic disciplines of informatics and mathematics lived on and was reinforced not only in general schooling, but also in extracurricular education. Because the discipline of informatics originated in university mathematics departments, and tertiary education and training initially focused primarily on mathematics students and student teachers, many of the instructors had an academic background in mathematics. The close ties between mathematics and informatics were reinforced through these instructors, by focusing on solving numerical problems and by targeting primarily the mathematically talented

---

[841] BBF PL 87–12–04, "Computerprogrammierung mit Schülern in der außerunterrichtlichen Tätigkeit," dated 1987, 11.
[842] BBF PL 87–12–04, "Computerprogrammierung mit Schülern in der außerunterrichtlichen Tätigkeit," dated 1987, 11.
[843] BBF PL 87–12–04, "Computerprogrammierung mit Schülern in der außerunterrichtlichen Tätigkeit," dated 1987, 11.
[844] BBF PL 87–12–04, "Computerprogrammierung mit Schülern in der außerunterrichtlichen Tätigkeit," dated 1987, 12.
[845] BBF PL 87–12–04, "Computerprogrammierung mit Schülern in der außerunterrichtlichen Tätigkeit," dated 1987, 12–13.

youth. In a similar fashion, *alpha* – a mathematics journal for students to stimulate their interest in mathematical problems and support their extracurricular learning – occasionally included articles on computers, programming, and informatics since the early 1980s. In 1986 and 1987, *alpha* published a seven-part series "Mini-BASIC for alpha readers" specifically catering to extracurricular computer classes and study groups.[846] This self-study course in BASIC programming was a valuable resource for educators and students, as they often exclusively had access to GDR microcomputers of the type KC 85, which could only be programmed in BASIC.

Numerous other so-called "Pupils' Societies" for computer technology and informatics, either with a mathematical or a technical orientation, were set up at universities and scientific institutions in the GDR, albeit often limited to larger cities where such research centers and universities were located. Examples of more technically oriented pupils' societies were the technical pupils' society in Berlin[847] or the pupils' society for electronic technology at the Technical College in Leipzig, where 20 selected 9th grade students from Leipzig were taught about microcomputers and involved in research projects since October 1986.[848]

A particularly prestigious institution offering extracurricular computer classes which served to promote particularly talented students was the "Schülerrechenzentrum Robotron" (eng.: Pupils' Computer Center) at the pioneer palace in Dresden,[849] established as the "Technisches Kabinett Mikroelektronik" (eng.: Technical Lab for Microelectronics) in 1982.[850] The initiative to create the Pupils' Computer Center stemmed from a decision by the SED district leadership in Dresden on 4 February 1983 to find and promote talent for the GDR computer combine Robotron from an early age.[851] The Robotron computing center for pupils offered courses in electronics, computer technology, and robotics, that often spanned over multiple years and included

---

846 "Mini-BASIC für alpha-Leser," *alpha*, no. 5 (1986): 102–3; "Mini-BASIC für alpha-Leser 2. Teil," *alpha*, no. 6 (1986): 126–27; "Mini-BASIC für alpha-Leser 3. Teil," *alpha*, no. 1 (1987): 17–18; "Mini-BASIC für alpha-Leser 4. Teil," *alpha*, no. 2 (1987): 44–45; "Mini-BASIC für alpha-Leser 5. Teil," *alpha*, no. 3 (1987): 62 and 67; "Mini-BASIC für alpha-Leser 6. Teil," *alpha*, no. 4 (1987): 84–85; "Mini-BASIC für alpha-Leser 7. Teil," *alpha*, no. 6 (1987): 134–35.
847 "Betriebe geben den Schülern Aufträge," *Berliner Zeitung*, February 28, 1987, 11.
848 "Bildungsvorlauf für technische Begabungen," *Neues Deutschland*, January 17, 1987, 7.
849 Steffi Heinicke and Michael Unger, "Das Schülerrechenzentrum Dresden von 1984 bis heute," in *Informatik in der DDR – Grundlagen und Anwendungen*, ed. Birgit Demuth (Bonn: Gesellschaft für Informatik, 2008), 220–30.
850 Meyer, *Computer in der DDR*, 104.
851 Heinicke and Unger, "Schülerrechenzentrum Dresden."

practical training and internships at Robotron as well as local research facilities.[852] The highest-performing pupils at the center received a support contract with Robotron to sponsor their higher education in computer technology, before they would return as highly skilled experts to work for Robotron.[853] This type of extracurricular computer education thus explicitly pursued the goal of preparing talented and high achieving students for a professional career in informatics or computer engineering. To be admitted to a study group at the Pupils' Computer Center pupils had to demonstrate a special aptitude for microelectronics and informatics in an entrance test.[854] A total of 250 pupils of grades 7 to 10 were members of one of the center's 20 study groups. In the 14 informatics study groups, they learned programming in BASIC and later PASCAL to solve various tasks on the computer. Pupils in one of the 6 microelectronics study groups learned about digital circuits and programming for computer-aided control of machines and assembly robots. The groups met at the center for two compulsory lessons per week and additional individual practice lessons. Computer specialists from the Robotron combine and scientists from the Technical University and the University of Education in Dresden, as well as from the APW research center, served as instructors at the Pupils' Computer Center.[855]

The promotion of special talents in the field of informatics and computing through extracurricular activities was clearly a strong political and economic motive for the SED to invest in providing children, teenagers, and young adults with access to computer technology in their spare time. To further fuel their zeal to learn, competitions were held for computer hardware tinkerers and software programmers. They were designed to encourage a spirit of innovation and strive for academic excellence among the younger generation, helping to create a pool of eager learners, innovators, and high performers from which the GDR's economy could recruit the skilled computer experts it needed to realize the SED's vision of a computerized future. In 1987, the GDR's first open Programming Olympiad was held by the Technical University of Dresden.[856] Participants were tasked with developing a BASIC program for data management on a KC 85/1 microcomputer as fast as possible, and then adapt their program in the final rounds to solve new challenges that were posed to the contestants. The first prize was a brand-new KC 87 microcomputer.[857] Countless other inventor competitions were held locally, for ex-

---

852 Weise, *Anlage 1 – Erläuterungen zu Begriffen*.
853 Meyer, *Computer in der DDR*, 104; "Begeistert für Chips und Computer," *Neues Deutschland*, July 16, 1988, 7.
854 "Begeistert für Chips und Computer," *Neues Deutschland*, July 16, 1988, 7.
855 "Begeistert für Chips und Computer," *Neues Deutschland*, July 16, 1988, 7.
856 Meyer, *Computer in der DDR*, 112–13.
857 Meyer, *Computer in der DDR*, 113.

ample at the FDJ "Pioneer Palace" (ger.: Pionierpalast) in Dresden, where young computer enthusiasts developed computer games, models, and teaching aids.[858]

Informal youth computer clubs

The establishment of private associations and independent social organizations was theoretically possible in the GDR, but they required state recognition in order to legally carry out their activities.[859] However, this was rarely put into practice because the idea of private associations conflicted with the ideology of state socialism. For this reason, computer clubs and societies were usually established under the auspices of officially recognized mass organizations such as the "Kulturbund," the GDR's Society for Sports and Technology (ger.: Gesellschaft für Sport und Technik, GST), and the FDJ. Through the SED's mass organization, extracurricular computer education in clubs or societies was thus directly integrated into the SED's system of power and state control over clubs, thereby securing a degree of supervision and regulation by the authoritarian regime over their members and activities. In contrast to the more school-like organized computer clubs associated with schools, universities and research institutions, the computer clubs established under the auspices of the GDR's mass organizations were open to anyone interested in computer technology, regardless of their academic achievements.

The GST, which was at its core a paramilitary sports organization, established numerous local sections for "Computer Sports" over the course of the 1980s.[860] The GST set up its own local computer labs or entered agreements with local institutions of higher education and vocational training to be allowed to use their computer labs at agreed-upon times. The GST also organized local and central programming competitions for the members of its computer sports sections and published the magazine *Funkamateur* which started to regularly feature articles on computer technology and programming.[861]

The FDJ targeted teenagers and young adults with its computer clubs, which were often set up at the FDJ's youth and leisure centers. One of the most popular FDJ computer clubs of the GDR was founded in January 1986 in East Berlin,

---

858 "Mikroelektroniker im Wettstreit," *Neues Deutschland*, November 21, 1987, 9.
859 SAPMO ZB 20049, "Verordnung über die Gründung und Tätigkeit von Vereinigungen vom 6. November 1975," Gesetzblatt der Deutschen Demokratischen Republik 1975, part 1, no. 44 (November 26, 1975): 723.
860 Schröder, *Auferstanden aus Platinen*, 92.
861 "Computersport in der GST bereits in 110 Sektionen," *Berliner Zeitung*, December 7, 1987, 3.

namely the computer club at the "House of Young Talents" (ger.: Haus der jungen Talente, HdjT).[862]

The club was led by Stefan Seeboldt, who was in his early 30s and had studied mechanical engineering. While the clubs' younger visitors were mostly interested in games, Seeboldt was enthusiastic about graphics programs. He gave lectures at the computer club on graphics software and programming languages.[863] The computer club met every Wednesday evening and was visited mostly by teenagers and young adults between the ages of 16 and early 20s. Although it was open for girls and women in theory, it was purely a boys' and men's club in practice.[864]

In comparison with the classroom-like instruction in the school-associated computer clubs and extracurricular classes, where small groups of students were instructed and closely guided by the club group leader, "open" computer clubs like the one at the pioneer palace resembled more an informal meetup of young people who shared a common interest in computers and software. In contrast to the fixed groups of school-associated computer clubs, the meetings at the pioneer palace were less exclusive, open to everyone and also allowed for irregular visits as it did not require a membership status.[865] Thus, it served as an ideal contact point for computer enthusiasts who were tinkering with hardware or programming their own software at home by themselves and occasionally wanted to share and exchange the results of their efforts with likeminded people, or needed technical support. Other regular visitors were in vocational education and training to become data processing specialists and wanted to explore computing outside the realm of workplace-related use. Apart from occasional lectures and presentations,[866] there was little structure to guide, control or constrain the activities of visitors who met up at the computer club. Initially, the club leader had offered programming courses in BASIC.[867] However, these were met by decreasing demand and were soon discontinued.

Upon Seeboldt's request and with the agreement of the HdjT's director, the computer club was equipped exclusively with Western home computers, produced by the American computer manufacturers Commodore and Atari.[868] Some

---

862 "Haus der jungen Talente hat jetzt Computerklub," *Berliner Zeitung*, January 23, 1986, 12.
863 Denis Gießler, "Video Games in East Germany: The Stasi Played Along," Zeit Online, November 21, 2018, https://www.zeit.de/digital/games/2018-11/computer-games-gdr-stasi-surveillance-gamer-crowd (accessed May 31, 2024).
864 Gießler, "Video Games in East Germany."
865 Meyer, *Computer in der DDR*, 109.
866 Meyer, *Computer in der DDR*, 108–9.
867 "Haus der jungen Talente hat jetzt Computerklub," *Berliner Zeitung*, January 23, 1986, 12.
868 Gießler, "Video Games in East Germany;" Erdogan, *Avantgarde der Computernutzung*, 276.

visitors also brought their own home computers to the meetings at the HdjT.[869] Unlike most computer labs at schools and combines which were equipped predominantly with domestic hardware, it offered new experiences by escaping the technical limitations of microcomputers produced in the GDR that could only run software specifically programmed for these devices. Western computers allowed young computer enthusiasts to, at least medially, transcend the borders of the GDR and try out games and software which had been developed in capitalist countries. However, there were simply not enough computers available at the pioneer palace in East Berlin to cater for the many visitors that showed up for the club's meetings. The focus was rather on the social aspect of meeting with likeminded young people and the formation of a collective identity, as well as being a part of a network for swapping computer games and programs.[870] In essence, as Erdogan (2021) describes in her book on hacker cultures in both Germanies, computer clubs and associations thus served as platforms for community building among young computer users with shared interests and experiences.[871]

The computer club at the HdjT seems to have met a real need and desire among the youth of the GDR, which was certainly not limited to East Berlin. In the second half of the 1980s, other clubs which focused exclusively on Western microcomputers were set up in larger cities, such as the C-16 Club in Dresden, the Atari Clubs in Berlin and Dresden, the Commodore Club Jena and the Atari Interest Group Rostock.[872] The clubs were often established in the form of "Kulturbundeinrichtungen," that is, as institutions under the umbrella of the "Kulturbund" which provided for the necessary legal and material basis for the official recognition and equipment of the clubs with computer technology.[873] The members of these clubs were usually private computer users who had managed to acquire an Atari or Commodore home computer.

As individual home computer users evaded ideological control by politically conformist pedagogues and leaders of computer clubs and study groups in schools and companies, they aroused the suspicion of the state security service. The Stasi was keen to find out, what kind of unsupervised activities private computer users were engaging in – especially if they preferred the use of Western hardware and software over domestic technology – and whether they posed a potential threat to

---

869 Erdogan, *Avantgarde der Computernutzung*, 274.
870 Gießler, "Video Games in East Germany."
871 Erdogan, *Avantgarde der Computernutzung*, 264–65.
872 Gießler, "Video Games in East Germany;" Meyer, *Computer in der DDR*, 78–79.
873 "Let's Go East – ATARI-Club in der DDR," *ST-Computer*, no. 7 (1990), https://www.stcarchiv.de/stc1990/07/atari-ddr (accessed May 31, 2024).

the SED's political hegemony.[874] The Stasi thus closely monitored the computing scene in the GDR and collected files with names, contact information and notes on the activities of some home computer owners and members of clubs dedicated to Western computer technology.[875]

6.2 Computer Technology Education in Popular Media

A number of popular science and technology magazines in the GDR regularly featured news articles and reports on the development and use of robots and computer technology in the GDR and other socialist countries, highlighting their potential to revolutionize economic production and bring prosperity for all.[876] In some cases, these articles were accompanied by reports on rising unemployment and deteriorating working conditions in Western capitalist countries, most prominently in the FRG.[877] The message was clear and simple: The computer was portrayed as an essentially neutral technology, the societal effects of which depended solely on the sociopolitical and economic system in which they were deployed. Automation and computerization were heralded as the decisive means to raise living standards and bring prosperity for all under socialism. But in the hands of monopoly capital, computer technology would only serve the purpose of profit maximization and run against the interests of the working population and society at large. An article in a popular youth magazine pointed out, that what is to be distrusted is not the technology itself, but the masters who hold control over the technology.[878] In the GDR, the article claimed, the SED party and the people were bound by a deep level of trust. In consequence, there would be no need to fear that a conflict between the SED party leadership and the working people would ever occur. The article went on to say that, in contrast to capitalist systems, in the socialist society of the GDR there was simply no sense of selfishness and egotism that could turn

---

874 Gießler, "Video Games in East Germany."
875 BArch, MfS, BV Halle, KD Weißenfels, no. 237, "Information zum Zusammenschluss privater Computerbesitzer," dated April 22, 1986; cf. Erdogan, *Avantgarde der Computernutzung*, 97–98 and 281; Gießler, "Video Games in East Germany."
876 For example, the youth magazines 'Jugend und Technik' and 'technikus' or the special interest magazines for a predominantly adult readership 'Funkamateur', 'Radio Fernsehen Elektronik', 'Mikroprozessortechnik', and 'Kleinstrechner TIPS'.
877 See, for example, "Neue Armut in altem System," *technikus*, no. 4 (1986): 38–39; "Vertrauen ins Morgen," *technikus*, no. 9 (1986): 1–2; "Eine Generation hat Angst," *technikus*, no. 6 (1987): 28–29; "Schlüsseltechnologien: Das Herz moderner Produktion," *technikus*, no. 6 (1987): 1–2.
878 "Vertrauen ins Morgen," *technikus*, no. 9 (1986): 2.

the achievements of scientific and technological development against the working people.[879]

In this sense, youth magazines with a technical focus such as *Jugend + Technik* or *technikus* served to popularize the SED's sociotechnical imaginary of computer technology among the young, reiterating the ideological narrative of economic and social progress that would be achieved with the help of new computer technology, and in contrast to the detrimental effects of computer use under capitalism.

The magazine *technikus* was a technical-scientific magazine for FDJ-"pioneers" aged 13 to 16.[880] Its aim was to foster the youth's interest in science and technology and contribute to the polytechnic instruction of pupils. The magazine frequently published articles on the topic of robots, computer technology and, programming throughout the 1980s. The articles covered new developments or innovative applications of computer technology in the GDR and other socialist countries and introduced readers to programming and the basics of microelectronics and computer technology. To assist the numerous computer clubs that had formed in the GDR over the course of the 1980s, the magazine also included suggestions and detailed instructions for projects in the field of microelectronics and computing, for example, to build an electronic dexterity game with an optical display.[881] It also regularly reported on the activities of computer clubs and study groups from all over the GDR. Readers were encouraged not to indulge in their hobby all on their own, but to join such a collective of young people who share the same interest.[882] On the one hand, this call corresponded to the socialist ethic of working in collectives rather than alone, emphasizing the value of working together in a team to achieve a common goal. On the other hand, the GDR's authoritarian leadership also had an interest in bringing individual tinkerers and computer enthusiasts under the supervision of official clubs and study groups, to have at least some degree of oversight and control over their activities. In addition, the pedagogical and ideological guidance provided by the club and group leaders and instructors was hoped to align the interests and activities of the young computer enthusiasts with broader societal and political goals and needs.[883] It was hoped that if these often very skilled computer users joined an official club, they could, for example, be encouraged to contribute to the solution of certain scientific and technical problems or to program software that could be used

---

879 "Vertrauen ins Morgen," *technikus*, no. 9 (1986): 2.
880 Pecher, "Kinderzeitschriften in der DDR," 36.
881 "Geschick zum Sehen," *technikus*, no. 10 (1986): 32–33.
882 "Unmögliches wird möglich," *technikus*, no. 1 (1987): 33. The article points out that scientific and technical problems can be solved better in collectives than by individuals, as the different knowledge and skills of the members complement each other.
883 Hutterer, "Orientierungen."

in schools or industry – and perhaps they could even be inspired to pursue a career in the field of new information technology and thereby actively contribute to the computerization of the socialist economy.

Frösi[884] was another children's magazine, but in contrast to *technikus* and *Jugend + Technik*, it was aimed at the Ernst Thälmann-pioneers, that is, younger children in grades 4 to 7.[885] It was also not exclusively focused on science and technology but covered a broad variety of topics and included inspiration for handicrafts and playing. The target group for *Frösi* were therefore children at an age when they had not yet encountered computers at school. Nevertheless, in 1986, the *Frösi* issue number 4 included a supplement, a so-called "Mini-Frösi," which was dedicated to introducing the young readers to computer technology in a playful manner. The small booklet was entitled "Tümico – Tütes Minicomputer," featuring *Frösi's* iconic character "Tüte," a black-haired boy, and his friend "Tümico," a microcomputer.[886] It contained puzzles loosely based on the concept of binary code. It was followed by two other "Mini-Frösi" supplements, which both featured a collaboration with the computer engineers Marina and Albert Jugel from the Robotron combine.[887] These supplements introduced the young readers in a child-friendly way to basic concepts of informatics and computing, such as binary code, programming commands and the inner components and workings of a microcomputer. Interspersed were short puzzles, in which the young readers would apply a simple concept such as the flow of electricity through a circuit with electric switches that were either on or off. Remarkably, one of the "Mini-Frösi" supplements used puzzles that were intended to introduce children to algorithmic thinking and computer graphics.[888] Each puzzle presented the reader with a table of numbers that encoded a particular location on a grid, a particular shape, and a particular color. If the correct shapes were drawn with pencils in the correct place and color, the solver was rewarded with a small picture of a light bulb, a pencil, or a space rocket (see Figure 9).

In addition to the playful introduction of the young readers to computer technology and informatics, the three "Mini-Frösi" supplements also used imagery to

---

[884] The name "Frösi" is derived from the first line of the popular GDR pioneer song "Fröhlich sein und singen" (eng.: "to be happy and sing"). "Fröhlich sein und singen" was originally the full title of the magazine. The abbreviated form 'Frösi' was introduced in 1965.
[885] Pecher, "Kinderzeitschriften in der DDR," 28–29; Wilkendorf, "Was bleibt?,"; Christine Lost, "Kinderzeitschriften und -zeitungen der DDR: Zwischen verschiedenen Betrachtungsweisen," in *Kinderzeitschriften in der DDR*, ed. Christoph Lüth and Klaus Pecher (Bad Heilbrunn: Verlag Julius Klinkhardt, 2007), 183.
[886] Mini-Frösi no. 27, supplement to Frösi issue no. 4 (1986).
[887] Mini-Frösi no. 28, supplement to Frösi issue no. 10 (1986), 1.
[888] Mini-Frösi no. 29, supplement to Frösi issue no. 1 (1987).

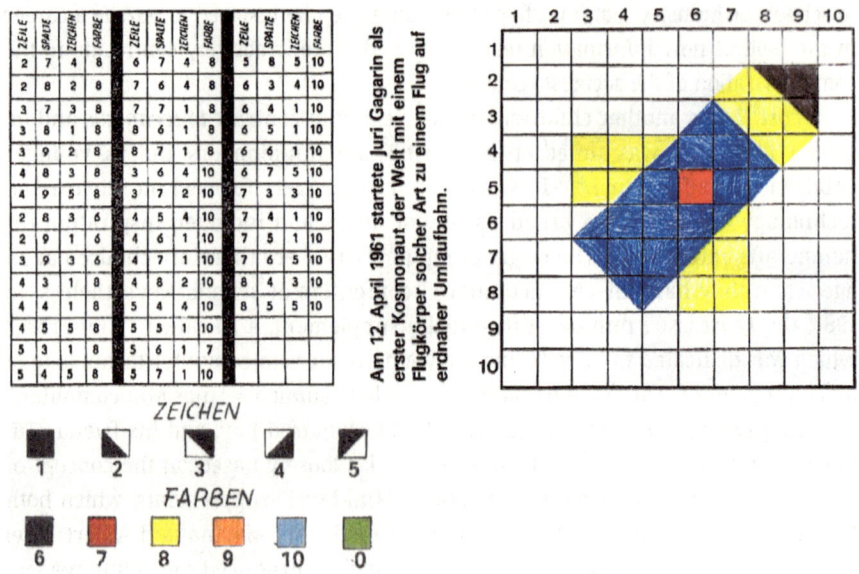

Figure 9: Number encoding puzzle in "Mini-Frösi" number 29.
Source: Mini-Frösi no. 29, supplement to Frösi issue no. 1 (1987), 41 and backpage.

convey a specific vision of the computer to the children, which corresponded to the SED's sociotechnical imaginary. Two prominent and reoccurring motifs are depictions of the "computer as a friend" and portrayals of "Tüte inside in the computer."

The motif of the "computer as a friend" (see Figure 10) can be seen as an attempt to evoke a positive attitude towards computing technology in the children, depicting the computer "Tümico" as a friends and obedient companion of the boy "Tüte." In this sense, children were presented with a vision of computers as a benign technology to accompany and assist people in their everyday tasks, dutifully relieving them of strenuous and repetitive work.

The motif of "Tüte in the computer" portrays the *Frösi* character "Tüte" inside or reaching through the computer screen or automat (see Figure 11). It can be interpreted as a vision of the computer as a mere machine or tool that required human input to create a meaningful output. In this sense, the pictures can be understood as an illustration of the fact that all that a computer could ever accomplish required a human that had "taught" the computer what to do. The illustrations were thus aimed at debunking the children's probable idea of the computer as a "magi-

Figure 10: Motif of the "computer as a friend" in the illustrations of the "Mini-Frösi" booklets on computer technology.
Sources: Mini-Frösi no. 27, supplement to Frösi issue no. 4 (1986), back page; Mini-Frösi no. 29, supplement to Frösi issue no. 1 (1987), 14; Mini-Frösi no. 29, supplement to Frösi issue no. 1 (1987), 48.

cal" device with its own intentions and agency.[889] Instead, it showed that there was a human mind and body – in the figurative sense – inside or behind every computer, which had built and programmed the computer to carry out the task it was instructed to do. The illustrations were accompanied by text explaining that "the computer is not a hand tool, but a thinking tool. [. . .] But don't forget: Everything it can think, it has previously learned from us. We programmed it. That means: We told it what to do and how to do it."[890] The motif of "Tüte inside the computer" can thus be understood as an illustration of a basic premise behind the SED's sociotechnical imaginary, which posited that computer technology as such was "neutral" in the sense that it would not itself bring about social change, be it positive or nega-

---

889 In fact, Mini-Frösi no. 28 explicitly proclaimed that the fast calculation power of the computer is not 'sorcery' ("Es ist keine Hexerei!") (Mini-Frösi no. 28, supplement to Frösi issue no. 10 (1986), 30). Moreover, the section explaining the basic principles of how a computer works was entitled "Zauberei (wird) Wirklichkeit" (eng.: Magic (becomes) reality) (Mini-Frösi no. 28, supplement to Frösi issue no. 10 (1986), 4).
890 Mini-Frösi no. 28, supplement to Frösi issue no. 10 (1986), 8 (translated from German).

Figure 11: Motif of "Tüte inside the computer" in the illustrations of the "Mini-Frösi" booklets on computer technology.
Sources: Mini-Frösi no. 29, supplement to Frösi issue no. 1 (1987), 8; Mini-Frösi no. 28, supplement to Frösi issue no. 10 (1986), 14; Cover of Mini-Frösi no. 29, supplement to Frösi issue no. 1 (1987); Mini-Frösi no. 29, supplement to Frösi issue no. 1 (1987), 36.

tive. Instead, the computer was to be considered merely as a tool under the complete control of the people who developed and used it, and the effects of its use were guided by human goals and visions – and thus dependent on the political and ideological system of the society as a whole.

6.3 Computer Games in the Context of Socialist Education

For many children in the GDR, the first contact with computer technology was through articles in the news press or magazines, as well as television and radio reports. Without even having seen or touched a computer, they were informed by the state media about how this new technology was allegedly revolutionizing the economy and bringing about economic prosperity and progress for all under socialism. Media representation of computer technology played a vital role in enforcing the SED's sociotechnical imaginaries among the young people, because computer technology was still largely absent from their real, everyday lives in the GDR – especially as schools were only equipped gradually and slowly with computers. The imaginary propagated by the state relied on a future vision of a society, where computer technology was available in abundance and accessible for all. This would allow the state to recruit workers that were keen on using computer technology in their future jobs and to acquire the necessary skills and knowledge to do so. But to lend more credibility to this imaginary, state propaganda could not merely rely on media reports about the possible innovative uses of computers. Rather, the people of the GDR needed to be able to experience the possibilities and benefits of computer technology outside of school and work, as citizens and consumers.

One possibility to inspire enthusiasm and an interest in computers among the youth was through a playful approach. The production of electronic entertainment games, thus, became a part of the SED's efforts to make more consumer goods available to the people of the GDR. In 1984, the "Poly-Play" was created, the GDR's only arcade machine.[891] Its design and development was inspired by Western computer hardware and popular games such as "Space Invaders" and "Pac Man," but with a twist, to render them ideologically appropriate for the socialist consumer market of the GDR.[892] "Poly-Play" machines were put up in youth centers and holiday camps, targeting in particular the young.[893] Computer games became a focal point of attention for many teenagers, which made learning about and engaging with microcomputers especially attractive to them. Once they had gained access to a computer and mastered the basics about how to use it, they could participate in a prolific gaming culture and engage in the practices of developing, copying, and sharing computer games within their communities. Informal computer clubs served as a popular platform for trading and swapping software, including self-written programs, but also games imported from the West.[894]

The SED's attitude to Western computer software was characterized by wariness, if not hostility. Computer hardware was seen as a "neutral" technology whose effects depended on the social, economic, and political system in which it was used. As a result of this social determinist stance, computer hardware was placed under little suspicion of reflecting certain ideological positions and values. However, software was an entirely different matter.

---

891 Meyer, *Computer in der DDR*, 125–26.
892 Jens Zirpins, "Geheimdienstspiele. Die Entwicklungsgeschichte des Poly-Play," *Retro*, no. 30 (2014): 10–14.
    2014; Erdogan, *Avantgarde der Computernutzung*, 156.
893 Schröder, *Auferstanden aus Platinen*, 99.
894 Erdogan, *Avantgarde der Computernutzung*, 281–82; Gießler, "Video Games in East Germany."

On the one hand, the GDR's Ministry for State Security feared the import of viruses and spying software from the West, which could – disguised as a Trojan horse, masquerading as a computer game – land on a computer within an educational institution, a research facility or an industrial plant, which were used outside of working hours by students and youth groups.[895] On the other hand, the possible influx of Western software raised political concerns regarding extracurricular computer activities of young people. In particular politicians and educators warned against the indoctrination of "capitalist" values, beliefs, and norms through the use of "capitalist" software and, most notably, computer games.[896] It was argued that computer games – just like other children's games and toys – fulfilled an educational function and were by no means "neutral" or "value-free" with regard to their content. The youth in capitalist countries, pedagogues in the GDR argued, was being ideologically manipulated by computer games to accept the behavioral patterns, values, and norms of capitalist society.[897] Such concerns were intensified by the fact that various games which originated from Western countries were played, copied, and distributed among the youth in the GDR. The import of computer games from the capitalist West into the GDR was declared illegal, but disks were simply smuggled across the border, copied, and swapped among young computer enthusiasts.[898]

This eventually triggered the Ministry for State Security (MfS), also known as the 'Stasi', to investigate whether the games that the young people in the GDR were playing needed to be considered a threat to the socialist cause, and what exactly the children and teenagers were doing when they met up in computer clubs. In July 1987, the Stasi put together a list of computer games available at the computer club in the HdjT in East Berlin,[899] highlighting the ones they deemed problematic for political and ideological reasons, or because they were "of a particularly militaristic and inhumane nature," such as "Raid Over Moscow," "G.I. Joe: Cobra Strike," "Blue Max," "Bomb Jack," and "Rambo."[900] These games were eventually deemed illegal and officially banned from the club, which, however, did

---

895 Gießler, "Video Games in East Germany."
896 Hutterer, "Orientierungen, 15.
897 Hutterer, "Orientierungen, 15.
898 BArch, MfS, BV Magdeburg, Abt. XX 3278, "Informationsbedarf der Abteilung XX, Gen. Doberstein," dated March, 5, 1986; cf. Erdogan, *Avantgarde der Computernutzung*, 75; Gießler, "Video Games in East Germany."
899 BArch, MfS, BV Berlin, Abt. XX, no. 3118, "Programmliste," undated; cf. Gießler, "Video Games in East Germany."
900 Erdogan, *Avantgarde der Computernutzung*, 77; Gießler, "Video Games in East Germany."

not prevent at least some if their fans in the GDR to keep playing and sharing them in secret.[901]

Nevertheless, some pedagogues recognized the educational potential of computer games which were in line with the socialist ideology and pleaded the case for integrating them into extracurricular education. They claimed that computer games could effectively contribute to the development of the learner's imagination and computer literacy, as well as to the reduction of individual fears and reservations about the computer.[902] While it was acknowledged that the games that had dominated the scene of computer gaming so far had focused predominantly on reaction speed and the ability to coordinate motor functions, the pedagogues pointed out need to develop a different kind of educational games. Specifically, they identified a need to develop games for individual or organized leisure-time activities which focused, for example, on solving construction tasks, or involved logical thinking, brain teasing or perhaps "scientific adventures," leaving scope for individual decision-making by the player who would be able to choose between various alternatives, as well as simultaneous competitions with adjustable difficulty levels.[903]

At the University of Teacher Education "N. K. Krupskaja" in Halle, a research collective was formed in the second half of the 1980s to investigate the role of computer use in extracurricular education. Under the lead of Gerd Hutterer, Professor of Pedagogy in Halle, the research collective published three brochures between 1987 and 1989 on the practical approaches and educational aims of computer use in computer study groups and youth clubs, programming courses, and computer camps. Amongst others, Hutterer's research group focused in detail on the issue of computer games in extracurricular education. In the first brochure published in 1987, pedagogue Frank Hille argued that games generally played an important role in moral and social education. Hille considered play activities as a necessary form of children's engagement with their environment to adopt socially appropriate and desirable behavior, such as cooperativeness, a sense of responsibility, and creativity. Hille was convinced that, although computer game situations are fictitious, the psychological changes which they triggered in the player are real and personality-forming.[904] However, not every computer game was deemed suitable to form the desired personality traits that a socialist education strove for. Many of the available

---

901 Erdogan, *Avantgarde der Computernutzung*, 282.
902 Hutterer, "Orientierungen," 14.
903 Hutterer, "Orientierungen," 14.
904 Frank Hille, "Computerspiele – Herangehen, Anforderungen, Klassifizierungen, Beispiele," in *Computer in der außerunterrichtlichen Tätigkeit. Standpunkte und Anregungen*, ed. Gerd Hutterer (Halle: Pädagogische Hochschule Halle N. K. Krupskaja, 1987), 51.

computer games were merely variations of "shooting a dangerous object" with low appeal to the mental abilities of players. Such games were concealing what Hille called the "essence of new information technology" instead of making it comprehensible to players and learners. Moreover, many of games were contradicting socialist values. Hence, he saw an urgent need to develop "better" games and make them available to the masses.[905] Hille declared that popular "adventure-games" to which Western hardware and software producers owed a large part of their sales, had so far been of "staggering stupidity." Instead, he suggested that new adventure games should be developed that require solidaristic actions to achieve goals, and in which the player would only be rewarded for taking ethical and moral decisions. In addition, educational games could be designed which would require the player to apply factual, normative, and value knowledge acquired during school lessons. Such computer games were to be developed jointly by experienced educators and computer experts.[906] Only by combining technical and pedagogical expertise, Hille claimed, would it be possible to create software that was technically, scientifically, and pedagogically sound and, thus, adequate for the general and moral education of "socialist personalities."

6.4 Pedagogical Control and Guidance in the Development of Software in Schools

The tasks of the instructors of the courses and study groups included not only teaching about the use of computers and programming, but also "pedagogically advising" the participants on the appropriate use and development of software. Self-developed games and programs in extracurricular courses, clubs, and study groups were supposed to reflect the values of socialism: partisanship, adherence to scientific standards, and relevance to users' everyday lives in the socialist society.[907]

To some educators, it seemed particularly important to correct "wrong" ideas of computers as game machines. A teacher from Templin argued in 1987, that the introduction of students to microtechnology had to focus on the use of this technology for solving scientific-technical problems, instead of unnecessary things such as

---

905 Hille, "Computerspiele," 50.
906 Hille, "Computerspiele," 55.
907 Hutterer, "Orientierungen," 16.

computer games.[908] Following this line of argument, the development of software in extracurricular clubs and study groups had to be guided by "serious" aims and goals rather than the joy of playing. Another teacher and leader of a school-based extracurricular computer club noted in 1987, that an uncontrolled and unguided use of computers was causing him problems. Pupils who had primarily encountered the computer as a playing machine before joining the club barely showed any interest in the "more serious side of the program," he lamented.[909] However, he still maintained that computer games served a purpose in getting students excited about this new technology and had a legitimate place in computer clubs. Under the guidance of instructors, extracurricular activities involving computer games would not focus on who could achieve the highest score, but on how the game could be altered and improved. For this purpose, it was considered necessary that students were able to understand the code of a program, and how it was related to the computer's output visible on the screen. Great importance was attached to these kinds of insights, as they supported an imaginary that envisioned computers not as magical machines with their own artificial intelligence on a par with humans – but as mere tools, subjected to the needs, ideas, and purposes of humans and ultimately under their complete control.

The main educational concern behind such an understanding was that humans should not feel inferior to computers. Pupils needed to understand that computer programs had been created by humans, and thus, that the computer's responses to the user's input were under the complete control of the programmer as the "teacher" of the computer, and not a creative product by the computer itself.[910] In consequence, the people living under a socialist regime would not need to fear any negative social effects from the use of computers, as this new technology was allegedly completely subjected to human and political control, which would ensure that it was deployed for the benefit of all members of the society. Therefore, recognizing the computer as a fully controllable technology, subjected to the needs and purposes of a socialist society, featured as an important aspect in reinforcing the sociotechnical imaginary of a desirable computerized future in the GDR.

To make sure that the games and other programs that students developed in extracurricular courses and clubs were ideologically sound and could be consid-

---

908 BBF PL 87–11–40, "Erfahrungen bei der Gestaltung des Prozesses der Einführung von Inhalten aus dem Bereich der Mikrorechentechnik und Informatik in die Allgemeinbildung," dated 1987, 7.
909 BBF PL 87–15–20, "Ergebnisse, Probleme und Folgerungen aus der Tätigkeit der Schülerarbeitsgemeinschaft 'Mikrorechentechnik' an der Otto-Grotewohl Oberschule," dated 1987, 9.
910 Hille, "Computerspiele," 52.

ered a pedagogically worthwhile endeavour, it was recommended that educators assessed the students' program ideas as to their feasibility and usefulness before implementation. In 1987, Gerd Hutterer, the leader of the Halle research collective on computers in extracurricular education, suggested the introduction of a set of requirements for software developed by students. According to Hutterer, the programs needed to be not only intelligible to the programmer, but had to be well-structured, easily comprehensible, accompanied by explanations, and an informative program documentation. It had to be designed in such a way that it could be expanded, improved, or changed at will by other interested learner.[911] His colleague Frank Hille made a similar argument: The real thrill of computer games, he claimed, lay in programming and in constantly adding to the code. The learners had to realize that the programming language was merely a tool and that a human being was the actual creator of any function a computer carried out.[912] Hille concluded that pedagogically meaningful computer programs must demonstrate the possibilities of computer technology, its broad field of application, and the "programmability" of computers as a central principle of interaction with a computer. The purpose of using computer games in extracurricular education, therefore, was to encourage players to want to "get behind their secret."[913] The code of the games should therefore not be protected, Hille argued, but must be freely accessible to their users.[914]

In essence, both calls by Hutterer and Hille for openly accessible and well-documented code were aimed at opening the "black box" of a program to learners, to give them the opportunity to comprehend the instructions the computer was following to perform certain functions and produce a specific output.[915] Consequently, any claims to intellectual property rights or author's rights over the self-developed programs had to be negated. Instead, games and other software needed to be openly accessible for analysis and adaptation by their users, to offer learners the opportunity to understand and make the program code their own. This rejection of copyright claims on software was further substantiated by a landmark decision of the Leipzig District Court in September 1979 that considered software to be neither a scientific product nor creative achievement.[916]

---

911 Hutterer, "Orientierungen," 16.
912 Hille, "Computerspiele," 54.
913 Hille, "Computerspiele," 57.
914 Hille, "Computerspiele," 57.
915 Hutterer, "Orientierungen," 16; Hille, "Computerspiele," 57.
916 Gießler, "Video Games in East Germany."

6.5 Chapter Conclusion

The various extracurricular opportunities to engage in computing was aimed at offering a diverse range of interesting and meaningful leisure activities for young people, not only to learn and master the use of computer technology, but also as a productive and creative pursuit. The main political aim behind offering and promoting extracurricular computer education was to broaden the impact of computer technology on the citizenry of the GDR by inspiring interest in computer technology among the youth and teach basic computer literacy to anyone who was willing to learn. Thus, extracurricular computer education was intended to both realize and reinforce the SED's dominant sociotechnical imaginary of a widely computerized socialist society and economy in the future, in which everyone would need a basic understanding of new information technology to contribute as a skilled worker and citizen to the thriving economy and prosperity of the GDR. The more selective and merit-based computer study groups and clubs in association with schools and universities helped to promote particularly interested and talented students in response to the urgent need for computer hard- and software specialists in industry and research. Competitions and extracurricular classes focused on skill and talent development while serving to recruit highly skilled computer experts to drive the technological progress of the GDR forward.

Creative computer use, where learners improved and modified existing software or developed their own programs, was put at the heart of many computer clubs and study groups. Over the span of a few years, an overwhelming abundance of computer programs was developed in extracurricular settings both by educators and learners. To harness the economic potential of this software, efforts were made to document and register promising and well-documented programs in centralized software libraries, as was already customary in companies. The pioneer palace Ernst Thälmann in East Berlin maintained such a database to make self-developed accessible to other students, educators, as well as companies. Such efforts highlight another important aspect regarding the use of computers in extracurricular education in the GDR, namely that the learner developed software was meant to have a social utility beyond the individual purposes and interests of its creators. Educational policymakers and educators promoted a close relationship of computer clubs and study groups with local companies and schools. Extracurricular computer projects were preferably meant to respond to the needs and requirements of the latter, by developing software for the use in economic production or school instruction. In this perspective, extracurricular computer clubs served not only as an opportunity for participants to develop their programming skills, but also to raise their awareness of the use of computers in solving collective tasks in socialist society and economy. Educators and club leaders were urged to closely relate the fields of application and goals of extracurricular computer activities to anticipated economic and social benefits in line with the SED's political priorities. In this way, they were meant to reinforce the dominant sociotechnical imaginary that envisioned computers – even when used as a leisure activity – primarily as a working tool to serve the socialist community rather than for the satisfaction of individual needs and creative self-expression. Similarly, popular children's and youth magazines under the control of the authoritarian leadership served not only to impart basic technical education and to stimulate the readers' interest in science and technology, but also to propagate and reinforce the SED's sociotechnical imaginary of computer technology among the younger generation.

Nevertheless, independent computer users as well as members of more informal and less pre-structured computer clubs found ways to develop their own ideas of computer technology, its social and political meanings – which were not always aligned with the state's sociotechnical imaginaries, or even outright subversive. By creating their own hard- and software in extracurricular settings, computer enthusiasts became "productive users" as opposed to the notion of merely passive consumers of technology that learned how to use pre-built hard- and software. Computer clubs, meet-ups, and hobby communities served as platforms for the circulation and shaping of beliefs, practices, and sense-making through social interaction with others who shared an interest in computers. Mediated by popular technology magazines and books on computing and programming, these communities helped to form collective identities as computer enthusiasts with the potential to create their own, alternative sociotechnical imaginaries, despite constant attempts of ideological influencing by the SED.

The question of how these alternative visions of computing related to the dominant and more powerful sociotechnical imaginary of the political leadership refers to the notion of contested and competing sociotechnical imaginaries at a local level. In the field of Science and Technology studies, this idea has recently lead to a novel approach, which "[. . .] resists treating imaginaries only as semiotic codes or cultural systems and instead embeds them in social fields that connect collective representations with the strategies of collective and individual actors."[917] Politicians and education policymakers in the GDR were trying to make sense of and act in the present through their sociotechnical imaginary, or in other words: They were devising a new computer curriculum by borrowing from

---

917 David J. Hess and Benjamin K. Sovacool, "Sociotechnical matters: Reviewing and integrating science and technology studies with energy social science," *Energy Research & Social Science*, no. 65 (2020): 7. For a study of conflicting imaginaries between government, industry leaders, and the public, see David J. Hess, "Publics as Threats? Integrating Science and Technology Studies and Social Movement Studies," *Science as Culture* 24 no. 1 (2014): 69–82.

an imagined emergent, volatile, and uncertain future. In contrast, individual computer users and learners were guided in their everyday activities by more short-term, subjectively meaningful goals and purposes.

However, as sociotechnical imaginaries are embedded in social practices centered around computers and in the technology itself, it must be assumed that they were affected both by the SED leadership's dominant sociotechnical imaginary, as well as less powerful imaginaries shared by the members of computer clubs and the more loosely connected community of individual home computer users. The notion of contested imaginaries invites us to look at how "differences in imaginaries (or other cultural meanings) are connected to differences in social positions" and to pay attention to "power in the sense of the capacity to change events."[918] Small social groups such as computer clubs and societies with very restricted political agency and influence can develop and enact alternative sociotechnical imaginaries, but have little power to take it up with the official imaginaries propagated by the SED outside of their small communities. Nevertheless, such extracurricular clubs and hobby groups opened spaces where alternative imaginaries could potentially challenge the dominant imaginary of the political leadership, especially if these social spaces functioned more or less autonomously and enjoyed a certain degree of freedom outside of strict political control and regulation.

The example of computer games and the historical debates around their rightful place within extracurricular education in the GDR are well suited to elaborate on the issue of contested imaginaries. Embedded within different sociotechnical imaginaries, computer games were either considered as a potential threat or an educational opportunity, which could either run against or support the SED's political goals and aims. According to the social-determinist stance of state authorities towards computer technology, the embedded values and norms, as well as the effects of computer games on society at large essentially depend on the political, social, and economic system in which they were produced, distributed, and played. Consequently, the political leadership and educators in the GDR were determined to prevent Western computer games and other software from imparting capitalist or militaristic values and ideas to young people in the GDR. Individual home computer users and their informal networks were suspected of engaging in and propagating uses of computer technology that were misaligned with the dominant imaginary of the political leadership, according to which technology needed to benefit the socialist society as envisioned by its economic strategy. Computer games produced in Western capitalist countries, however, were accused of serving a hedonistic consumer culture and profit-maximizing com-

---

918 Hess, "Publics as Threats?," 77.

puter companies. Moreover, the sociotechnical imaginary shared by GDR pedagogues attributed software and especially computer games the power to exert considerable influence on the moral values and attitudes of its users and players.

The SED's efforts to promote extracurricular activities for the young to learn about and use computer technology in their free time essentially pursued the purpose to align young people's interest in computers and games with political and economic aims. By inspiring a willingness in them to learn how to develop their own computer games, young people would be motivated to acquire valuable skills that could eventually be harnessed by the socialist state to cover the demand for computer experts and software developers. For many young computer enthusiasts, however, playing computer games was an end in itself. In youth computer clubs and through informal networks, they developed and enacted a collectively shared sociotechnical imaginary that envisioned computers first and foremost as a source of pleasure and self-realization.

The individual beliefs, intentions, and practices which computer users developed in the process of learning about computer technology on their own terms were sometimes at odds with state authorities' attempts to provide a particular motivation for learning about and using computers – that is, to computerize economic production. In the eyes of the SED leadership, computer technology was a vital success factor for the GDR to compete in a world economy. But in addition, the GDR's microelectronics industry was given a political mandate to develop and produce a certain share of consumer products for the domestic market. This was to give the people of the GDR access to coveted products and experiences over which Westerners in capitalist countries seemed to have access in abundance. State-mandated production of computer games and arcade machines for recreational use served to prove to both its own people and the outside world that the socialist economy of the GDR was capable of providing the same amenities and satisfying the desires of its citizens. At the same time, by highlighting the essential role of key technologies such as microelectronics in enabling the domestic industry to do so, it also served to reinvigorate public credence to the sociotechnical imaginaries propagated by media and state authorities.

7 Conclusion and Outlook

In 1986, three philosophers of technology and science at the Technical University in Dresden published an article on the impact of the development of new computer technology on humans, society, and socialist ideology. In the article, they wrote:

> Hardly any other scientific-technological development has been accompanied by such a wealth of ideological opinions, hypotheses, and visions of the future as the emergence of modern computer technology and robotics – and as is well known, even before they came into being. In this respect, computer technology can generate dreams and nightmares of a new world. There is no doubt that it is becoming an influential force that will significantly alter the human condition.[919]

The development of microelectronics and computer technology in the GDR was indeed accompanied by a "dream of a new world," of an ideologically loaded vision of a computerized future under socialism. The dominant imaginary of a sociotechnical future promoted by the SED sought to link technological advances with socialist ideology and put it at the service of the party's claim to power. The imaginary of a desirable computerized future itself was embedded in and shaped by the social, economic, political, and ideological context of the time. In turn, it also sought to shape the way in which computer technology was perceived, adopted, and deployed in the GDR, by creating a normative shared understanding of what the new technology was, how it was to be used, and what purposes it was to serve. In this sense, scientific and technological innovations reflect social interests and power relations, and the effects of technological change are shaped by the dynamic relations within the sociotechnical system in which they occur.

To investigate the entangled relationship between sociotechnical imagination and education policymaking in the GDR between 1960 and 1990, this study followed a two-fold approach. The first research question pertained to the specific vision of a desirable computerized future endorsed and advocated by the socialist political leadership of the GDR, in the form of a dominant sociotechnical imaginary. This entailed an examination of the anticipated and societal and economic effects and promises linked to the development and use of computer technology within the GDR. The second research question centered around the educational

---

[919] Lothar Striebing, Karin Zänker, and Bernd Zschaler, "Neue Computergenerationen und die Perspektive des Menschen," *Deutsche Zeitschrift für Philosophie* 34, no. 1 (1986): 13–14 (translated from German).

policy response of the SED to the emergence of modern computer technology, and how it related to the party leadership's vision of a desirable sociotechnical future.

By looking at developments in different sectors of education in the GDR, the study has highlighted that the introduction of computers at different levels of the education system cannot be fully understood in isolation. It must be considered in its entanglement with the visions, concerns and concrete policies related to computer education as it was introduced elsewhere in the system. In this sense, for example, computer education in schools was modelled on the disciplinary development of computer science in higher education, rooted in both technical and mathematical disciplines, and therefore included in both mathematics and polytechnic education. At the same time, it was intended to lay the foundation for further computer education as part of vocational training, preparing pupils for their future work in a computerized socialist economy. Extracurricular computer clubs and activities complemented school-based instruction by providing ample opportunities to nurture talent and interest in computing outside the scope of the uniform comprehensive polytechnic school.

7.1 Introducing Computer Education to Navigate Technological Change

Over the course of the three decades under study, the gradual introduction of computer technology into higher education, vocational education, general and extracurricular education in the GDR was guided by a dominant sociotechnical imaginary brought forward and disseminated by the country's socialist political leadership. In line with socialist ideology, this positive "dream of a new world" encompassed the promise of both economic and social progress for the socialist society of the GDR, which would be brought about by the development and use of new computer technology, under the political guidance of the SED's authoritarian leadership. The imaginary provided a frame of reference for education policymakers in terms of ideological alignment with the SED's political strategy, and for anticipating the skills and knowledge that would be required of people in the GDR to thrive in a desired computerized future economy and society.

The origins of the SED's imaginary of socialist computerization can be traced back to an earlier imaginary of cybernetics in the 1960s, which envisioned the possibility of more effective control and regulation of the socialist economy and society through the management of information processes, thereby making the planned economy more efficient. Initially, however, the SED's stance on cybernetics, and by extension computing, as a theory and as technology that had been decisively shaped in the capitalist USA, was ambivalent. In analogy to the rejection of cybernetic visions of technocratic governance of society, the computer could be perceived as a potential threat to the political authority of the SED in organizing the society and economy, subjugating humans to its power.

To be deemed ideologically and politically acceptable by the socialist political leadership, cybernetics and new information technology thus needed to function as tools in the hands and under the control of the regime, and thus furthering its power rather than threatening it. In this sense, modern computers needed to be envisioned as a "neutral" technology with no meaningful agency on its own, the societal effects of which depended solely on how it would be used by those who hold power over it. It was the specific social and political conditions under which a technology was deployed, which were considered key in setting a certain trajectory for its development and use, and thus promoting social change in a positive or negative sense. In the process of establishing cybernetics and computing as ideologically and politically legitimate technologies, the GDR philosopher Georg Klaus and his students played a key role. They convincingly argued that cybernetics and computing were indeed compatible with Marxist-Leninist philosophy and could even be used to advance the socialist cause.

Accordingly, the dominant imaginary of computer technology was based on the SED leadership's conviction that, above all, it was the relations of production that the economic and social effects of computer technology depended on, namely whether its development and use benefitted all people, or followed the desires and needs of few. Scientific-technological progress in the field of computer technology was thus, since its beginnings, closely tied to the task of furthering the socialist cause. The SED's imaginary posited that any scientific-technological progress under socialism would necessarily include progress of the intellectual culture and was to be deemed desirable, as it would be deployed for the common good. Under the socialist leadership of the SED, computer technology was not to replace human decision making, but rather to support and facilitate it by modelling complex systems and simulating the likely outcomes of policy interventions. It was therefore envisioned as a powerful tool to further consolidate the power of the authoritarian regime by providing it with vast amounts of data that can be processed by computers to gain even greater control and oversight of the socialist society and economy.

For computer technology to be perceived as fully controllable, and to function under the absolute authority of the government, it was deemed crucial to control its *use* in the GDR, namely through a powerful imaginary that would convince future developers and users in the GDR of the SED's envisioned role, purposes, and applications of this technology. In this sense, education was identified as a crucial means of "taming" the computer. The GDR's educational system was to shape people's understanding of the computer in accordance with the SED's imaginary, and thereby set clear priorities for the further development and use of modern computer technology in the socialist GDR. In the process of introducing computers into education, and in developing new curricula for computer education, the SED's dominant sociotechnical imaginary helped policymakers to establish, what the computer was to be used for, by whom, and how – by relating its development and use by the people of the GDR to the country's economic and social welfare, and to the struggle of socialist systems against capitalist systems.

During the 1960s and early 1970s, the first digital computers in the GDR were developed and used for scientific computation, as well as in business and state administration. In this phase, a cybernetic aspect was particularly prominent in the SED's sociotechnical imaginary, which foregrounded applications of computer technology for managing the planned economy, by envisioning the computerized control of machines to automate industrial production and administration. The imaginary posited that computers would make state planning more efficient, industrial production more cost-effective, and work in manufacturing more humane. It was also proclaimed as being crucial for the well-being of the GDR's society and the strengthening of the socialist system and ideology in the struggle against capitalism.

During this early phase, a new field of scientific research and academic education concerned with computer technology formed in institutions of higher education, in connection with the establishment of computing centers associated with university departments of mathematics or engineering. In mathematics, research and teaching focused on complex mathematical calculations and the modelling of social and economic systems that could then be managed, controlled, and regulated by powerful information processing machines. In engineering, the focus was on technical machines as cybernetic systems, the development of computer systems and computerized machines. Especially the latter gained in importance following the SED leadership's decision on an extensive microelectronics program in 1977 to build up a domestic microelectronics industry to computerize economic production in the GDR and secure the competitiveness of its export products, in particular through the development of computerized machine tools. In addition to the higher education and training of development-oriented computer specialists in technical disciplines and mathematics, computer education was also introduced into the study curricula of economists, to prepare them for computer-aided planning and management, as well as the implementation of computerization projects in combines and state bureaucracy. With further advances in computer technology which allowed for smaller and more powerful computers, the SED's imaginary was extended by including the envisioned future use of computers in virtually all fields of science and the economy. Consequently, an application-oriented computer education and training was also introduced into study courses in other disciplines, especially in agricultural and natural sciences over the course of the 1980s. While all students needed to learn a basic level of computer-user education, only a smaller fraction received a more in-depth training to enable them to develop hardware and software systems for applications in their specific field of expertise.

The process of institutionalization and the search for a name for the newly emerging discipline reflected the struggles to reach a common understanding of the scope of the problems and phenomena it was concerned with, defining its boundaries, and establishing its own scientific paradigms and methods. This process of the genesis of a new discipline also influenced the design and choice of curricular content of computing education and training in higher education, which differed between technical, mathematical, or economic faculties. In the GDR, the new discipline was commonly referred to as "information processing," emphasizing the focus on information processes in a wide variety of applications, rather than on the device of the computer itself, as suggested by the designation of "computer science." In the second half of the 1980s, the name of institutions and courses concerned with information processing gradually changed to "informatics," reflecting the consolidation of the process of institutionalization in which informatics was increasingly released from its close affiliation with mathematics and engineering departments and established as an academic science and discipline in its own right.

The operation of the GDR's first large mainframe computers in the 1960s required a considerable number of skilled workers. Their development and deployment in the GDR's industry, state administration, and computing centers, thus triggering a need to train skilled personnel for the operation, maintenance, and repair of these machines. Computer technology no longer remained the exclusive domain of highly educated experts, but gradually became part of new skilled worker professions in the GDR's vocational training system. With the development of smaller and more powerful computers in the 1970s, which allowed for a broader range of applications, the computer increasingly came to be imagined as a virtually universal tool that would fundamentally change a wide range of jobs, particularly in industry. A milestone on this path was the development of the microprocessor. Its main applications were in electronic data processing, industrial automation technology, and information and communication technology. The microprocessor served as the prototype and symbol of far-fetching changes in technology, the relationship between man and technology, and in the conditions of human labor. It became a focal point in the SED's sociotechnical imaginary of the potential and desired effects of modern computer technology on society and the economy.

The microprocessor laid the foundation of the development of the GDR's first microcomputers towards the mid-1980s, which were used for computation for research, planning and administration, as well as modelling and simulation in industrial production. It also enabled the development of more sophisticated computerized machines, industrial robots, and automated systems and production lines. This development gave new momentum to that aspect of the SED's imaginary that envisioned computer technology as the key to increasing economic competitiveness to gain the upper hand in the economic and political power struggle between the systems. The dominant sociotechnical imaginary posited that the development and use of modern computer technology would prove the GDR's technological prowess, both domestically and in the world, thereby legitimizing the authoritarian rule of the party. In this sense, the SED's sociotechnical imaginary not only helped to justify new investments in science and technology, and far-fetching curricular reforms in education. Through the imaginary, expected and realized advances in science and technology also served to reaffirm the popular belief in the SED regime's "capacity to act as responsible stewards of the public good."[920]

Computer technology was expected to lead to economic growth through intensification and rationalization, as well as to help alleviate the GDR's permanent problem of severe labor shortages. Computers were intended to replace certain work tasks or entire jobs done by humans, setting them free for new job tasks. The SED's imaginary promised that the tasks to be automated would concern monotonous, repetitive, or physically demanding jobs. People whose jobs were automated, thus, were promised a new assignment in their company or elsewhere in the socialist economy, where they would be able to engage in more mentally stimulating, creative, and appealing work activities. This, in turn, presupposed a high degree of professional mobility. Workers thus needed to be provided with a solid and broad foundation of education and training, which would allow them to quickly get settled into a new work position. A solid basic education in computer technology and a willingness to undergo retraining and further education were therefore considered essential assets for workers in preparation for the expected widespread computerization of the GDR economy.

From the mid-1980s, the vision of widespread computerization affecting all workers in the GDR gained traction within the SED's imaginary of computer technology. With the development of microcomputers and the increased political salience of CAD/CAM technology, the envisioned future user base of computer technology broadened significantly. Computers were no longer imagined as the exclusive domain of experts, but as a key technology that would soon be used by a large part of the socialist workforce.

---

[920] "The Sociotechnical Imaginaries Project," Harvard STS Research Platform, https://sts.hks.harvard.edu/research/platforms/imaginaries/ (accessed May 31, 2024).

Therefore, education policymakers in the GDR were tasked with revising all VET curricula. The training of skilled machine operators was to include the application of CAM technology, technical draughtsmen were to be trained in the use of CAD technology, and clerical trainees were to be prepared for the anticipated office automation and computerization of administration. As the curricular innovations were geared to anticipated technological changes in the workplace, newly trained workers were in some cases still being employed at traditional workplaces and machines that had not yet been reached by the planned widespread introduction of computer technology. For them, the computer education they had acquired signified a promise of a still unattained sociotechnical future. At the same time, it was a source of frustration not to be able to apply what they had learned, and to find that the high hopes and expectations raised by their training in computer technology were disappointed, at least for the time being.

The technological innovation of microcomputers was accompanied with the belief of the GDR's political leadership in the possibility and necessity for all workers to participate in the "scientific-technological revolution" of socialist computerization, as envisioned by the SED's imaginary. It was therefore considered essential that everyone in the GDR gained a basic understanding of the new technology to play an active role in the computerization of the country as envisaged by the political leadership.

Consequently, the second half of the 1980s saw the introduction of mandatory basic computer education in both VET and general education, in the form of the courses "Basics of Automation" and "Basic course in Informatics" for all pupils and apprentices. The aim of this basic computer education "for all" was to empower people to become masters of computer technology, rather than being controlled and passively subjugated by the process of computerization. It served to counteract fears and reservations among the young generation and future workforce of the GDR, in line with the dominant sociotechnical imaginary, which framed the process of socialist computerization as a democratic process with strong worker participation in shaping it. This aspect of the dominant sociotechnical imaginary was also promoted by children's and youth magazines, which portrayed the computer as a friend and helper, and crucially, as a technical tool controlled by the people.

However, not everyone was seen as in need of the same level of computer education. A basic computer education "for all" served to foster positive attitudes towards the new technology among the people and a common understanding of the purpose and importance assigned to computer technology as part of the SED's political and economic strategy for the coming decades. But in the eyes of the SED leadership and economic policymakers, not everyone needed to become a computer expert. Thus, computer education was included in curricula in higher education, VET, and general education in a differentiated manner, both in terms of quality and quantity of instruction. This was done by distinguishing between computer education for the envisioned roles of technology developers on the one hand, and mere users of the new technology on the other. The mandatory basic computer courses for users focused on teaching "algorithmic thinking" and problem-solving skills, programming in BASIC and the use of ready-made software. A more in-depth computer instruction was provided for pupils in the EOS, and through additional elective courses for particularly talented, high-achieving pupils with a strong interest in computers. This advanced general computer education was aimed at preparing pupils for a higher education or professional role specializing in computing, enabling them to develop their own software and engaging in more sophisticated use of computers for problem-solving.

In both cases, the inculcation and reaffirmation of the SED's dominant sociotechnical imaginary played a key role. Computer education in the GDR generally included instruction on the anticipated social and economic effects of the development and use of computers under the political leadership of the SED, and the desired future of socialist computerization with the associated promises of social progress and economic prosperity which would be felt by all people in the GDR. In this sense, a basic computer education in general education not only served to prepare pupils for their further education in VET and higher education, or a possible career in the field of computer technology and informatics. Crucially, it also served to align individual interests in using computer technology with societal requirements as envisioned in the SED's sociotechnical imaginary.

Regarding the teaching content, the computer instruction in general education in some sense mirrored the institutionalization of the field of computing in higher education in the disciplines of mathematics and engineering. The inclusion of computer education into general education in the form of both mandatory and elective courses followed a double-track approach of more mathematically oriented informatics courses on the one hand, focusing on algorithms and programming, and more technically oriented courses on the other hand, with a stronger focus on hardware, the use of computers for the control and regulation of technical processes, and applications in industrial automation technology.

Particularly in general education, educators saw the computer not only as a work tool for industrial production and enterprises, which a modern polytechnic education needed to include in its curriculum. The computer was also seen as an educational technology with the potential to develop human mental abilities and to promote learning. However, this aspect played a minor role in the SED's sociotechnical imaginary of computerization and was largely side-lined by the SED's prioritization of the economic aspect, which emphasized the role of computer technology as a means of increasing productivity. Thus, the introduction of computer education and training into schools in the GDR primarily followed the instrumental purpose of preparing a skilled workforce for an envisioned computerized future rather than highlighting its potential to support processes of teaching and learning. Consequently, the introduction of computers into schools during the 1980s was geared first and foremost towards the computer as a subject of instruction, rather than a means to support teaching and learning. Nevertheless, the practice-oriented teaching of informatics crucially involved the computer as a tool of instruction that was deemed indispensable by pedagogues for pupils to acquire the desired skills of algorithmic thinking and problem-solving. In this sense, the computer also took on the function of an educational technology, although its use as such was largely limited to teaching in computer courses and was more a side effect than the result of a conscious political decision based on the dominant sociotechnical imaginary of computer technology of the SED.

However, the potential of the microcomputer as a tool of instruction had triggered the interest of pedagogues and educational scientists to further explore this imaginary of the computer as an educational technology in a broader range of subject. But these efforts had to be subordinated to the sociotechnical imaginary and the corresponding educational strategy of the SED leadership, and were hence postponed until the 1990s, when the introduction of computers into schools as a subject of teaching and learning was expected to be completed.

The prioritization of the introduction of computers into schools as a subject of instruction over their use as an educational technology highlights the power-exerting and normative nature of sociotechnical imaginaries. A vision of a "desirable" future, crucially, is a normative rather than a speculative vision, that does not detail how things will *likely* turn out, but how they *ought* to turn out. As a result, the imaginary of the SED leadership also excluded reflection on possible negative and undesirable consequences, such as the paradoxical effects of computerization in the workplace, which ran counter to the promised improvement of working conditions in the GDR. Within the SED's political rhetoric and propaganda, a reflection on the negative sides of computerization only occurred on the basis of setting the imaginary of positive socialist computerization apart from a contrasting, dystopian counter-imaginary of capitalist computerization construed by the SED. This rationale was underlined by repeated reports in newspaper and youth magazines on the negative effects of computerization on the working people in capitalist countries, most prominently in West Germany. These reports of rising unemployment and deteriorating working conditions resulting from the use of computer technology under capitalism, as well as the alleged use of computer technology as a means of political manipulation and surveillance, and as a weaponized technology, were used by the SED to construct a stark contrast to its own sociotechnical imaginary of socialist computerization, which promised social progress, peace, and prosperity for all. However, as this study has pointed out, computerization in the socialist GDR had rather mixed effects on working conditions and social well-being. In fact, it seems that the social and economic challenges brought about by the emergence of new information technology under socialism did not differ from those under capitalism all that much.[921] The search for an adequate response to the rapid changes in society and the economy arising from the development and use of new information technologies also confronted education policymakers with similar issues and problems on both sides of the Iron Curtain.

Nevertheless, the SED's sociotechnical imaginary posited that the social effects of technological change depend on in whose hands the technology was, and who decided over its development and use in society. A positive vision of a desirable computerized future was thus also aimed at building popular trust in the party leadership to act in the interest of the people. Through its reaffirmation in education, it sought to mobilize support for the SED's political strategy with regard to the development and use of computer technology in the GDR society and economy.

But the case of the GDR, as a socialist one-party state, also highlights the "contested nature" of sociotechnical imaginaries and their function within power structures.[922] The SED used its political power to keep emerging alternative imaginaries in check, by capturing them and attempting to align them with its own dominant sociotechnical imaginary of computer technology.

Alternative imaginaries emerged, for example, among young computer enthusiasts, who envisioned the computer primarily as a tool for individual playful exploration. Extracurricular computer clubs were officially supported, monitored, and controlled by the SED, in an effort to align the youth's visions and ideas of computers with the SED's dominant sociotechnical imaginary. Pedagogical supervision and guidance of young people's activities involving computers were seen as a way to channel their activities and talents towards the SED's economic goals, by interpreting playful computing as a means to stimulate learning

---

[921] A systematic comparative study of computerization in socialist and capitalist countries during the Cold War remains a desideratum. Nevertheless, existing publications on the introduction of computers in the two Germanies offer some insights into this matter: Danyel and Schuhmann, "Wege in die digitale Moderne"; Erdogan, *Avantgarde der Computernutzung;* Schmitt, *Digitalisierung der Kreditwirtschaft;* Pieper, *Hochschulinformatik;* Frank Bösch, ed., *Wege in die Digitale Gesellschaft: Computernutzung in der Bundesrepublik 1955–1990* (Göttingen: Wallstein Verlag, 2018); Thomas Kasper, *Wie der Sozialstaat digital wurde. Die Computerisierung der Rentenversicherung im geteilten Deutschland* (Göttingen: Wallstein Verlag, 2020).

[922] Jessica Smith and Abraham Tidwell, "The everyday lives of energy transitions: Contested sociotechnical imaginaries in the American West," *Social Studies of Science* 46, no. 3 (2016): 327–50.

for their further professional career and guiding their programming efforts towards tackling economic and social challenges for the benefit of the socialist society.

In this sense, supervised and guided forms of extracurricular computer education served to prevent and correct "wrong" uses of computers that did not correspond to the SED's dominant imaginary of computer technology, or involved programming practices and habits that were deemed "inferior" to the standard set by pedagogues and computer scientists.

7.2 A Multifaceted and Fluid Imaginary of Socialist Computerization

The main aim of this case study on the introduction of computer technology into education in the GDR was to historically reconstruct the SED's sociotechnical imaginary of socialist computerization, and to explore its role in guiding the educational policy response to the advent of modern computer technology. To fulfill its intended function of promoting a common understanding of the value and purpose of the new computer technology, and to provide effective guidance for policymaking, this imaginary had to be both compelling and durable in order to persuade people, but at the same time also dynamic enough to adapt to social and technological developments over the years.

The SED's imaginary incorporated several different facets, which came to the fore in response to different political, economic, and social challenges and opportunities that computer technology was intended to address. As a result of both scientific-technological and societal change in the GDR over the years, the SED's sociotechnical imaginary was amended and updated with new facets. This points to the temporal dynamic of the SED's imaginary of socialist computerization. In order to be able to preserve its dominance over the years, the SED's imaginary required continuous reworking and reaffirmation in face of rapid scientific and technological change – for example, the developments of cybernetics, modern digital computers, the integrated circuit, the microchip, and microprocessor – as well as societal and economic developments – for example, the challenges of global market competition, a growing desire for consumer electronics, or fundamental changes in workplace organization. This somewhat fluid and dynamic character of the SED's sociotechnical imaginary, taking on different shapes over time to adapt to scientific, technological, and societal change, and highlighting different aspects in addressing different target groups, corresponds to Sergio Sismondo's characterization of sociotechnical imaginaries as typically "changeable, flexible and loose around the edges."[923]

Over the course of the three decades under study, different aspects of envisioning a desirable computerized future were bound together to form a single, but fluid dominant imaginary of socialist computerization. The SED's imaginary was intended to legitimize the authoritarian regime's political and economic strategy with regard to the introduction of new computer technology, and to convince the people of the GDR to bring individual sacrifices and commit to the project of working towards a desirable future of a computerized society and economy. Crucially, it served to guide educational policymaking and curricular reforms, by envisioning not just a future computerized world, but also the future computer users.

Tasked with responding to the technological, political, and economic challenge of the emergence of new computer technology, education policymakers in the GDR tried to make sense and act in their present by borrowing from an imagined emergent future. The SED's dominant multifaceted sociotechnical imaginary served as a reference framework for the design of new curricula for technology education, by creating and affirming a common, normative understanding of what the new technology was, how and what it should be used for, and by whom. In turn, education also served to reinforce the dominant sociotechnical imaginary by propagating a certain conception of the technology and its anticipated social, political, and economic effects. The case of the GDR highlights the potentially fragile and inconsistent character of popular belief in the SED's imaginary. The authoritarian regime made continuous and rigorous efforts to counter disappointed promises and hopes through computer courses, popular science and technology books and journals, and political propaganda to foster and reinvigorate popular support and belief in its vision of a desirable computerized future under socialism.

With the fall of the Berlin Wall and the subsequent demise of the authoritarian SED regime, the power base on which the hegemony of the sociotechnical imaginary of socialist computerization had rested was lost. This created space for the emergence of new sociotechnical imaginaries that envisioned alternative purposes, uses, and lines of development for modern computer technology. The German reunification would thus make for a fascinating period to study in terms of whether certain aspects of the SED's imaginary lived on in one form or another, and how new imaginaries came to supplement or replace earlier visions of a desirable computerized future. It also raises the question of how sociotechnical imaginaries come to dominate and are stabilized in pluralistic and democratic so-

---

[923] Sergio Sismondo, "Sociotechnical imaginaries: An accidental themed issue," *Social Studies of Science* 50, no. 4 (2020): 505.

cieties in order to achieve an (at least temporary) political consensus on the role of new computer technology in education.

While this study focused on a single, albeit multifaceted sociotechnical imaginary in a socialist one-party state, further research can draw upon the findings of the case of the GDR and compare them with the entanglements of sociotechnical imaginaries and education policymaking in pluralist societies of capitalist countries, or among different states of the Socialist Bloc. Furthermore, international comparative or transnational research perspectives could prove fruitful in shedding light on the phenomenon of "traveling imaginaries"[924] in relation to computer technology instruction and the rise of computers as an educational technology. The GDR's involvement in international organizations in the field of computing, such as the IFIP, and the participation of education policymakers and pedagogues from the GDR in various international congresses on computers in education, raises the question of how "national" imaginaries of computers in education were shaped by developments and imaginaries in other countries, or by the sociotechnical imaginaries promoted by powerful transnational actors in education. In the context of the Cold War, it would be particularly interesting to explore what kinds of sociotechnical imaginaries were able to cross the Iron Curtain or were shared on both sides, if any. Further research is therefore needed to investigate whether these "travelling," or shared imaginaries have potentially served to facilitate (or impede) the international transfer of educational policies and national strategies in the design of computer education curricula and the use of computers as tools of teaching and learning.

In her work on the "Soviet Information Age," Ksenia Tatarchenko reminds us that in order to do accurately reflect the historical memory of the Soviet version of the information age, it is essential to "suspend the received notions on the inevitable implosion of socialism."[925] This consideration is equally pertinent to the case of the GDR, since the fall of the Berlin Wall in 1989 and the subsequent dissolution of the GDR came as somewhat unexpected events at the time. The utopian vision of a computerized future under socialism, therefore, should not be lightly dismissed as mere pipe dreams of a crumbling political system with a faltering planned economy. As this study has shown, the SED's imaginary of socialist computerization informed a comprehensive policy agenda to prepare the GDR's society and economy for the future. It set the stage for an elaborate educational policy program and carefully crafted curricula from the comprehensive school to

---

924 Sebastian Pfotenhauer and Sheila Jasanoff, "Traveling imaginaries: The 'practice turn' in innovation policy and the global circulation of innovation models," in *The Routledge Handbook of the Political Economy of Science*, ed. David Tyfield et al. (London: Routledge, 2017), 416–28.
925 Ksenia Tatarchenko, "Why We Should Remember the Soviet Information Age," 264.

post-secondary and continued education and training that could guide the people of the GDR through the rapidly changing world of new information technology. In this process, the courses provided them with the necessary updates in knowledge and skills on an ongoing basis. In this sense, this study presents a counter-narrative to the alleged supremacy of capitalist computerization that looks beyond the short-sighted verdict of technological backwardness, to acknowledge the effort to devise a coherent system of computer education and training that would not just respond to short-term qualification needs but was set in place to follow a vision for the next decades. This clearly shows that the SED's ambitious plans to computerize the GDR were not an act of desperation in the face of its slow, seemingly inevitable demise – but rather an indication of a vigorous and long-term aspiration for the future.

List of Abbreviations

AdW	Akademie der Wissenschaften der DDR (eng.: Academy of Sciences of the GDR)
APW	Akademie der Pädagogischen Wissenschaften der DDR (eng.: Academy of Pedagogical Sciences of the GDR)
BASIC	Beginner's All-purpose Symbolic Instruction Code
BIC	Bildungscomputer (eng.: Educational Computer)
BmA	Berufsausbildung mit Abitur (eng.: Vocational Training with Abitur)
BMSR	Betriebsmess-, Steuerungs- und Regelungstechnik (eng.: Operational Measurement, Control and Regulation Technology)
BUW	Bezirkskabinett für Unterricht und Weiterbildung (eng.: Distric Cabinet for Instruction and Further Training of Teachers and Educators)
CAD	Computer-Aided Design
CAM	Computer-Aided Manufacturing
CIM	Computer-Integrated Manufacturing
CNC	Computerized Numerical Control
COCOM	Coordinating Committee on Multilateral Export Controls
COMECON	Council for Mutual Economic Assistance (also abbreviated as CMEA)
CP/M	Control Program for Microcomputers
DPZI	Deutsches Pädagogisches Zentralinstitut (eng.: German Pedagogical Central Institute)
ECCE	European Conference on Computers in Education
EOS	Erweiterte Oberschule (eng.: Extended Secondary School)
ESER	Einheitliches System Elektronischer Rechentechnik (eng.: Unified System of Electronic Computing Machines)
ESP	Einführung in die Sozialistische Produktion (eng.: Introduction to Socialist Production)
FDGB	Freier Deutscher Gewerkschaftsbund (eng.: Free German Trade Union Federation)
FDJ	Freie Deutsche Jugend (eng.: Free German Youth)
FRG	Federal Republic of Germany (ger.: Bundesrepublik Deutschland, BRD)
GDR	German Democratic Republic (ger.: Deutsche Demokratische Republik, DDR)
GIDDR	Gesellschaft für Informatik der DDR (eng.: Society for Informatics of the GDR)
GST	Gesellschaft für Sport und Technik (eng.: Society for Sports and Technology)
HdjT	Haus der jungen Talente (eng.: House of Young Talents)
IBM	International Business Machines Corporation
IDV	Institut für Datenverarbeitung (eng.: Institute for Data Processing)
IED	Institut für Elektronik Dresden (eng.: Dresden Institute for Electronics)
IFIP	International Federation for Information Processing
IMN	Institut für Mathematischen und Naturwissenschaftlichen Unterricht (eng.: Institute for Mathematics and Science Education)
IMR	Institut für Maschinelle Rechentechnik (eng.: Institute for Computing Machinery)
IPB	Institut für Polytechnische Bildung (eng.: Institute for Polytechnic Education)
KdT	Kammer der Technik (eng.: Chamber of Technology)
MfS	Ministerium für Staatssicherheit ('Stasi') (eng.: Ministry for State Security)
MMM	Messe der Meister von Morgen
NATO	North Atlantic Treaty Organization

NÖSPL	Neues Ökonomisches System der Planung und Leitung (eng.: New Economic System of Planning and Management)
OECD	Organisation for Economic Co-operation and Development
PA	Produktive Arbeit (eng.: Productive Work)
PAP	Programmablaufplan (eng.: programming flowchart)
PKJM	Potsdamer Klub der Jungen Mathematiker (eng.: Potsdam Club of Young Mathematicians)
POS	Polytechnische Oberschule (eng.: Polytechnic Secondary School)
RBASIC	Robotron-BASIC
SCP	Single User Control Program (commonly used operating system in the GDR)
SED	Sozialistische Einheitspartei Deutschlands (eng.: Socialist Unity Party of Germany)
SEG	Schülerexperimentiergerät (eng.: Pupils' Experiment Device)
STS	Science and Technology Studies
USA	United States of America
VEB	Volkseigener Betrieb (eng.: Publicly Owned Enterprise)
VET	Vocational Education and Training
VVB	Vereinigung Volkseigener Betriebe (eng.: Association of Publicly Owned Enterprises)
WCCE	World Conference on Computers in Education
WPA	Wissenschaftlich-Praktische Arbeit (eng.: Scientific-Practical Work)
ZIB	Zentralinstitut für Berufsbildung (eng.: Central Institute for Vocational Education and Training)
ZIA	Zentralinstitut für Automatisierung (eng.: Central Institute for Automation)
ZIW	Zentrales Institut für Weiterbildung der Lehrer und Erzieher (eng.: Central Institute for the Continuing Education of Teachers and Educators)
ZKI	Zentralinstitut für Kybernetik und Informationsprozesse (eng.: Central Institute for Cybernetics and Information Processes)

List of Figures

Figure 1   The delegation of the district of Erfurt during the big parade through the city center of Berlin to celebrate the 750th anniversary of Berlin on 4th July 1987 —— **29**

Figure 2   Four-tier model of computer technology education in higher education in the GDR —— **84**

Figure 3   Magnetic tape units (top) and control panel (bottom) of the 'R 300' —— **103**

Figure 4   Article in the newspaper *Berliner Zeitung* dated July 16, 1988 entitled "Guided by a computer, the blouses float through the room – Berlin ladies' fashion relies on new technology / Over 7200 more products per year" —— **113**

Figure 5   Simplified governance structure for the determination of curricular content in VET —— **122**

Figure 6   Advertisements in the *Berliner Zeitung* from the early 1990s for a "Basic course for computer users" and a training course in "Clerical administration and EDP accounting" —— **148**

Figure 7   The system of education in the GDR since 1965, with the comprehensive ten-class general educational Polytechnic Secondary School (POS) at its core —— **157**

Figure 8   Example of a programme flowchart for a phone call —— **180**

Figure 9   Number encoding puzzle in "Mini-Frösi" number 29 —— **240**

Figure 10 Motif of the "computer as a friend" in the illustrations of the "Mini-Frösi" booklets on computer technology —— **241**

Figure 11 Motif of "Tüte inside the computer" in the illustrations of the "Mini-Frösi" booklets on computer technology —— **242**

List of Tables

Table 1 Higher education computer centers in the GDR (1964) —— **67**
Table 2 Introduction of higher education study courses on computing and informatics in the GDR during the 1960s —— **71**
Table 3 University graduates with a higher education in informatics and/or CAD/CAM, differentiated according to the level of computer technology education they received —— **86**
Table 4 Generational shifts in the GDR's domestically produced computer technology in comparison to international developments —— **111**
Table 5 Overview of computer education in general schooling in the GDR by the end of the 1980s —— **193**

Historical Sources

Archival Documents

BArch | Bundesarchiv Lichterfelde Berlin
DR 2 Ministerium für Volksbildung
DR 3 Ministerium für Hoch- und Fachschulwesen
DQ 4 Staatssekretariat für Berufsbildung
DR 201 Zentralinstitut für Weiterbildung der Lehrer und Erzieher
DQ 400 Zentralinstitut für Berufsausbildung
MfS Stasi-Unterlagen-Archiv

SAPMO | Stiftung Archiv der Parteien und Massenorganisationen der DDR im Bundesarchiv
DY 30 Sozialistische Einheitspartei Deutschlands
ZB 20049 Gesetzblatt der Deutschen Demokratischen Republik

DIPF | Leibniz-Institut für Bildungsforschung und Bildungsinformation, BBF | Bibliothek für Bildungsgeschichtliche Forschung
BBF Archive: Akademie der Pädagogischen Wissenschaften der DDR (APW)
BBF Library: Sammlung der Pädagogischen Lesungen (PL)
BBF Library: Sammlung der Lehrpläne der DDR

Periodicals and Magazines

alpha
Berliner Zeitung (zefys.staatsbibliothek-berlin.de)
Einheit – Zeitschrift für Theorie und Praxis des Wissenschaftlichen Sozialismus
Funkamateur
Mathematik in der Schule
Mini-Frösi (supplements to the children's magazine *Frösi*)
Neues Deutschland (zefys.staatsbibliothek-berlin.de)
Neue Zeit (zefys.staatsbibliothek-berlin.de)
Polytechnische Bildung und Erziehung
practic
technikus

Literature

Afinogenov, Gregory. "Andrei Ershov and the Soviet Information Age." *Kritika: Explorations in Russian and Eurasian History* 14, no. 3 (2013): 561–84.
Allinson, Mark. "More from Less: Ideological Gambling with the Unity of Economic and Social Policy in Honecker's GDR." *Central European History* 45, no. 1 (2012): 102–27. https://doi.org/10.1017/S0008938911001002.
Ammer, Thomas, 1995. "Die Machthierarchie der SED." In *Materialien der Enquete-Kommission "Aufarbeitung von Geschichte und Folgen der SED-Diktatur in Deutschland" (12. Wahlperiode des Deutschen Bundestages)*, vol. II, 2, 803–67, https://enquete-online.de/pdf?pdf=wp12b2_2_15-79 (accessed May 31, 2024).
Ansorg, Leonore. *Kinder im Klassenkampf: Die Geschichte der Pionierorganisation von 1948 bis Ende der fünfziger Jahre*. Berlin: Akademie Verlag, 1997.
Anweiler, Oskar. "Berufsbildung in der Deutschen Demokratischen Republik unter vergleichenden Aspekten." *Berufsbildung in Wissenschaft und Praxis* 16, special issue 'Berufsbildung in der DDR' (1987): 3–7.
Anweiler, Oskar. *Schulpolitik und Schulsystem in der DDR*. Opladen: Leske+Budrich, 1988.
Anweiler, Oskar. "Berufsbildung in der Deutschen Demokratischen Republik unter vergleichenden Aspekten." In *Wissenschaftliches Interesse und politische Verantwortung: Dimensionen vergleichender Bildungsforschung*, edited by Jürgen Henze, Wolfgang Hörner, and Gerhard Schreier, 83–92. Wiesbaden: VS Verlag für Sozialwissenschaften, 1990. https://doi.org/10.1007/978-3-322-95936-2_6
Anweiler, Oskar, Wolfgang Mitter, Hansgert Peisert, Hans-Peter Schäfer, and Wolfgang Stratenwerth. *Vergleich von Bildung und Erziehung in der Bundesrepublik Deutschland und in der Deutschen Demokratischen Republik*. Köln: Verlag Wissenschaft und Politik, 1990.
APW der DDR. *Lehrbuchergänzung ESP Klasse 9*. Berlin: Volk und Wissen, 1988.
APW der DDR. *Experimentallehrplan für den fakultativen Unterricht in Informatik in der erweiterten Oberschule: Lehrgang Entwickeln von Programmen*. Berlin: Volk und Wissen, 1988.
Arnold, Thomas. "Lehrkräfte(aus)bildung im Deutschland des 20. Jahrhunderts – Kontinuitäten und Brüche." *Seminar*, no. 3 (2021): 5–17. https://bak-lehrerbildung.de/wp-content/uploads/seminar-3-21-arnold.pdf (accessed May 31, 2024).
Augustine, Dolores L. "Berufliches Selbstbild. Arbeitshabitus und Mentalitätsstrukturen von Software-Experten in der DDR." In *Eliten im Sozialismus. Beiträge zur Sozialgeschichte der DDR*, edited by Peter Hübner, 405–33. Köln: Böhlau, 1999.
Augustine, Dolores L. (1999b). "The Socialist 'Silicon Ceiling': East German Women in Computer Science." In *International Symposium on Technology and Society – Women and Technology: Historical, Societal, and Professional Perspectives, New Brunswick NJ, USA, 29–31 July 1999. Proceedings*, 347–355. https://doi.org/10.1109/ISTAS.1999.787357
Bacchi, Carol. "Sociotechnical Imaginaries and WPR: Exploring connections." Posted November 29, 2022. https://carolbacchi.com/2022/11/29/sociotechnical-imaginaries-and-wpr-exploring-connections (accessed May 31, 2024).
Barkleit, Gerhard. *Mikroelektronik in der DDR: SED, Staatsapparat und Staatssicherheit im Wettstreit der Systeme*. Dresden: Hannah-Arendt-Institut für Totalitarismusforschung, 2000.
Barth, Bernd-Rainer and Helmut Müller-Enbergs. "Hager, (Leonhard) Kurt." In *Wer war wer in der DDR? Ein Lexikon ostdeutscher Biographien*, edited by Helmut Müller-Enbergs, Jan Wielgohs, Dieter Hoffmann, Andreas Herbst, and Ingrid Kirschey-Feix. Berlin: Christoph Links Verlag, 2010.

https://www.bundesstiftung-aufarbeitung.de/de/recherche/kataloge-datenbanken/biographi sche-datenbanken (accessed May 31, 2024).

Bergien, Rüdiger. "Programmieren mit dem Klassenfeind: Die Stasi, Siemens und der Transfer von EDV-Wissen im Kalten Krieg." *Vierteljahrshefte für Zeitgeschichte* 67, no.1 (2019): 1–30. https://doi.org/10.1515/vfzg-2019-0001

Biener, Klaus. "Karl Steinbruch – Informatiker der ersten Stunde." *RZ-Mitteilungen*, no. 15 (1997): 53–54.

Blechinger, Doris, Alfred Kleinknecht, Georg Licht, and Friedhelm Pfeiffer. *The impact of innovation on employment in Europe: An analysis using CIS data*. Mannheim: Zentrum für Europäische Wirtschaftsforschung, 1998.

Blyth, Tilly. *The Legacy of the BBC Micro: Effecting Change in the UK's Cultures of Computing*. London: Nesta, 2012.

Blyth, Tilly. "Computing for the Masses? Constructing a British Culture of Computing in the Home." In *Reflections on the History of Computing*, edited by Arthur Tatnall, 231–42. Berlin/Heidelberg: Springer, 2012.

Böhme, Hans-Joachim. "Aus- und Weiterbildung der Ingenieure und Ökonomen." *Das Hochschulwesen* 33, no. 3 (1985): 57–83.

Bohnsack, Siegfried. "Aufbau, Arbeitsweise, Programmierung des Kleincomputers KC 85/1." *Polytechnische Bildung und Erziehung* 29, no. 2/3 (1987): 57–60.

Bösch, Frank. ed. *Wege in Die Digitale Gesellschaft: Computernutzung in der Bundesrepublik 1955–1990*. Göttingen: Wallstein Verlag, 2018.

Böse, Viktor and Friederike Krump. *Computer in der DDR-Presse: eine ZEFYS gestützte Mikrostudie*. Berlin, 2016. https://nbn-resolving.org/urn:nbn:de:0168-ssoar-47244-8 (accessed May 31, 2024).

Breton, Philippe, Alain-Marc Rieu, and Franck Tinland. *La Techno-Science en Question. Éléments pour une Archéologie du XXe Siècle*. Seyssel: Editions Champ Vallon, 1990.

Brock, Angela. "The Making of the Socialist Personality: Education and Socialisation in the German Democratic Republic 1958–1978." PhD diss., University College London, 2005. https://discovery.ucl.ac.uk/id/eprint/1363641/ (accessed May 31, 2024).

Brock, Angela. "Producing the 'Socialist Personality'? Socialisation, education, and the Emergence of New Patterns of Behaviour." In *Power and Society in the GDR, 1961–1979*, edited by Mary Fulbrook, 220–52. New York/Oxford: Berghahn Books, 2009. https://doi.org/10.1515/9781845459130-012

Brömmel, Klaus-Dieter. "Schüler an die moderne Technik – Aspekt meiner Leitungstätigkeit." *Polytechnische Bildung und Erziehung* 29, no. 1 (1987): 10–12.

Büchler, Grit. *Mythos Gleichberechtigung in der DDR: Politische Partizipation von Frauen am Beispiel des Demokratischen Frauenbunds*. Frankfurt/New York: Campus, 1997.

Buchwald, Angela. "Die Ausbildung von Informatikern in Dresden – frühe Anfänge." In *Informatik in der DDR – Grundlagen und Anwendungen*, edited by Birgit Demuth, 129–41. Bonn: Gesellschaft für Informatik, 2008.

Cain, Frank. "Computers and the Cold War: United States Restrictions on the Export of Computers to the Soviet Union and Communist China." *Journal of Contemporary History* 40, no. 1 (2005): 131–47. https://doi.org/10.1177/0022009405049270

Cain, Victoria. *Schools and Screens. A Watchful History*. Cambridge, MA: MIT Press, 2021.

Campbell-Kelly, Martin. "Origin of Computing." *Scientific American* 301, no. 3 (2009): 62–69.

Campbell-Kelly, Martin, William Aspray, Nathan Ensmenger, and Jeffrey R. Yost. *Computer: A History of the Information Machine*. Boulder, CO: Westview Press, 2014.

Caruso, Marcelo. *Geschichte der Bildung und Erziehung: Medienentwicklung und Medienwandel*. Stuttgart: utb, 2019.

Cope, Bill and Mary Kalantzis. "A little history of e-learning: finding new ways to learn in the PLATO computer education system, 1959–1976." *History of Education* 52, no. 6 (2023): 905–36. https://doi.org/10.1080/0046760X.2022.2141353

Cortada, James W. "Information Technologies in the German Democratic Republic (GDR), 1949–1989." *IEEE Annals if the History of Computing* 34, no. 2 (2012): 34–48. https://doi.org/10.1109/MAHC.2012.27

Cortada, James W. *IBM: The Rise and Fall and Reinvention of a Global Icon*. Cambridge, MA: MIT Press, 2019.

Coy, Wolfgang. "Was ist Informatik? Zur Entstehung des Faches an den deutschen Universitäten." In *Geschichten der Informatik: Visionen, Paradigmen, Leitmotive*, edited by Hans Dieter Hellige, 473–98. Berlin/Heidelberg: Springer, 2004. https://doi.org/10.1007/978-3-642-18631-8_17

Coy, Wolfgang and Peter Schirmbacher. *Informatik in der DDR – Tagung Berlin 2010. Tagungsband zum 4. Symposium 'Informatik in der DDR' am 16. und 17. September 2010 in Berlin*. Berlin: Humboldt-Universität zu Berlin, 2010. https://edoc.hu-berlin.de/handle/18452/18526 (accessed May 31, 2024).

Criblez, Lucien (2022). "Zur Entwicklung der Weiterbildung/Erwachsenenbildung während der Bildungsexpansionsphase der 1960er- und 1970er-Jahre." *Education Permanente*, no. 2 (2022): 8–19.

Cuban, Larry. *Oversold and Underused: Computers in the Classroom*. Cambridge, MA: Harvard University Press 2001.

Danyel, Jürgen and Annette Schuhmann. "Wege in die digitale Moderne: Computerisierung als gesellschaftlicher Wandel." In *Geteilte Geschichte: Ost- und Westdeutschland 1970–2000*, edited by Frank Bösch, 283–319. Göttingen: Vandenhoeck & Ruprecht, 2015.

Demuth, Birgit, ed. *Informatik in der DDR: Grundlagen und Anwendungen. Drittes Symposium 'Informatik in der DDR,' 15. und 16. Mai 2008 in Dresden, Deutschland*. Bonn: Gesellschaft für Informatik, 2008.

Demuth, Birgit, Frank Rohde, and Uwe Aßmann. "50 Jahre universitärer Informatik-Studiengang an der TU Dresden aus der Sicht von Zeitzeugen in einem Zeitstrahl." *Informatik Spektrum* 45, no. 3 (2022): 183–91. https://doi.org/10.1007/s00287-022-01457-0

Dennis, Mike. "Working under Hammer and Sickle: Vietnamese Workers in the German Democratic Republic, 1980–89." *German Politics* 16, no. 3 (2007): 339–57. https://doi.org/10.1080/09644000701532700

Denzer, Horst. "Kybernetische Planung und Politische Ordnungsform: Ein Aspekt der Kybernetikdiskussion in der DDR." *Zeitschrift Für Politik* 15, no. 1 (1968): 65–86.

Diesel, Harald. "Integration der Informatik und der informationsverarbeitenden Technik in den polytechnischen Unterricht." *Polytechnische Bildung und Erziehung* 30, no. 5 (1988): 154–57.

Dittmann, Frank. "Microelectronics under Socialism." *Icon*, no. 8 (2002): 43–54.

Dittmann, Frank and Rudolf Seising. *Kybernetik steckt den Osten an. Aufstieg und Schwierigkeiten einer interdisziplinären Wissenschaft in der DDR*. Berlin: trafo, 2007.

Donig, Simon. "Die DDR-Computertechnik und das COCOM-Embargo 1958–1973. Technologietransfer und institutioneller Wandel im Spannungsverhältnis zwischen Sicherheit und Modernisierung." In *Informatik in der DDR – eine Bilanz*, edited by Friedrich Naumann and Gabriele Schade, 251–72. Bonn: Gesellschaft für Informatik, 2006.

Donig, Simon. "Informatik im Systemkonflikt – Der Technik- und Wissenschaftsdiskurs in der DDR." In *Informatik in der DDR – eine Bilanz*, edited by Friedrich Naumann and Gabriele Schade, 462–78. Bonn: Gesellschaft für Informatik, 2006.

Donig, Simon. "Appropriating American technology in the 1960s: Cold War politics and the GDR computer industry." *IEEE Annals of the History of Computing* 32, no. 2 (2010): 32–45. https://doi.org/10.1109/MAHC.2010.6

Enders, Horst. "CAD-Technologie meistern – eine Herausforderung an die politisch-ideologische Arbeit. *Einheit*, no. 7 (1986): 605–10.

Engelmann, Lutz. *Zur Vermittlung und Aneignung informatischen Wissens und Könnens als Bestandteil sozialistischer Allgemeinbildung – dargestellt am Beispiel eines fakultativen Kurses 'Informatik' für die Klassen 9 und 10 (Dissertation A), Band 1.* Berlin: Akademie der Pädagogischen Wissenschaften der DDR, 1988.

Erdogan, Julia G. "Computerkids, Freaks, Hacker: Deutsche Hackerkulturen in internationaler Perspektive." In *Let's Historize it! Jugendmedien im 20. Jahrhundert*, ed. Aline Maldener and Clemens Zimmermann, 61–96. Köln: Böhlau Verlag, 2018. https://doi.org/10.7788/9783412512286.61

Erdogan, Julia G. *Avantgarde der Computernutzung: Hackerkulturen der Bundesrepublik und der DDR.* Göttingen: Wallstein, 2021.

Erdogan, Julia G. "Wie die Fertigung (zunächst) nicht in den Computer kam. Der schwierige Prozess der Umsetzung von CIM." *Technikgeschichte* 90, no. 1 (2023): 3–24.

Flade, Lothar. "Zum fakultativen Kurs 'Informatik' in den Klassen 9 und 10." *Mathematik in der Schule* 27, no. 7/8 (1989): 541–43.

Flamm, Kenneth. *Creating the Computer: Government, Industry, and High Technology.* Washington, DC: Brookings Institution, 1988.

Fleischer, Wolfgang. "Die Zukunft meistern." *technikus*, no. 10 (1985): 1–2.

Fleischer, Wolfgang. "Vertrauen ins Morgen." *technikus*, no. 9 (1986): 1–2.

Flury, Carmen and Michael Geiss, eds. *How Computers Entered the Classroom, 1960–2000: Historical Perspectives.* Berlin/Boston: De Gruyter Oldenbourg, 2023. https://doi.org/10.1515/9783110780147

Flury, Carmen and Michael Geiss. "Computers in Europe's Classrooms: An Introduction." In *How Computers Entered the Classroom, 1960–2000: Historical Perspectives* edited by Carmen Flury and Michael Geiss, 1–12. Berlin/Boston: De Gruyter Oldenbourg, 2023. https://doi.org/10.1515/9783110780147-001

Fraunholz, Uwe "'Revolutionäres Ringen für den gesellschaftlichen Fortschritt': Automatisierungsvisionen in der DDR." In *Technology Fiction: Technische Visionen und Utopien in der Hochmoderne*, edited by Uwe Fraunholz and Anke Woschech, 195–219. Bielefeld: transcript, 2012.

Friedrich, Gerd. "Beschleunigung des Reproduktionsprozesses in den Kombinaten." *Einheit*, no. 7 (1986): 598–604.

Friedrich, Steffen and Bettina Timmermann. "Lehrerbildung Informatik – Basis für die Informatik als Allgemeinbildung." In *Informatik in der DDR – Grundlagen und Anwendungen*, edited by Birgit Demuth, 197–208. Bonn: Gesellschaft für Informatik, 2008.

Fritsche, Detlev. "Mit Prototyprekonstruktion zum Welthöchststand? PC-Software in den letzten Jahren der DDR." *Dresdener Beiträge zur Geschichte der Technikwissenschaften*, no. 30 (2005): 105–23.

Fuchs, Hans-Jürgen and Eberhard Petermann. *Bildungspolitik in der DDR 1966–1990: Dokumente.* Berlin: Osteuropa-Institut der Freien Universität Berlin, 1991.

Fuchs-Kittowski, Klaus. "Grundlinien des Einsatzes der modernen Informations- und Kommunikationstechnologien in der DDR. Wechsel der Sichtweisen zu einer am Menschen orientierten Informationssystemgestaltung." In *Informatik in der DDR – eine Bilanz*, edited by Friedrich Naumann and Gabriele Schade, 55–70. Bonn: Gesellschaft für Informatik, 2006.

Fuchs-Kittowski, Klaus. "Zur Herausbildung von Sichtweisen der Informatik in der DDR unter Einfluss der Kybernetik I. und II. Ordnung." In *Kybernetik steckt den Osten an. Aufstieg und Schwierigkeiten einer interdisziplinären Wissenschaft in der DDR*, edited by Frank Dittmann and Rudolf Seising, 323–80. Berlin: Trafo, 2007.

Fuchs-Kittowski, Klaus, Edo Albrecht, Erich Langner, and Dieter Schulze. "Gründung, Entwicklung und Abwicklung der Sektion Ökonomische Kybernetik und Operationsforschung/Wissenschaftstheorie und Wissenschaftsorganisation an der Humboldt-Universität zu Berlin." In *Die Humboldt- Universität Unter den Linden 1945 bis 1990. Zeitzeugen – Einblicke – Analysen*, edited by Wolfgang Girnus and Klaus Meier, 155–97. Leipzig: Leipziger Universitätsverlag, 2010.

Fuchs-Kittowski, Frank and Werner Kriesel. "Biografie von Klaus Fuchs-Kittowski." In *Informatik und Gesellschaft: Festschrift zum 80. Geburtstag von Klaus Fuchs-Kittowski*, edited by Frank Fuchs-Kittowski and Werner Kriesel, 479–84. Frankfurt am Main: Peter Lang, 2016.

Fulbrook, Mary. "Popular Discontent and Political Activism in the GDR." *Contemporary European History* 2, no. 3 (1993): 265–82. https://doi.org/10.1017/S0960777300000527

Fulbrook, Mary. "Democratic Centralism and Regionalism in the GDR." In *German Federalism*, edited by Maiken Umbach, 146–71. London: Palgrave Macmillan, 2002. https://doi.org/10.1057/9780230505797_7

Gartz, Manfred. "Die Urania im Bildungssystem der DDR 1980/90." *REPORT Literatur- und Forschungsreport Weiterbildung*, no. 30 (1992): 99–108.

Gazzard, Alison. *Now the Chips Are Down: The BBC Micro*. Cambridge, MA: MIT Press, 2016.

Geiss, Michael. "Die Politik des lebenslangen Lernens in Europa nach dem Boom." *Zeitschrift für Weiterbildungsforschung* 40, no. 2 (2017): 211–28. https://doi.org/10.1007/s40955-017-0093-1

Gerovitch, Slava. "'Mathematical Machines' of the Cold War: Soviet Computing, American Cybernetics and Ideological Disputes in the Early 1950s." *Social Studies of Science* 31, no. 2 (2001): 253–87. https://doi.org/10.1177/0306312701031002

Gerovitch, Slava. *From Newspeak to Cyberspeak: A History of Soviet Cybernetics*. Cambridge, MA: MIT Press, 2002.

Gerth, Werner. "Jugend und wissenschaftlich-technische Revolution." *Einheit*, no. 6 (1987): 527–32.

Gießler, Denis. "Video Games in East Germany: The Stasi Played Along," Zeit Online, November 21, 2018. https://www.zeit.de/digital/games/2018-11/computer-games-gdr-stasi-surveillance-gamer-crowd (accessed May 31, 2024).

Gill, Ulrich. "Betriebsgewerkschaftsleitung (BGL)," in *FDGB-Lexikon: Funktion, Struktur, Kader und Entwicklung einer Massenorganisation der SED (1945–1990)*, edited by Dieter Dowe, Karlheinz Kuba, and Manfred Wilke. Berlin, 2009. http://library.fes.de/FDGB-Lexikon (accessed May 31, 2024).

Good, Katie Day. *Bring the World to the Child. Technologies of Global Citizenship in American Education*. Cambridge, MA: MIT Press, 2020.

Goodson, Ivor F. and J. Marshall Mangan. "Computers in Schools as Symbolic and Ideological Action: The Genealogy of the ICON." *The Curriculum Journal* 3, no. 3 (1992): 261–76. https://doi.org/10.1080/0958517920030305

Grote, Ursula and Horst Völz. *BASIC: Einmaleins des Programmierens. Eine Einführung in die Programmiersprache BASIC*. Leipzig/Jena: Urania-Verlag, 1987.

Grottker, Dieter. "Ungleiche Partner – Das Dogma der sozialistischen Allgemeinbildung und die berufliche Bildung in der DDR." *Syllabus. Gesammelte Aufsätze zur Berufs- und Bildungswissenschaft*, no. 1 (January 2020). https://syllabus-dresden.de/2020/01/18/223/ (accessed May 31, 2024).

Grütter, Fabian. "The Smaky School Computer. Technology and Education in the Ruins of Switzerland's Watch Industry, 1973-1997." *Learning, Media and Technology* 49, no. 1 (May 24, 2023): 49-62. https://doi.org/10.1080/17439884.2023.2216463

Guder, Michael. (1987). "Die Einstellung der beruflichen Bildung in der DDR auf neue Technologien." *Berufsbildung in Wissenschaft und Praxis* 16, special issue 'Berufsbildung in der DDR', 12 -19.

Guerrero Cantarell, Rosalía and Carmen Flury. "Making the Computer Fit for School: Efforts to Develop a State-Mandated Educational Computer in Sweden and East Germany (1980s-1990s)." *Historical Studies in Education / Revue d'histoire De l'éducation* 35, no. 2 (2023): 5-29. https://doi.org/10.32316/hse-rhe.2023.5141

Hacker, Winfried. "Software-Ergonomie; Gestalten rechnergestützter geistiger Arbeit?!." In *Software-Ergonomie '87: Nützen Informationssysteme dem Benutzer?*, edited by Wolfgang Schönpflug and Marion Wittstock, 31–54. Stuttgart: B. G. Teubner, 1987.

Haigh, Thomas and Paul E. Ceruzzi. *A New History of Modern Computing*. Cambridge, MA: MIT Press, 2021.

Hally, Mike. *Electronic Brains: Stories from the Dawn of the Computer Age*. Washington, DC: Joseph Henry Press, 2005.

Hartmann, Karl. "Schlüssel für kräftiges Wachstum zum Wohle des Volkes." *Einheit*, no. 7 (1986): 591-97.

Harvard STS Research Platform, "The Sociotechnical Imaginaries Project." https://sts.hks.harvard.edu/research/platforms/imaginaries/ (accessed May 31, 2024).

Hausten, Hans-Joachim. *Allgemeinbildung und Persönlichkeitsentwicklung: Ein Beitrag zur Aufarbeitung der DDR-Pädagogik*. Frankfurt am Main: Peter Lang, 2003.

Heinicke, Steffi and Michael Unger. "Das Schülerrechenzentrum Dresden von 1984 bis heute." In *Informatik in der DDR – Grundlagen und Anwendungen*, edited by Birgit Demuth 220–30. Bonn: Gesellschaft für Informatik, 2008.

Herrlich, Ottomar. "Gründung und Wirken der Sektion Informationsverarbeitung der Technischen Universität Dresden – ein Gedächtnisbericht." In: *Informatik in der DDR – eine Bilanz*, edited by Friedrich Naumann and Gabriele Schade, 319–30. Bonn: Gesellschaft für Informatik, 2006.

Hess, David J. "Publics as Threats? Integrating Science and Technology Studies and Social Movement Studies." *Science as Culture* 24, no. 1 (2014): 69–82. https://doi.org/10.1080/09505431.2014.986319

Hess, David J. and Benjamin K. Sovacool. "Sociotechnical matters: Reviewing and integrating science and technology studies with energy social science." *Energy Research & Social Science*, no. 65 (2020). https://doi.org/10.1016/j.erss.2020.101462

Heylighen, Francis and Cliff Joslyn. "Cybernetics and Second Order Cybernetics." In *Encyclopedia of Physical Science and Technology (3rd Edition), Vol. 4*, edited by Robert A. Meyers, 155–70. New York: Academic Press, 2003.

Hille, Frank. "Computerspiele – Herangehen, Anforderungen, Klassifizierungen, Beispiele." In *Computer in der außerunterrichtlichen Tätigkeit. Standpunkte und Anregungen*, edited by Gerd Hutterer, 50–57. Halle: Pädagogische Hochschule Halle N. K. Krupskaja, 1987.

Hof, Barbara. "From Harvard via Moscow to West Berlin: Educational Technology, Programmed Instruction and the Commercialisation of Learning after 1957." *History of Education* 47, no. 4 (2018): 445–65. https://doi.org/10.1080/0046760X.2017.1401125

Hof, Barbara. "The Turtle and the Mouse: How Constructivist Learning Theory Shaped Artificial Intelligence and Educational Technology in the 1960s." *History of Education* 50, no. 1 (2021): 93–111. https://doi.org/10.1080/0046760X.2020.1826053

Hof, Barbara and Regula Bürgi. "The OECD as an arena for debate on the future uses of computers in schools." *Globalisation, Societies and Education* 19, no. 2 (2021): 154–66. https://doi.org/10.1080/14767724.2021.1878015

Holloway, David. "Innovation in Science – The Case of Cybernetics in the Soviet Union." *Science Studies* 4, no. 4 (1974): 299–337. https://doi.org/10.1177/030631277400400401

Honecker, Margot. *Der gesellschaftliche Auftrag unserer Schule. Referat von Margot Honecker, Minister für Volksbildung, auf dem VIII. Pädagogischen Kongress*. Berlin: Verlag Zeit im Bild, 1978.

Hörner, Wolfgang. "Technisch-ökonomische Entwicklung und Reformen im Bildungswesen der DDR." *Bildung und Erziehung* 40, no. 1 (1987): 19–33.

Hörner, Wolfgang. "Informationstechnische Bildung." In *Vergleich von Bildung und Erziehung in der Bundesrepublik Deutschland und in der Deutschen Demokratischen Republik*, edited by Anweiler, Oskar, Wolfgang Mitter, Hansgert Peisert, Hans-Peter Schäfer, and Wolfgang Stratenwerth, 620–37. Köln: Verlag Wissenschaft und Politik, 1990.

Hoyer, Helmut and Norbert Klotz. *TINY, der kleine Selbstbau-Computer*. Berlin: Verlag Junge Welt, 1989.

Hübner, Peter. *Arbeit, Arbeiter und Technik in der DDR 1971 bis 1989. Zwischen Fordismus und digitaler Revolution*. Bonn: J.H.W. Dietz Nachf., 2014.

Hummel, Lothar and Rudi Rosenkranz. "CAD/CAM im Dienste des Menschen." *Einheit*, no. 12 (1986): 1103–8.

Hunger, Francis. "Sozialistische Co-Innovation – Wie in der DDR die relationale Datenbank DABA-1600 entwickelt wurde." *Zeitschrift für Medienwissenschaft* 14, no. 2 (2022): 65–78. https://doi.org/10.25969/mediarep/18945

Hutterer, Gerd. "Orientierungen für eine erziehungswirksame Gestaltung des inhaltlichen und methodischen Herangehens an die Beschäftigung mit dem Computer in der außerunterrichtlichen Tätigkeit." In *Computer in der außerunterrichtlichen Tätigkeit. Standpunkte und Anregungen*, edited by Gerd Hutterer, 5–20. Halle: Pädagogische Hochschule Halle N. K. Krupskaja, 1987.

Impagliazzo, John and John A. N. Lee. *History of Computing in Education*. New York: Springer, 2004.

Jasanoff, Sheila and Sang-Hyun Kim. "Containing the Atom: Sociotechnical Imaginaries and Nuclear Power in the United States and South Korea." *Minerva* 47, no. 2 (2009): 119–46. https://doi.org/10.1007/s11024-009-9124-4

Jasanoff, Sheila. "Future Imperfect: Science, Technology, and the Imaginations of Modernity." In *Dreamscapes of Modernity: Sociotechnical Imaginaries and the Fabrication of Power*, edited by Sheila Jasanoff and Sang-Hyun Kim, 1–33. Chicago/London: University of Chicago Press, 2015. https://doi.org/10.7208/chicago/9780226276663.003.0001

Jones, Robert J.D. "Shaping Educational Technology: Ontario's Educational Computing Initiative." *Educational & Training Technology International* 28, no. 2 (1991): 129–34. https://doi.org/10.1080/0954730910280207

Judy, Richard W. and Jane M. Lommel. "The New Soviet Computer Literacy Campaign." *Educational Communication and Technology* 34, no. 2 (1986): 108–23. https://doi.org/10.1007/BF02802584

Kaiser, Beate. *Die Pionierorganisation Ernst Thälmann. Pädagogik, Ideologie und Politik: eine Regionalstudie zu Dresden 1945–1957 und 1980–1990*. Frankfurt am Main: Peter Lang, 2013.

Kapplusch, Silvia. "Geschichte der Informatik an der technischen Universität Dresden." Technische Universität Dresden. Last modified September 6, 2016. https://tu-dresden.de/ing/informatik/die-fakultaet/geschichte/geschichte-der-informatik-an-der-technischen-universitaet-dresden (accessed May 31, 2024).

Kasper, Thomas. *Wie der Sozialstaat digital wurde. Die Computerisierung der Rentenversicherung im geteilten Deutschland*. Göttingen: Wallstein Verlag, 2020.

Keller, Gert and Gunter Kleinmichel. "Bildungscomputer robotron A 5105." *Mikroprozessortechnik* 2, no. 10 (1988): 292.

Keller, Gert and Gunter Kleinmichel. [Bildungscomputer (BIC) A 5105]. *Neue Technik im Büro* 33, no. 2 (1989): 62–64.

Kerner, Immo O. "Internationaler Trend des Computereinsatzes in allgemeinbildenden Schulen und Bemerkungen zur Computerdidaktik." In *1. Seminar 'Computer als Mittel und Gegenstand der Ausbildung', Tharandt 1982. Seminarberichte Teil 1*, 4–17. Dresden: Pädagogischen Hochschule Dresden, 1982.

Kerner, Immo O. *Studienmaterial zur Informatikausbildung für Lehrerstudenten*. Dresden: Pädagogische Hochschule K. F. W. Wander Dresden, 1988.

Kerner, Immo O. "Vorbereitung des Informatik-Unterrichts an den Schulen der DDR." In *Informatik in der DDR – eine Bilanz*, edited by Friedrich Naumann and Gabriele Schade, 422–31. Bonn: Gesellschaft für Informatik, 2006.

Kerner, Immo O. and Ulrich Winkler. *Taschenrechner in der Schule. Lehrmaterial zur Aus- und Weiterbildung von Mathematiklehrern an der PH Dresden*. Dresden: Pädagogische Hochschule K. F. W. Wander Dresden, 1985.

Kestere, Iveta and Katrina Elizabete Purina-Bieza. "Computers in the Classrooms of an Authoritarian Country: The Case of Soviet Latvia (1980s–1991)." In *How Computers Entered the Classroom, 1960–2000: Historical Perspectives*, edited by Carmen Flury and Michael Geiss, 75–98. Berlin/Boston: De Gruyter Oldenbourg, 2023. https://doi.org/10.1515/9783110780147-004

Klaus, Georg. "Zu einigen Problemen der Kybernetik." *Einheit* 13, no. 7 (1958): 1026–40.

Klaus, Georg. *Kybernetik in Philosophischer Sicht*. Berlin: Dietz Verlag, 1961.

Klaus, Georg. "Die Kybernetik, das Programm der SED und die Aufgaben der Philosophen." *Deutsche Zeitschrift für Philosophie* 11, no. 6 (1963): 693–707.

Klenke, Olaf. *Ist die DDR an der Globalisierung gescheitert? Autarke Wirtschaftspolitik versus internationale Weltwirtschaft – Das Beispiel Mikroelektronik*. Frankfurt am Main: Peter Lang, 2001.

Klenke, Olaf. "Globalisierung, Mikroelektronik und das Scheitern der DDR-Wirtschaft." *Deutschland Archiv*, 35, no. 3 (2002): 421–28.

Klenke, Olaf. *Kampfauftrag Mikrochip: Rationalisierung und sozialer Konflikt in der DDR*. Hamburg: VSA-Verlag, 2008.

Kline, Ronald R. *The Cybernetics Moment: Or Why We Call Our Age the Information Age*. Baltimore: Johns Hopkins University Press, 2015.

Kober, Wolfram. "Das neue Jahrtausend." *technikus*, no. 1 (1986): 10–11.

Koch, Hartmut. "Stand und Probleme der Informatikausbildung an landwirtschaftlichen Fachschulen der DDR." In *Agrarinformatik, Band 19: Referate der 11. GIL-Jahrestagung in Nürtingen, September 1990*, edited by Hans Geidel, Reiner Mohn, and Gerhard Schiefer, 353–55. Stuttgart: Ulmer, 1990.

Koch, Katja and Felix Linström. "Die Pädagogischen Lesungen im Rahmen der DDR-Lehrer*innenweiterbildung – Eine Systematisierung." In *DDR-Unterricht im Spiegel der Pädagogischen Lesungen*, edited by Katja Koch and Tilman von Brand, 35–54. Baltmannsweiler: Schneider Verlag Hohengehren, 2021. https://doi.org/10.18453/rosdok_id00002809

Kodron, Christoph. "Polytechnische Bildung und Erziehung in der Deutschen Demokratischen Republik." *International Review of Education* 24, no. 2 (1978): 207–16. https://doi.org/10.1007/BF00598982

Köhler, Helmut and Manfred Stock. *Bildung nach Plan? Bildungs- und Beschäftigungssystem in der DDR 1949 bis 1989*. Wiesbaden: Springer Fachmedien, 2004.

Koopmann, Kurt and Karl-Heinz Thieme. Moderne Technologien als hoher moralischer Anspruch. *Einheit*, no. 8 (1987): 708–12.
Kopstein, Jeffrey. "Ulbricht Embattled: The Quest for Socialist Modernity in the Light of New Sources." *Europe-Asia Studies* 46, no. 4 (1994): 597–615. https://doi.org/10.1080/09668139408412185
Körner, Helge. Berufsfachkommissionen. In *Aspekte der beruflichen Bildung in der ehemaligen DDR*, edited by Arbeitsgemeinschaft Qualifikations-Entwicklungs-Management Berlin, 13–58. Münster: Waxmann, 1996.
Krakat, Klaus. *Schlussbilanz der elektronischen Datenverarbeitung in der früheren* DDR (FS-Analysen, no. 5). Berlin: Forschungsstelle für Gesamtdeutsche Wirtschaftliche und Soziale Fragen, 1990.
Kranz, Susanne. "Women's Role in the German Democratic Republic and the State's Policy Toward Women." *Journal of International Women's Studies* 7, no. 1 (2005): 69–83.
Kreis, Reinhild. "A 'Call to Tools': DIY between State Building and Consumption Practices in the GDR." *International Journal for History, Culture and Modernity* 6, no. 1 (2018): 49–75. https://doi.org/10.18352/hcm.539
Kretzschmar, Albrecht. "Wissenschaftlich-technische Revolution und Persönlichkeit." *Einheit*, no. 2 (1987): 137–42.
Krisch, Henry. *The German Democratic Republic: The Search for Identity*. Boulder, CO: Westview Press, 1985.
Kröber, Günter. "Individualität und Kollektivität bei der Meisterung der wissenschaftlich-technischen Revolution." *Einheit*, no. 9 (1987): 815–20.
Krömke, Claus. "Ökonomische Strategie weist Wege zur Steigerung der Arbeitsproduktivität." *Einheit*, no. 9 (1987): 781–89.
Krückeberg, Fritz. *Die Geschichte der GI*. Bonn: Gesellschaft für Informatik, 2001.
Laitko, Hubert. "Wissenschaftlich-technische Revolution: Akzente des Konzepts in Wissenschaft und Ideologie der DDR." *Utopie kreativ*, no. 73–74 (1996): 33–50.
Lambrecht, Wolfgang. "Neuparzellierung einer gesamten Hochschullandschaft. Die III. Hochschulreform in der DDR (1965–1971)." *die hochschule*, no. 2 (2007): 171–89. https://doi.org/10.25656/01:16414
Lambrecht, Wolfgang. *Wissenschaftspolitik zwischen Ideologie und Pragmatismus: Die III. Hochschulreform (1965–71) am Beispiel der TH Karl-Marx-Stadt*. Münster: Waxmann, 2007.
Laux, Günter and Fritz Scholz. *Der Kybernetik-Report* (ZKI-Informationen, special issue no. 2). Berlin: Zentralinstitut für Kybernetik und Informationsprozesse der AdW der DDR, 1989.
Lean, Tom. "Mediating the Microcomputer: The Educational Character of the 1980s British Popular Computing Boom." *Public Understanding of Science* 22, no. 5 (2012): 546–58. https://doi.org/10.1177/0963662512457904
Lean, Tom. *Electronic Dreams: How 1980s Britain Learned to Love the Computer*. London/New York: Bloomsbury Sigma, 2016.
Lehmann, Nikolaus J. "Zur Geschichte des 'Instituts für maschinelle Rechentechnik' der Technischen Hochschule/Technischen Universität Dresden." In: *Zur Geschichte von Rechentechnik und Datenverarbeitung in der DDR 1946–1968*, edited by Erich Sobeslavsky and Nikolaus J. Lehmann, 123–57. Dresden: Hannah-Arendt-Institut für Totalitarismusforschung, 1996.
Lemke, Christiane. "Frauen, Technik und Fortschritt. Zur Bedeutung neuer Technologien für die Berufssituation von Frauen in der DDR." In *Die DDR in der Ära Honecker*, edited by Gert-Joachim Glaeßner, 481–98. Opladen: Westdeutscher Verlag.

Leonel Da Silva, Renan G. and Larry Au. "The Blind Spots of Sociotechnical Imaginaries: COVID-19 Scepticism in Brazil, the United Kingdom and the United States." *Science, Technology and Society* 27, no. 4 (2022): 611–29. https://doi.org/10.1177/09717218221125217

Leppin, Lars and Tom Schnabel. "Informatik und Rechentechnik in der DDR [Studienarbeit]," Humboldt-Universität zu Berlin, April 28, 1999. https://archive.org/details/Leppin-Schnabel_1999_Informatik-in-der-DDR (accessed May 31, 2024).

Leslie, Christopher. "From CoCom to Dot-Com: Technological Determinisms in Computing Blockades, 1949 to 1994." In *Histories of Computing in Eastern Europe*, edited by Christopher Leslie and Martin Schmitt, 196–225. Cham: Springer, 2019.

Lewis, Arthur J. "Education: Bridging Past, Present, and Future." *Journal of Thought* 16, no. 3 (1981): 61–71.

Liebscher, Heinz. *Kybernetik und Leitungstätigkeit*. Berlin: Dietz Verlag, 1966.

Lodahl, Hans-Jürgen. "Das Softwarehaus des VEB Kombinat Robotron." In *Informatik in der DDR – Grundlagen und Anwendungen*, edited by Birgit Demuth, 231–42. Bonn: Gesellschaft für Informatik, 2008.

Loebl, Eugen. "Computer socialism." *Studies in Soviet Thought* 11, no. 4 (1971): 294–300.

Lokatis, Siegfried. "Falsche Fragen an das Orakel? Die Einheit der SED." In *Zwischen 'Mosaik' und 'Einheit': Zeitschriften in der DDR*, edited by Simone Barck, Martina Langermann, and Siegfried Lokatis, 592–601. Berlin: Christoph Links Verlag, 1999.

Loos, Gottfried. "Produktive Arbeit an informationsverarbeitender Technik." *Polytechnische Bildung und Erziehung* 29, no. 4 (1987): 124–25.

Lorenzo, Mark J. *Endless Loop: The History of the BASIC Programming Language*. Philadelphia/Pittsburgh: SE Books, 2017.

Lösch, Andreas, Armin Grunwald, Martin Meister, and Ingo Schulz-Schaeffer, eds. *Socio-Technical Futures Shaping the Present*. Wiesbaden: Springer VS, 2019.

Lost, Christine. "Kinderzeitschriften und -zeitungen der DDR: Zwischen verschiedenen Betrachtungsweisen." In *Kinderzeitschriften in der DDR*, edited by Christoph Lüth and Klaus Pecher, 180–204. Bad Heilbrunn: Verlag Julius Klinkhardt, 2007.

MacKenzie, Donald and Judy Wajcman. "Introductory essay: the social shaping of technology." In *The social shaping of technology*, edited by Donald MacKenzie and Judy Wajcman, 3–27. Buckingham/Philadelphia: Open University Press, 1999.

Mahoney, Michael S. "The History of Computing in the History of Technology." *Annals of the History of Computing*, no. 10 (1988): 113–25. https://doi.org/10.1109/MAHC.1988.10011

Mahrad, Christa. "Jugendpolitik in der DDR." In: *Jugend im doppelten Deutschland*, edited by Walter Jaide and Barbara Hille, 195–225. Opladen: Westdeutscher Verlag, 1977.

Malycha, Andreas. "Im Zeichen von Reform und Modernisierung (1961 bis 1971)." *Informationen zur politischen Bildung*, no. 3 (2011): 37–48. https://www.bpb.de/shop/zeitschriften/izpb/geschichte-der-ddr-312/48537/im-zeichen-von-reform-und-modernisierung-1961-bis-1971/ (accessed May 31, 2024).

Malycha, Andreas. *Die SED in der Ära Honecker: Machtstrukturen, Entscheidungsmechanismen und Konfliktfelder in der Staatspartei 1971 bis 1989*. München: De Gruyter Oldenbourg, 2014.

Marx, Leo and Merritt R Smith. "Introduction." In *Does Technology Drive History? The Dilemma of Technological Determinism*, edited by Merritt R. Smith & Leo Marx, ix–xv. Cambridge, MA/London: MIT Press, 1994.

Medina, Eden. "Designing Freedom, Regulating a Nation: Socialist Cybernetics in Allende's Chile." *Journal of Latin American Studies* 38, no. 3 (2006): 571–606. https://doi.org/10.1017/S0022216X06001179

Medina, Eden. *Cybernetic Revolutionaries: Technology and Politics in Allende's Chile*. Cambridge, MA: MIT Press, 2011.

Merkel, Gerhard. *VEB Kombinat Robotron, Sitz Dresden: Ein Kombinat des Ministeriums für Elektrotechnik und Elektronik der DDR*. Dresden: AG Rechentechnik der Technischen Sammlungen Dresden, 2005. http://robotron.foerderverein-tsd.de/111/robotron111a.pdf (accessed May 31, 2024).

Merkel, Gerhard. "Computerentwicklungen in der DDR – Rahmenbedingungen und Ergebnisse." In *Informatik in der DDR – eine Bilanz*, edited by Friedrich Naumann and Gabriele Schade, 40–54. Bonn: Gesellschaft für Informatik, 2006.

Merkel, Gerhard. "Bildung und Wirken der Gesellschaft für Informatik der DDR." In *Informatik in der DDR – eine Bilanz*, edited by Friedrich Naumann and Gabriele Schade, 451–61. Bonn: Gesellschaft für Informatik, 2006.

Meyer, René. "Computer in der DDR." Erfurt: Landeszentrale für politische Bildung Thüringen, 2019.

Meyer, Uli. "The Institutionalization of an Envisioned Future. Sensemaking and Field Formation in the Case of 'Industrie 4.0' in Germany." In *Socio-Technical Futures Shaping the Present*, edited by Andreas Lösch, Armin Grunwald, Martin Meister, and Ingo Schulz-Schaeffer, 111–38). Wiesbaden: Springer VS, 2019.

Migdalek, Jürgen. "Zur Einführung von Elementen der Informatik in die Allgemeinbildung." *Polytechnische Bildung und Erziehung* 28, no. 4 (1986): 105–10.

Migdalek, Jürgen. "Grundlagen der Informatik im polytechnischen Unterricht." *Polytechnische Bildung und Erziehung* 28, no. 12 (1986): 429–31.

Migdalek, Jürgen. "Herausbildung eines Grundverständnisses der Informatik als fester Bestandteil des polytechnischen Unterrichts in den Klassen 9 und 10." *Polytechnische Bildung und Erziehung* 30, no. 5 (1988): 157–59.

Ministerium für Hoch- und Fachschulwesen. *Studienplan für die Grundstudienrichtung Wirtschaftswissenschaften. Berlin: Ministerium für Hoch- und Fachschulwesen*. Berlin: Ministerium für Hoch- und Fachschulwesen, 1972.

Ministerium für Hoch- und Fachschulwesen. *Studienplan für die Grundstudienrichtung Informationsverarbeitung zur Ausbildung an Universitäten und Hochschulen der DDR*. Berlin: Ministerium für Hoch- und Fachschulwesen, 1976.

Ministerium für Hoch- und Fachschulwesen. *Studienplan für die Grundstudienrichtung Mathematik zur Ausbildung an Universitäten und Hochschulen der DDR*. Berlin: Ministerium für Hoch- und Fachschulwesen, 1976.

Ministerium für Hoch- und Fachschulwesen. *Studienplan für die Grundstudienrichtung Mathematik zur Ausbildung an Universitäten und Hochschulen der DDR*. Berlin: Ministerium für Hoch- und Fachschulwesen, 1982.

Ministerium für Hoch- und Fachschulwesen. *Ergänzung zum Studienplan für die Grundstudienrichtung Mathematik. Mathematische Informatik*. Berlin: Ministerium für Hoch- und Fachschulwesen, 1982.

Ministerium für Hoch- und Fachschulwesen. *Studienplan für die Grundstudienrichtung Informationsverarbeitung zur Erprobung der Ausbildung in Verwirklichung der Konzeption für die Gestaltung der Aus- und Weiterbildung der Ingenieure und Ökonomen in der Deutschen Demokratischen Republik an der Technischen Universität Dresden, der Technischen Hochschule Karl-Marx-Stadt, der Technischen Hochschule Magdeburg, der Wilhelm-Pieck-Universität Rostock und der Ingenieurhochschule Dresden*. Berlin: Ministerium für Hoch- und Fachschulwesen, 1986.

Ministerrat der DDR/Ministerium für Volksbildung. *VIII. Pädagogischer Kongress der Deutschen Demokratischen Republik vom 18. bis 20. Oktober 1978: Protokoll*. Berlin: Volk und Wissen, 1979.

Ministerrat der DDR/Ministerium für Volksbildung. *Arbeitsbereich "Informationsverarbeitung und Rechentechnik" für die produktive Arbeit der Schüler der Klassen 9 und 10*. Berlin: Volk und Wissen, 1989.
Ministerrat der DDR/Ministerium für Volksbildung. *Experimentallehrplan Informatik Klasse 11*. Berlin: Volk und Wissen, 1989.
Ministerrat der DDR/Ministerium für Volksbildung. *Rahmenprogramm für den fakultativen Kurs Informatik in den Klassen 9 und 10*. Berlin: Volk und Wissen, 1989.
Ministerrat der DDR/Ministerium für Volksbildung. *Rahmenprogramm für den fakultativen Kurs Informationsverarbeitung – Prozessautomatisierung in den Klassen 9 und 10*. Berlin: Volk und Wissen, 1989.
Ministerrat der DDR/Ministerium für Volksbildung. *Lehrplan der zehnklassigen allgemeinbildenden polytechnischen Oberschule: Technik, Variante mit Stoffgebiet 'Grundkurs Informatik', Klassen 9 und 10*. Berlin: Volk und Wissen, 1990.
Ministerrat der DDR/Ministerium für Volksbildung/Ministerium für Hoch- und Fachschulwesen. *Studienplan für das postgraduale Studium zur Qualifizierung von Diplomlehrern auf dem Gebiet der Informatik an Universitäten und Hochschulen der DDR*. Berlin: Volk und Wissen, 1987.
Ministerrat der DDR/Ministerium für Volksbildung/Ministerium für Hoch- und Fachschulwesen. *Studienplan für die Ausbildung von Diplomlehrern der allgemeinbildenden polytechnischen Oberschulen in der Fachkombination Mathematik/Informatik an Universitäten und Hochschulen der DDR*. Berlin, 1989.
Ministerrat der DDR/Ministerium für Volksbildung/Ministerium für Hoch- und Fachschulwesen. *Studienplan für die Ausbildung von Diplomlehrern der allgemeinbildenden polytechnischen Oberschulen in Polytechnik – Informatik an Universitäten und Hochschulen der DDR*. Berlin, 1989.
Ministerrat der DDR/Ministerium für Volksbildung/Ministerium für Hoch- und Fachschulwesen. *Lehrprogramme für die Ausbildung von Diplomlehrern der allgemeinbildenden polytechnischen Oberschulen im Fach Polytechnik – Informatik an Universitäten und Hochschulen der DDR*. Berlin, 1989.
Modrow, Hans. "Schlüsseltechnologie – politische Bewährung auf dem Hauptkampffeld." *Einheit*, no. 8 (1987): 733–38.
Möhle, Horst. *Aus- und Weiterbildung Erwachsener auf Hochschulebene in der DDR*. Hagen: Zentrales Institut für Fernstudienforschung, 1986.
Morandi, Pietro. "Die ordnungspolitische Gegenrevolution in der DDR der 60er Jahre. Die Absage an den Militarismus und die Verwirtschaftlichung der Nationalen Mission der DDR." In *Kommunikation und Revolution*, edited by Kurt Imhof & Peter Schulz, 263–83. Zürich: Seismo, 1998.
Müller, Dieter. "Zur Einführung des obligatorischen Unterrichts im Fach Informatik an den erweiterten Oberschulen." *Mathematik in der Schule* 26, no. 9 (1988): 577–84.
Müller, Wolfgang. "Hat die Mikroelektronik-Industrie in der DDR noch eine Zukunft?." *Wechselwirkung* 12, no. 44 (1990): 35.
Müller-Enbergs, Helmut. "Kleiber, Günther". In *Wer war wer in der DDR? Ein Lexikon ostdeutscher Biographien*, edited by Helmut Müller-Enbergs, Jan Wielgohs, Dieter Hoffmann, Andreas Herbst, and Ingrid Kirschey-Feix. Berlin: Christoph Links Verlag, 2010. https://www.bundesstiftung-aufarbeitung.de/de/recherche/kataloge-datenbanken/biographische-datenbanken (accessed May 31, 2024).
Naumann, Friedrich and Gabriele Schade. *Informatik in der DDR – eine Bilanz. Tagungsband zu den Symposien 7. bis 9. Oktober 2004 in Chemnitz und 11. bis 12. Mai 2006 in Erfurt*. Bonn: Gesellschaft für Informatik, 2006.

Nick, Harry. "Informationsverarbeitende Technik erschliesst neue Quellen des ökonomischen und sozialen Fortschritts." *Einheit*, no. 7 (1986): 611–17.

Nick, Harry. "Über Wesen, Effekte und Tragweite flexibler Automatisierung." *Einheit*, no. 9 (1987): 790–95.

Nooney, Laine. *The Apple II Age: How the Computer Became Personal*. Chicago: The University of Chicago Press, 2023.

Oppermann, Lothar and Rudi Oelschlägel. "Ideologisch-theoretische Probleme der weiteren Ausprägung des polytechnischen Charakters unserer Oberschule." Einheit 31, no. 11 (1976): 1241–48.

O'Regan, Gerard. *A Brief History of Computing*. London: Springer, 2012.

Organisationsbüro IX. Pädagogischer Kongress. *Bulletin no. 3 IX. Pädagogischer Kongress: Diskussionsbeiträge vom zweiten Konferenztag (14. Juni 1989)*. Berlin, 1989.

Pecher, Klaus. "Kinderzeitschriften in der DDR – erziehungsstaatliche Okkupation der Kindheit." In *Kinderzeitschriften in der DDR*, edited by Christoph Lüth and Klaus Pecher, 12–43. Bad Heilbronn: Verlag Julius Klinkhardt, 2007.

Petrov, Victor. *Balkan Cyberia*. Cambridge, MA: MIT Press, 2023.

Pfotenhauer, Sebastian and Sheila Jasanoff. "Traveling imaginaries: The 'practice turn' in innovation policy and the global circulation of innovation models." In *The Routledge Handbook of the Political Economy of Science*, edited by David Tyfield, Rebecca Lave, Samuel Randalls, and Charles Thorpe, 416–28. London: Routledge, 2017.

Pieper, Christine. "Wechselbeziehungen zwischen Wissenschaft, Politik und Wirtschaft in der Hochschulinformatik der DDR (1960er Jahre)." In *Informatik in der DDR – eine Bilanz*, edited by Friedrich Naumann and Gabriele Schade, 351–68. Bonn: Gesellschaft für Informatik, 2006.

Pieper, Christine. *Hochschulinformatik in der Bundesrepublik und der DDR bis 1989/90*. Stuttgart: Franz Steiner Verlag, 2009.

Pohlers, Volker. *Dr. Helmut Hoyer: JU+TE Computer selbst gebaut*. 2004. https://hc-ddr.hucki.net/wiki/lib/exe/fetch.php/tiny/jutecomp1.pdf (accessed May 31, 2024).

Prager, Eberhard and Evelyn Richer. *Software – Was ist das?*. Berlin: Verlag die Wirtschaft, 1986.

Radkau, Joachim. *Geschichte der Zukunft: Prognosen, Visionen, Irrungen in Deutschland von 1945 bis heute*. München: Carl Hanser Verlag, 2017.

Rahm, Lina. "Educational Imaginaries: Governance at the Intersection of Technology and Education." *Journal of Education Policy* 38, no. 1 (2021): 46–68. https://doi.org/10.1080/02680939.2021.1970233

Rankin, Joy Lisi. *A People's History of Computing in the United States*. Cambridge, MA: Harvard University Press, 2018.

Rensfeldt, Annika B. and Catarina Player-Koro. "'Back to the future': Socio-technical imaginaries in 50 years of school digitalization curriculum reforms." *Seminar.net* 16, no. 2 (2020). https://doi.org/10.7577/seminar.4048

Rieder, Gernot. "Tracing Big Data Imaginaries through Public Policy: The Case of the European Commission." In *The Politics and Policies of Big Data: Big Data, Big Brother?*, edited by Ann R. Sætnan, Ingrid Schneider, and Nicola Green, 89–109. New York/London: Routledge, 2018.

robotrontechnik.de (2023). "Integrierte Schaltkreise." Robotrontechnik. Last modified September 21, 2023. https://www.robotrontechnik.de/index.htm?/html/komponenten/ic.htm (accessed May 31, 2024).

Röhr, Rita. "Die Beschäftigung polnischer Arbeitskräfte in der DDR 1966–1990." *Archiv für Sozialgeschichte*, no. 42 (2002): 211–36.

Rossow, Gerd. "Wissenschaftlich-technischer Fortschritt und Kulturniveau der Arbeiterklasse." *Einheit*, no. 8 (1987): 759–62.

Rudolph, Wolfgang. "Weiterbildung als ein erstrangiges Erfordernis." *Einheit*, no. 12 (1986): 1109–14.

Rummler, Günter. "Heranführen der Schüler an die automatisierte Produktion." *Polytechnische Bildung und Erziehung* 28, no. 11 (1986): 402–6.

Sabrow, Martin. "Zukunftspathos als Legitimationsressource: Zu Charakter und Wandel des Fortschrittsparadigmas in der DDR." In: *Aufbruch in die Zukunft. Die 1960er Jahre zwischen Planungseuphorie und kulturellem Wandel. DDR, CSSR und Bundesrepublik Deutschland im Vergleich*, edited by Heinz-Gerhard Haupt and Jörg Requate, 165–84. Weilerswist: Velbrück Wissenschaft, 2004.

Saeltzer, Gerhard. *Kollege Personalcomputer*. Leipzig: VEB Fachbuchverlag Leipzig, 1988.

Salecker, Wolfgang. "Aufgaben und Erfahrungen beschleunigter Einführung von CAD/CAM in unseren Kombinaten." *Einheit*, no. 7 (1986): 618–21.

Schilling, Alfred and Wolfgang Töpfer. *Informatik: Lehrbuch für das strukturierte Programmieren*. Berlin: Volk und Wissen Volkseigener Verlag, 1988.

Schmidt, Erwin. "Zur Geschichte und Lehre der Informatik an der Technischen Universität Dresden bis zur Gründung des Informatik-Zentrum 1986." In *Informatik in der DDR – Grundlagen und Anwendungen*, edited by Birgit Demuth, 142–53. Bonn: Gesellschaft für Informatik, 2008.

Schmidt, Werner H. "Informatikausbildung an Schulen der DDR." In *Informatik in der DDR – eine Bilanz*, edited by Friedrich Naumann and Gabriele Schade, 378–81. Bonn: Gesellschaft für Informatik, 2006.

Schmitt, Martin, Julia Erdogan, Thomas Kasper, and Janine Funke. "Digitalgeschichte Deutschlands: Ein Forschungsbericht." *Technikgeschichte* 83, no. 1 (2016): 33–70.

Schmitt, Martin. "Socialist Life of a U.S. Army Computer in the GDR's Financial Sector: Import of Western Information Technology into Eastern Europe in the Early 1960s." In *Histories of Computing in Eastern Europe*, edited by Christopher Leslie and Martin Schmitt, 139–64. Cham: Springer, 2019.

Schmitt, Martin. *Die Digitalisierung der Kreditwirtschaft: Computereinsatz in den Sparkassen der Bundesrepublik und der DDR, 1957–1991*. Göttingen: Wallstein, 2021.

Schnabel, Tom. "Der Einsatz der Taschenrechner," in *Kleincomputer in der DDR*. Thesis, Humboldt-Universität zu Berlin, 1999. http://waste.informatik.hu-berlin.de/Diplom/robotron/diplom/texte/einsatz/taschenrechner.html (accessed May 31, 2024).

Schröder, Jens. *Auferstanden aus Platinen: Die Kulturgeschichte der Computer- und Videospiele unter besonderer Berücksichtigung der ehemaligen DDR*. Stuttgart: ibidem, 2010.

Schuhmann, Annette. "Die Zukunft der Arbeit in der Übergangsgesellschaft: Überlegungen zur Produktion von (Zukunfts-)Erwartungen in der DDR." In *Vergangene Zukünfte von Arbeit: Aussichten, Ängste und Aneignungen im 20. Jahrhundert*, edited by Franziska Rehlinghaus and Ulf Teichmann, 157–78. Bonn: Dietz, 2019.

Schuster, Regine. *Welche Anforderungen werden an einen Facharbeiter für Schreibtechnik gestellt?* [Hausarbeit zur Prüfung als Facharbeiter für Schreibtechnik]. Berlin, 1990. Available at the library of the federal archive in Berlin-Lichterfelde, shelfmark 05 C 1307.

Schwarz, Martin. "'Zauberschlüssel zu einem Zukunftsparadies der Menschheit': Automatisierungsdiskurse der 1950er- und 1960er-Jahre im deutsch-deutschen Vergleich." PhD diss., TU Dresden, 2015. https://nbn-resolving.org/urn:nbn:de:bsz:14-qucosa-191680 (accessed May 31, 2024).

Schwenkel, Christina. "Rethinking Asian Mobilities." *Critical Asian Studies* 46, no. 2 (2014): 235–58. https://doi.org/10.1080/14672715.2014.898453

SED. *Direktive des X. Parteitages der SED zum Fünfjahrplan für die Entwicklung der Volkswirtschaft der DDR in den Jahren 1981–1985*. Berlin: Dietz Verlag, 1981.

Segal, Jérôme. "Kybernetik in der DDR: Begegnung mit der marxistischen Ideologie." *Dresdener Beiträge zur Geschichte der Technikwissenschaften*, no. 27 (2001): 47–75.

Selwyn, Neil. "Learning to Love the Micro: The Discursive Construction of 'Educational' Computing in the UK, 1979–89." *British Journal of Sociology of Education* 23, no. 3 (2002): 427–43. https://doi.org/10.1080/0142569022000015454

Selwyn, Neil. "Making the Most of the 'Micro': Revisiting the Social Shaping of Microcomputing in UK Schools," *Oxford Review of Education* 40, no. 2 (2014): 170–88. https://doi.org/10.1080/03054985.2014.889601

Sismondo, Sergio. "Sociotechnical imaginaries: An accidental themed issue." *Social Studies of Science* 50, no. 4 (2020): 505–7. https://doi.org/10.1177/0306312720944753

Smith, Jessica M. and Abraham S. D. Tidwell. "The everyday lives of energy transitions: Contested sociotechnical imaginaries in the American West." *Social Studies of Science* 46, no. 3 (2016): 327–50. https://doi.org/10.1177/03063127166445

Sobeslavsky, Erich and Nikolaus J. Lehmann. *Zur Geschichte von Rechentechnik und Datenverarbeitung in der DDR 1946–1968*. Dresden: Hannah-Arendt-Institut für Totalitarismusforschung, 1996.

Somogyvári, Lajos, Máté Szabó, and Gábor Képes. "How Computers Entered the Classroom in Hungary: A Long Journey from the Late 1950s into the 1980s." In *How Computers Entered the Classroom, 1960–2000: Historical Perspectives*, edited by Carmen Flury and Michael Geiss, 39–74. Berlin/Boston: De Gruyter Oldenbourg, 2023. https://doi.org/10.1515/9783110780147-003

Spangenberg, Sabine. *The Institutionalised Transformation of the East German Economy*. Heidelberg/New York: Physica-Verlag, 1998.

Spitzky, Norbert. *Die 30. Zentrale Messe der Meister von morgen (ZMMM) im Urteil der Besucher*. Leipzig: Zentralinstitut für Jugendforschung, 1988. https://nbn-resolving.org/urn:nbn:de:0168-ssoar-401358 (accessed May 31, 2024).

Spitzky, Norbert, Leonhard Kasek, and Michael Chrapa. *Information zur 31. Zentralen Messe der Meister von Morgen*. Leipzig: Zentralinstitut für Jugendforschung, 1989. https://nbn-resolving.org/urn:nbn:de:0168-ssoar-402897 (accessed May 31, 2024).

Sprengel, Hans-Jürgen. "Kleincomputer und Programmierung (Teil 1)." *Mathematik in der Schule* 24, no. 12 (1986): 835–39.

Sprengel, Hans-Jürgen. "Kleincomputer und Programmierung (Teil 2)." *Mathematik in der Schule* 25, no. 1 (1987): 8–13.

Sprengel, Hans-Jürgen "Kleincomputer und Programmierung (Teil 3)." *Mathematik in der Schule* 25, no. 2/3 (1987): 91–97.

Staatliche Zentralverwaltung für Statistik. *Statistisches Jahrbuch 1985 der Deutschen Demokratischen Republik*. Berlin: Staatsverlag der Deutschen Demokratischen Republik, 1985.

Stachniak, Zbigniew. "Red Clones: The Soviet Computer Hobby Movement of the 1980s." *IEEE Annals of the History of Computing* 37, no. 1 (2015): 12–23. https://doi.org/10.1109/MAHC.2015.11

Stephan, Helga and Eberhard Wiedemann. "Lohnstruktur und Lohndifferenzierung in der DDR." Sonderdruck aus: *Mitteilungen aus der Arbeitsmarkt- und Berufsforschung* 23, no. 4 (1990): 550–62.

Stokes, Raymond G. *Constructing Socialism: Technology and Change in East Germany 1945–1990*. Baltimore/London: Johns Hopkins University Press, 2000.

Striebing, Lothar, Karin Zänker, and Bernd Zschaler. "Neue Computergenerationen und die Perspektive des Menschen." *Deutsche Zeitschrift für Philosophie* 34, no. 1 (1986): 12–20.

Švelch, Jaroslav. *Gaming the Iron Curtain: How Teenagers and Amateurs in Communist Czechoslovakia Claimed the Medium of Computer Games*. Cambridge, MA: MIT Press, 2018.

Sysło, Maciej M. "The First 25 Years of Computers in Education in Poland: 1965 – 1990." In *Reflections on the History of Computers in Education*, edited by Arthur Tatnall and Bill Davey, 266–90. Berlin/ Heidelberg: Springer, 2014. https://doi.org/10.1007/978-3-642-55119-2_18

Tatarchenko, Ksenia. "Why We Should Remember the Soviet Information Age." In *Vertical atlas*, edited by Leonardo Dellanoce, Amal Khalaf, Klaas Kuitenbrouwer, Nanjala Nyabola, Renée Roukens, Arthur Steiner, and Mi You, 264–68. Arnhem: ArtEZ Press, 2022. Available at: https://ink.library.smu.edu.sg/cis_research/71 (accessed June 11, 2024).

Tatnall, Arthur. "The Australian educational computer that never was." *IEEE Annals of the History of Computing* 35, no. 1 (2013): 35–47.

Tatnall, Arthur and Bill Davey. *Reflections on the History of Computers in Education: Early Use of Computers and Teaching about Computing in Schools*. Berlin/Heidelberg: Springer, 2014. https://doi.org/10.1007/978-3-642-55119-2

Ulbricht, Walter. *Das Programm des Sozialismus und die geschichtliche Aufgabe der Sozialistischen Einheitspartei Deutschlands. Schlusswort des Genossen Walter Ulbricht*. Berlin: Dietz Verlag, 1963.

Venske, Benjamin. *Das Rechenzentrum der Universität Rostock 1964–2010* (Rostocker Studien zur Universitätsgeschichte, Band 19). Rostock: Universität Rostock, 2012. https://doi.org/10.18453/rosdok_id00002238

Vidal, Matt. "Work and Exploitation in Capitalism." In *The Oxford Handbook of Karl Marx*, edited by Matt Vidal, Tony Smith, Tomás Rotta, and Paul Prew, 241–60. New York: Oxford University Press, 2019. https://doi.org/10.1093/oxfordhb/9780190695545.013.11

Vollmer, Uwe. "Vollbeschäftigungspolitik, Arbeitsplatzplanung und Entlohnung." In *Die Endzeit der DDR-Wirtschaft – Analysen zur Wirtschafts-, Sozial- und Umweltpolitik*, edited by Eberhard Kuhrt, 323–73. Opladen: Leske+Budrich, 1999.

Völz, Horst. "Persönliches zur Rechentechnik in der DDR." 2015. https://docplayer.org/68357290-Persoenliches-zur-rechentechnik-in-der-ddr.html (accessed May 31, 2024).

Wähler, Josefine and Maria-Annabel Hanke. "'Pacemakers Report': GDR Pedagogical Innovators and the Collection of Pädagogische Lesungen, 1952–1989." *Paedagogica Historica* 58, no. 1 (2022): 66–83. https://doi.org/10.1080/00309230.2020.1796720.

Waller, Paul. *Nightmare of the Imaginaries. A Critique of Socio-technical Imaginaries Commonly Applied to Governance*. London, 2020. https://doi.org/10.2139/ssrn.3605494

Wasiak, Patryk and Jaroslav Švelch. "Designing Educational and Home Computers in State Socialism: The Polish and Czechoslovak Experience." *Journal of Design History* 36, no. 4 (December 30, 2023): 377–93. https://doi.org/10.1093/jdh/epad027

Weise, Klaus-Dieter. *Erzeugnislinie Heimcomputer, Kleincomputer und Bildungscomputer des VEB Kombinat Robotron*. Dresden: Förderverein für die Technischen Sammlungen Dresden, 2005. http://robotron.foerderverein-tsd.de/322/robotron322a.pdf (accessed May 31, 2024).

Weise, Klaus-Dieter. *Anlage 1 – Erläuterungen zu Begriffen im Dokument 'Erzeugnislinie Heimcomputer, Kleincomputer und Bildungscomputer des VEB Kombinat Robotron' der UAG Historie Robotron der Arbeitsgruppe Rechentechnik in den Technischen Sammlungen Dresden*. Dresden: Förderverein für die Technischen Sammlungen Dresden, 2005. http://robotron.foerderverein-tsd.de/322/robotron322b.pdf (accessed May 31, 2024).

Wellisch, Hans. "From Information Science to Informatics: a terminological investigation." *Journal of Librarianship* 4, no. 3 (1972): 157–87. https://doi.org/10.1177/096100067200400302

Wentker, Hermann. *Außenpolitik in engen Grenzen: Die DDR im internationalen System 1949–1989. Veröffentlichungen zur SBZ-/DDR-Forschung im Institut für Zeitgeschichte*. München: Oldenbourg Wissenschaftsverlag, 2007. https://doi.org/10.1524/9783486707380

Wiener, Norbert. *Cybernetics: or Control and Communication in the Animal and the Machine.* Cambridge, MA: MIT Press, 1948.

Wilkendorf, Dieter. "Was bleibt? Die Kinderzeitschrift 'Fröhlich sein und singen – Frösi' im Erinnern und Nachdenken ihres Chefredakteurs." In: *Kinderzeitschriften in der DDR*, edited by Christoph Lüth and Klaus Pecher, 139–51. Bad Heilbrunn: Verlag Julius Klinkhardt, 2007.

Winkler, Gunnar, ed. *Frauenreport '90.* Berlin: Verlag Die Wirtschaft, 1990.

Winkler, Jürgen F. H. (2019). "Oprema – The Relay Computer of Carl Zeiss Jena." August 26, 2019. https://arxiv.org/pdf/1908.09549.pdf (accessed May 31, 2024).

Wissenschaftsrat. *Stellungnahmen zu den außeruniversitären Forschungseinrichtungen der ehemaligen Akademie der Wissenschaften der DDR in den Fachgebieten Mathematik, Informatik, Automatisierung und Mechanik.* Köln: Wissenschaftsrat, 1992. http://www.adw-zki.de/docs/evaluierung_B051_4-92_AdW.pdf (accessed May 31, 2024).

Wüstneck, Klaus. "Der kybernetische Charakter des neuen ökonomischen Systems und die Modellstruktur der Perspektivplanung als zielstrebiger, kybernetischer Prozeß." *Deutsche Zeitschrift für Philosophie* 13, no. 1 (1965): 5–31.

Yatsko, Viatcheslav A. "Informatics, Informations Science, and Computer Science." *Scientific and Technical Information Processing* 45, no. 4 (2018): 235–40. https://doi.org/10.3103/S0147688218040081

Zabel, Nicole. "Die Lehrmaschinen und der Programmierte Unterricht – Chancen und Grenzen im Bildungswesen der DDR in den 1960er und 1970er Jahren." In *Jahrbuch für Historische Bildungsforschung 2014 (Vol. 20),* edited by Sektion Historische Bildungsforschung der DGfE and Bibliothek für Bildungsgeschichtliche Forschung des DIPF, 123–52. Bad Heilbrunn: Verlag Julius Klinkhardt, 2015. https://doi.org/10.25656/01:14577

Zentralinstitut für Berufsbildung der DDR. *Facharbeiter für Schreibtechnik. Berufsbild für die Berufsberatung.* Berlin: Staatsverlag der DDR, 1985.

Zentralkomitee der SED. *Bericht des Zentralkomitees der Sozialistischen Einheitspartei Deutschlands an den X. Parteitag der SED, Berichterstatter: Genosse Erich Honecker.* Berlin: Dietz Verlag, 1981.

Zentralkomitee der SED. *Bericht des Zentralkomitees der Sozialistischen Einheitspartei Deutschlands an den XI. Parteitag der SED, Berichterstatter: Genosse Erich Honecker.* Berlin: Dietz Verlag, 1986.

Zirpins, Jens. "Geheimdienstspiele. Die Entwicklungsgeschichte des Poly-Play." *Retro,* no. 30 (2014): 10–14.

Zschockelt, Peter. "Wirtschaftsinformatik an der HfÖ Berlin." In *Informatik in der DDR – Tagung Berlin 2010. Tagungsband zum 4. Symposium 'Informatik in der DDR' am 16. und 17. September 2010 in Berlin,* edited by Wolfgang Coy and Peter Schirmbacher, 145–53. Berlin: Humboldt-Universität zu Berlin, 2010.